Photographic Science

Photographic Science

Advances in Nanoparticles, J-Aggregates, Dye Sensitization, and Organic Devices

TADAAKI TANI

OXFORD
UNIVERSITY PRESS

OXFORD
UNIVERSITY PRESS

Great Clarendon Street, Oxford OX2 6DP

Oxford University Press is a department of the University of Oxford.
It furthers the University's objective of excellence in research, scholarship,
and education by publishing worldwide in

Oxford New York

Auckland Cape Town Dar es Salaam Hong Kong Karachi
Kuala Lumpur Madrid Melbourne Mexico City Nairobi
New Delhi Shanghai Taipei Toronto

With offices in

Argentina Austria Brazil Chile Czech Republic France Greece
Guatemala Hungary Italy Japan Poland Portugal Singapore
South Korea Switzerland Thailand Turkey Ukraine Vietnam

Published in the United States
by Oxford University Press Inc., New York

First published 2011

British Library Cataloguing in Publication Data

Data available

Library of Congress Cataloging in Publication Data

Data available

Typeset by SPI Publisher Services, Pondicherry, India
Printed in Great Britain
on acid-free paper by
CPI Antony Rowe, Chippenham, Wiltshire

ISBN 978–0–19–957295–3

1 3 5 7 9 10 8 6 4 2

Contents

Preface

Since its birth in 1839, silver halide (AgX) photography has achieved marvelous progress and variety on the basis of photographic science and technology. Photographic science is composed of the knowledge accumulated during the development of AgX photographic products over such a long period. Since photographic products have required the development of various kinds of technologies for their progress, they have accumulated unique and fruitful scientific knowledge over many years. Since basic research has been very active and effective in the development of photographic products, it has made a great contribution to the progress of photographic science. However, further progress in photographic science is unlikely owing to the decline in the production of AgX photographic products. The first part of this book describes the final and highest level that photographic science has ever achieved.

Photographic science, which this book describes, includes the formation, properties, and behaviors of AgX and Ag nanoparticles, and dye sensitization with J-aggregated sensitizing dye layers as photosensitive p-channel organic semiconductors for AgX grains acting as n-channel substrates. It is therefore expected that photographic science provides advanced knowledge for developing new areas related to nanoparticles, J-aggregates, dye sensitization, and organic semiconductors. The second part of this book describes the results of extending photographic science into these new areas.

Although photographic science has declined recently, I have made efforts to extend the results of photographic science to the above-stated new areas. Meanwhile, it turns out that synergetic interaction between photographic science and sciences in the new areas can take place and create new ideas. This book also describes the ideas arising from these synergetic interactions, not only for the new areas described in Part II, but also for the photographic science described in Part I.

This book is regarded as the second edition of *Photographic Sensitivity: Theory and Mechanisms*, which was published by Oxford University Press (OUP) in 1995 and mostly read by scientists and engineers in the field of AgX photography. On the other hand, this book is expected to be read by scientists and engineers in new areas such as digital photography with CCD and CMOS as photo-sensors, nuclear physics with AgX nanoparticles as sensors for dark matter, metal nanoparticles with localized surface plasmons, dye-sensitized solar cells, and organic semiconductors. This book is entitled *Photographic Science* instead of *Photographic Sensitivity* in order to include the expected readers of this new edition.

I have been involved in basic research on functional materials for a half of a century, and have been influenced by many people, to whom I am sincerely grateful. Among them, several persons were standing at my crossroads. Professor Shin-ichi Kikuchi guided me to AgX photographic science at the University of Tokyo, Mr. Yasuo Koseki laid the systematic ground for me to carry out basic research throughout my tenure in Fuji Photo Film Co., and Professor Junichi Hanna invited me to study organic semiconductors at the

Tokyo Institute of Technology. Collaborating with us for many years, Professor Kazuhiko Seki and Professor Keitaro Yoshihara made it possible to apply related fundamental sciences to photographic science and technology, enabling photographic science to be attractive to scientists in areas other than photography.

I have been happy to study photographic science in atmosphere and community with many active and eminent scientists and engineers. Many of them were at Kikuchi's Laboratory at the University of Tokyo, including Professor Ken-ichi Honda and Professor Akira Fujishima, at Ashigara Research Laboratories of Fuji Photo Film Co., including Professor Hiroshi Kanzaki, Dr. Shunji Takada, and Dr. Akihiro Hirano, and at Research Laboratories of Eastman Kodak Co., including Dr. Paul Gilman, Dr. Joe Maskasky, Dr. R. Hailstone, and Dr. Annabel Muenter. I am sincerely grateful to them for their cooperation, encouragement, and stimulation in addition to their great contributions to the progress of photographic science, which made this book attractive.

It was on January 17 in 2008 that I met Dr. Sonke Adlung, Senior Editor in physical sciences in OUP, after I enjoyed staying in London with my family for several days. We talked about the possibility of publishing a second edition of *Photographic Sensitivity* under the condition that AgX photography was declining rapidly. I would like to thank him and others at OUP for accepting my proposal, and encouraging and supporting me to publish this book.

I have ever enjoyed photographic science as a harmonized part of my life with my dear wife, Aoi, for a half of a century. In particular, it has left a strong impression on me that I could really enjoy many scientific meetings with Aoi in various countries, including Japan. Recalling many pleasant memories, I dedicate this book to her.

July 5, 2011
Tadaaki Tani

Part I

Progress in Photographic Science

1

Introduction to Photographic Science

1.1 Silver Halide Photographic Materials

Photography has a long history [1–3]. The oldest existing photograph was taken by Nicephore Niepse in 1826 in France by means of heliography with asphalt (natural photopolymer) as a photosensitive material. J. L. Marignier succeeded in reproducing the heliographic process by exposing a photosensitive plate of heliography to sunlight for 5 days to take a picture [4].

It is generally accepted that silver halide (AgX) photography was born in 1839 with the daguerreotype invented by L. Jacques M. Daguerre in France and the calotype by W. H. Fox Talbot in England. In the former case, a thin layer of AgI was the photosensitive material, and a latent image was revealed by a development process involving mercury vapor. However, it took many minutes to take a picture. Furthermore, producing a daguerreotype was very laborious and expensive [1–3]. Nevertheless, many people enthusiastically accepted photography. Such enthusiasm indicates the strong attraction of human beings to photography as a medium for the exact recording and depiction of visual images.

Based on the literature, it is judged from the exposure time required to take a picture by daguerreotype that photographic sensitivity has increased by about 400,000 times [5]. In addition, improvement in image quality and the invention of color photography have also been achieved, making it much easier and cheaper to take pictures now than in the early days of photographic history.

A color negative film contains AgX photographic materials. Figure 1.1 is a scanning electron micrograph of the photosensitive layers of a current color negative film with ISO 1600 sensitivity. The white lines are the sections of thin tabular AgX grains shown in Figure 1.2. In the photosensitive layers, there are blue-, green-, and red-sensitive layers. Moreover, each color-sensitive layer is composed of high-, medium-, and low-speed sublayers.

In addition, a fourth layer has been introduced into color negative films, as shown in Figure 1.1. It was theoretically predicted that the red-sensitive layer should have considerable negative sensitivity in the blue–green region in order to achieve faithful color reproduction by a color negative film with layers sensitive to the three primary colors only. In order to fulfill this prediction, the fourth layer depresses the development in the red-sensitive layer according to the exposure to blue–green light [6,7].

Thus, in a typical color negative film there are about fourteen photosensitive layers of different functions with a total thickness of 20 μm, coated on a triacetate cellulose base that is about 100 μm thick. There are some twenty million AgX grains per square millimeter, and more than one hundred kinds of functional organic compounds in a film.

The remarkable progress in AgX photography over the past 170 years is evident when we compare the composition of a daguerreotype and that of a current color film. The former is a black-and-white photograph with a thin layer of AgI as a photosensitive

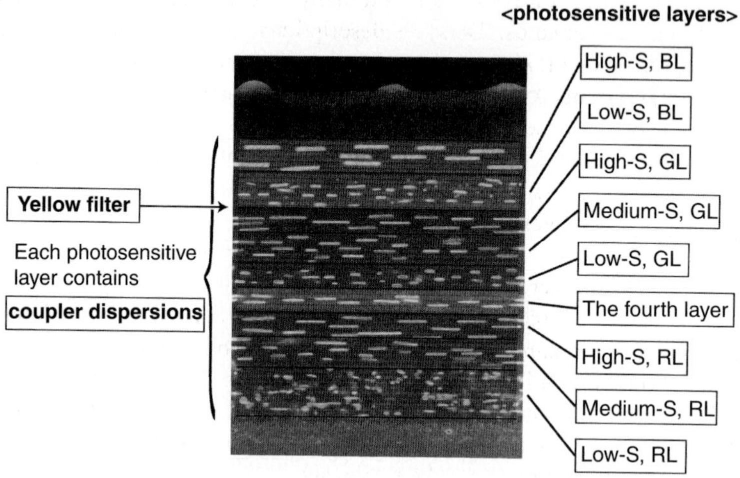

Figure 1.1 Scanning electron micrograph of the photosensitive layers of a color negative film with three major layers sensitive to blue, green, and red light (BL, GL, and RL, respectively), and sub-layers with high, medium, and low sensitivities (high-S, medium-S, and low-S, respectively). White lines and spots are the sections of AgX grains. A yellow filter between Low-S, BL and High-S, GL is composed of Ag nanoparticles, which absorb blue light to bring about surface plasmon resonance as described in Section 8.3.1. Each photosensitive layer contains the dispersion of a coupler, which takes part in the formation of a dye during development as exemplified in Figure 1.5.

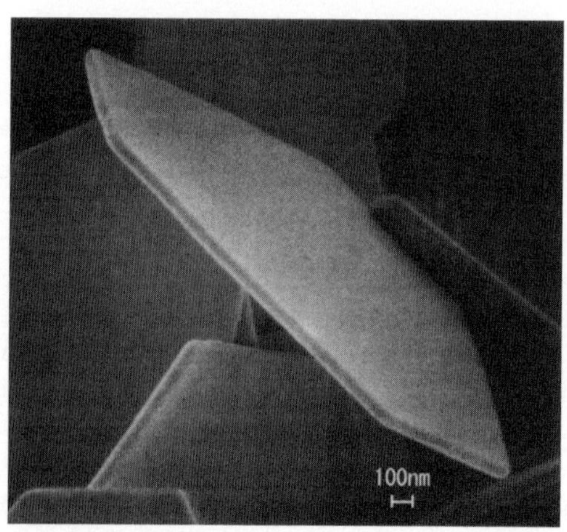

Figure 1.2 Scanning electron micrograph of tabular AgX grains used for a color negative film.

material, while the latter is a color film with many well-designed AgX grains and many functional organic compounds. Detailed descriptions are made in Chapter 2 for the structure and properties of AgX grains, showing the sophistication in the design and preparation of these grains. It is also remarkable that so many kinds of functional organic compounds are used in concert with each other to take color pictures.

As described above, during its long history AgX photography has progressed enormously in capability and changed in structure. The distance between the actual sensitivity of AgX photographic materials and its ultimate limit is decreasing. However, this situation does not mean that little room remains for further progress of AgX photographic materials [8]. On the contrary, AgX photographic systems have continuously expanded the variety of its capabilities, including sensitivity, image quality, color reproduction, easy handling, maintaining of stability, environmental protection, and cost performance along with its applications, which include color negative film, color reversal film, color paper, diagnostic X-ray film, microfilm, and films for printing by incorporating various technologies and functional materials [9]. These technologies and functional materials have in turn expanded the area and room for AgX to grow.

Figure 1.3 shows typical photographic processes. When exposed to light, an AgX grain in a photographic emulsion (i.e. a suspension of AgX grains in gelatin) absorbs incident photons to create photoelectrons, which form on the grain surface a latent image center composed of a silver nanocluster by combining with mobile silver ions in the grain. An AgX grain with a latent image center on its surface is reduced to a mass of silver (i.e. developed silver) by a developer, whereas a grain without a latent image center does not change in a developer and is dissolved by a fixing solution. Thus, the exposure to light of an AgX emulsion layer forms a black-and-white negative-working image composed of developed silver.

As shown in Figure 1.1, a color film needs AgX grains, which are sensitive separately to blue, green, and red light, while AgX grains by themselves are sensitive to only blue light in visible region. Thus, dye sensitization (i.e. spectral sensitization by sensitizing dyes) is used to render AgX grains sensitive to green and red light in addition to blue. Namely,

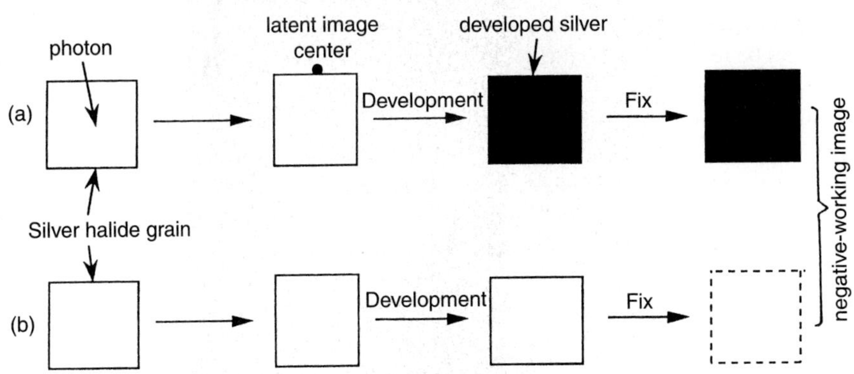

Figure 1.3 An illustration showing fundamental photographic processes with (a) an exposed AgX grain and (b) an unexposed one. On exposure to light, an AgX grain forms a latent image center composed of a silver cluster according to the mechanism described in Sections 4.1 and 4.2.

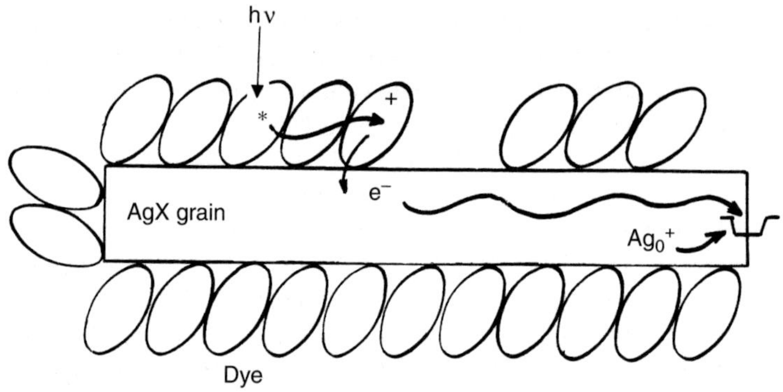

Figure 1.4 Dye sensitization of an AgX grain, where * and + are an excited dye molecule and a dye positive hole, respectively, while e^- and Ag_o^+ are a free electron and an interstitial silver ion in a grain, respectively, for latent image formation. The mechanism of dye sensitization is described in Section 5.9.1.

excited dye molecules inject electrons into the conduction band of a AgX grain. The injected electrons then collaborate with mobile silver ions in the grain to form the latent image, as illustrated in Figure 1.4. As shown in Figure 1.2, very thin tabular AgX grains are used in color films. Since they have very large specific surface area, they can adsorb sensitizing dyes, whose amount is so large that incident light is solely absorbed by sensitizing dyes on the grain.

According to the above-stated processes, photographic sensitivity depends on light absorption, efficiency of latent image formation, and size of the smallest latent image center, as described in detail in Chapter 6 [8]. The light absorbance of color films, which had been as small as one third, became to be one-half owing to the development of thin tabular AgX grains, and still leaves the room for improvement in light absorption to double. The efficiency of latent image formation in color films, which had been also small, became to be about one-third owing to several technologies that have developed over many years, and still leaves the room for improvement in efficiency of latent image formation to triple. The number of atoms in the smallest latent image center is three, and will not be reduced in the future. In total, there is room for increases in sensitivity by several times in the future.

As stated in detail in Chapter 6, there are several technologies that have contributed to the increase in photographic sensitivity and that are expected to contribute to it further. It is possible to increase the light absorbance of a color film by increasing the amount of sensitizing dyes adsorbed on AgX grains. This can be achieved by increasing the aspect ratio (i.e. the ratio of equivalent diameter of main surface to thickness) of a tabular AgX grain and by antenna dye sensitization, according to which a dye molecule separate from an AgX grain absorbs and transfers the energy of an incident photon to a sensitizing dye molecule on the grain. It is possible to improve the yield of free electrons by the two-electron sensitization, according to which one absorbed photon creates two free electrons in an AgX grain by utilizing a positive hole for the creation of an additional free electron. It is possible to improve the sites for latent image formation, which typically for many years have been sulfur sensitization centers, by introducing selenium sensitization centers and tellurium sensitization ones instead. It is possible to depress the recombination between an electron and a positive hole by the above-stated two-electron sensitization. Dislocations introduced in the area along the edge of a tabular AgX grain bind electrons and prevent them from recombining with positive holes localized by sensitizing dyes on the main surfaces of the grain. These technologies are described in detail in Chapters 4, 5, and 6.

Figure 1.5 shows an example of the chemical reactions taking place during the development of a color film, where each color-sensitive layer with a corresponding color coupler (1) and AgX grains is developed in a color developer with p-phenylenediamine compounds (PPD) as developing agents [10, 11]. Namely, a PPD molecule reduces two silver ions and is converted into a quinonediimine (QDI). Then, a QDI takes part in the formation of an image composed of a dye (3) through an intermediate compound (2) by its reaction with a coupler in each layer.

An AgX grain with a latent image center on its surface can be discriminated from a grain without any latent image center through the development process. As described in Section 4.4.1, the smallest latent image center is composed of three or four atoms. Figure 1.6 shows a cubic AgBr grain with an edge length of 1 μm, in which twenty billion silver ions exist. The photolytic reduction and clustering of only four silver ions are needed to change the fate of twenty billion silver ions in a grain. The degree of amplification is as large as five billion times in this case, and is one of the most important reasons why AgX photographic materials achieve high sensitivity [12].

Figure 1.7 illustrates the steps for the formation of a latent image center. A single, dimer, and tetramer of silver atoms are called a latent pre-image center, a latent sub-image center, and a latent image center, respectively. A latent pre-image center is unstable, whereas a latent sub-image center and a latent image center are stable. A latent image

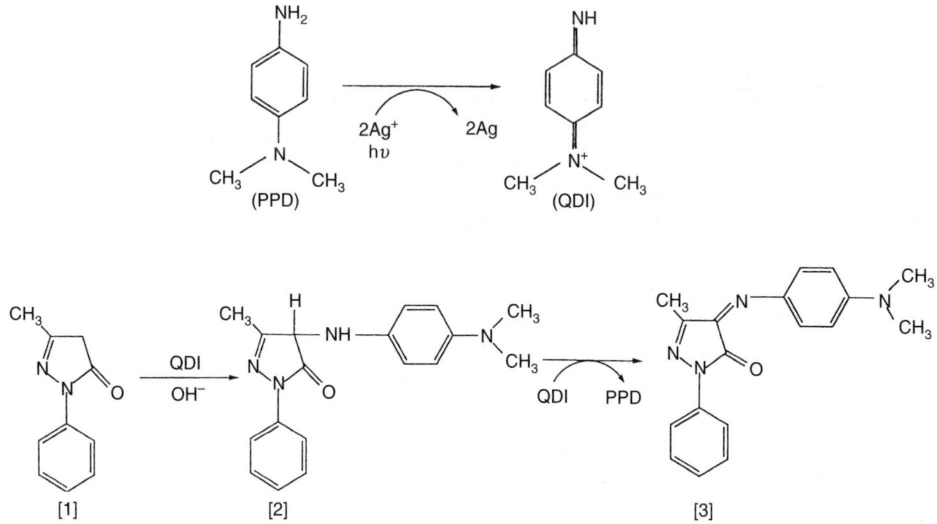

Figure 1.5 An example of coupling reactions for color image formation in AgX photographic films, where PPD is a p-phenylenediamine developing agent, QDI is a quinonediimine, and [1], [2], and [3] are a magenta coupler, a leuco dye, and a magenta dye, respectively.

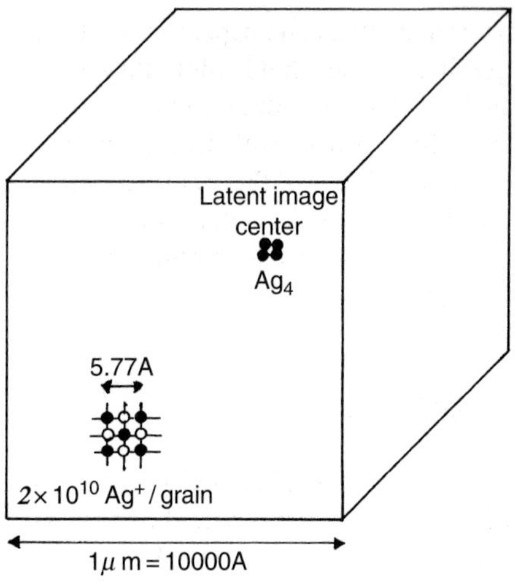

Figure 1.6 A cubic AgBr grain with edge length of 1 μm and a latent image center composed of four silver atoms, demonstrating the very large degree of amplification brought about by development.

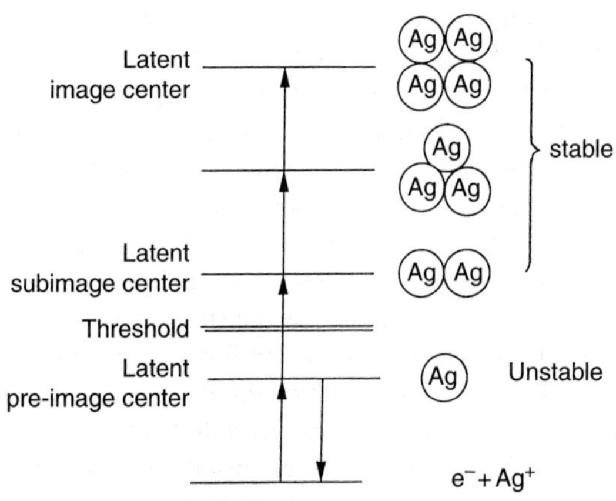

Figure 1.7 The steps and threshold for latent image formation, including a latent pre-image center without developability, a latent sub-image center without developability, and a latent image center with developability.

center can initiate development, whereas a latent sub-image center cannot. Since a latent sub-image center is composed of two silver atoms, its rate of the formation is proportional to the square of the number of absorbed photons per second and thus to the square of light intensity. Therefore a threshold is said to exist between a latent pre-image center and sub-image center in the steps for formation of a latent image center in terms of light intensity, and any illumination whose intensity is not strong enough to form a latent sub-image center does not change AgX grains even when the formation and dissociation of a latent pre-image center are repeated many times over a long time period. It is therefore considered that AgX photographic films and papers are quite sensitive to capturing a light image with intensity larger than the threshold, while they are insensitive to a light image with intensity weaker than the threshold. A threshold is also said to exist in the steps for the formation of a fog center, which occurs with free electrons formed owing to thermal excitation instead of optical excitation, and that photographic emulsions are stable during their production and storage processes for a long time owing to the existence of this threshold. This is the reason why AgX photographic films and papers can achieve both high sensitivity and stability [12,13].

1.2 Changes of Photographic Science

The development of AgX photographic films and papers was achieved on the basis of photographic science. The present author reviewed the progresses in photographic science while focusing on the theory and mechanisms of the phenomena in relation to photographic sensitivity [13–15]. It is noted that progress in photographic science has always been driven by progress in the fundamental sciences on which photographic science is based. Since photographic sensitivity heavily depends on the processes of latent image formation and dye sensitization, their mechanisms have always been the center of discussions in photographic science, and are outlined hereafter in relation to the chapters that follow.

In 1925, Sheppard, Trivelli, and Loveland proposed the mechanism of latent image formation [16], according to which light absorption of AgX grains creates a latent image with silver specks. The specks, which are named latent image centers, are too small to be observed and can be revealed only by development. Sheppard et al. also proposed the concept of the concentration principle, according to which the actions of absorbed photons are concentrated to form a silver speck in an AgX grain. Although the mechanism of their proposal was rejected, it has been improved upon by investigators over the years.

In the absence of knowledge about solid-state physics, Sheppard, Trivelli, and Loveland considered that incident photons were absorbed to form silver atoms at different places in an AgX grain, and that silver atoms thus formed migrated and coagulated with each other to give silver specks. Based on the progress of solid-state physics, Gurney and Mott proposed in 1938 that incident photons were absorbed to create free electrons (i.e. photoelectrons) and positive holes, and that photoelectrons migrated and combined with mobile silver ions (i.e. interstitial silver ions) to form a silver speck under the assumptions that the number of interstitial silver ions in each AgX grain was large enough for latent image formation and that the mobility of positive holes was too small to allow their recombination with electrons [17]. Based on the progress of solid-state physics of large, single AgX crystals, Michell denied the above-stated assumptions and proposed the so-called Mitchell mechanism [18–21]. Based on the progress of solid-state physics of AgX microcrystals, Hamilton and colleagues revised the Mitchell mechanism to reveal that the number of interstitial silver ions in each grain was so large that electrons should combine with them before they recombine with positive holes [22,23]. The present author considers that studies on the principal mechanism of latent image formation was thus complete in the mid-1960s. Details of the mechanism of latent image formation are described in Chapter 4.

From the present author's viewpoint, the mechanism of dye sensitization was the next to debut in the field of photographic science, with several scientists, including the present author, initiating extensive studies on it in the late 1960s. Two mechanisms for dye sensitization were proposed: electron transfer and energy transfer. The former was put forward by Gurney and Mott in 1938 with the idea that an electron optically excited in a dye molecule should transfer to the conduction band of an AgX grain [17]. The proposal was later denied by Mott in 1948 [24] with the consideration that the transfer of an electron from an excited dye molecule to the conduction band of AgX was impossible, based on the estimated electronic energy levels of dye molecules according to Coulson's evaluation with molecular orbital theory [24]. Coulson's evaluation was supported by Terenin and Akimov with ultraviolet photoelectron spectroscopy (UPS) [25,26]. Mott then proposed the energy transfer mechanism [24], according to which energy transfer should take place between the excited dye molecule and an AgX grain by the Foerster-type resonance transfer of energy [27–29] to bring about the electronic transition from an intermediate energy level in the forbidden band to the conduction band of AgX.

The present author initiated the investigation by estimating the electronic energy level diagram of sensitizing dye molecules and AgX grains [30], while Kuhn and colleagues tried to prove the energy transfer mechanism by means of Langmuir–Blodget membranes [31–36]. The present author pointed out the difference in the electronic energy levels of dye molecules in vacuum and in solid [30], and introduced the idea of the vacuum level shift between dye molecules and AgX [37,38] into the then-available data [26,39,40]. As a result of revising the electronic energy level diagram in order to solve these problems, the possibility of an excited electron transferring to the conduction band of AgX could be proved [30,38] and supported by others [41,42].

Although Kuhn and colleagues confirmed that the energy transfer actually occurred, its rate was an order of magnitude smaller than that of the electron transfer [36]. Concluding that dye sensitization takes place with the electron transfer mechanism completed the investigation by the 1980s. Dye sensitization is now used, not only in AgX photographic films, but also in dye-sensitized solar cells [43,44] in the light of the electron transfer mechanism, as described in Chapters 5 and 9.

Progress in photographic science and technology has been supported by unique and useful materials and methods, including mono-dispersed AgX grains such as cubic and octahedral ones [45–49], thin tabular AgX grains [50–53], and tremendous numbers of J-aggregated cyanine dyes [54]. The preparation of mono-dispersed AgX grains was initiated independently by Berry and Moisar in 1962 by means of the controlled double-jet method. Photographic emulsions with thin tabular AgX grains only were developed in 1981. J-aggregates of dyes were discovered by Jelly [55] and Scheibe [56,57] in 1936, and have become indispensable to color films due to their very sharp absorption band and attractive interaction among dye molecules in the aggregates. Details of the structure and preparation of AgX grains are described in Chapter 2, while the structure and behavior of J-aggregated cyanine dyes on AgX grains are given in Chapter 5.

Studies on the mechanisms of latent image formation and dye sensitization have been followed and have been reinforced by quantitative measurements of photographic processes and properties of materials used in photographic products. These quantitative studies have made the above-stated progresses in photographic science relevant and useful for the development of photographic materials and attractive to scientists in other fields.

Methods used for the above-stated purpose include the following. Microwave and radio frequency photoconductivity methods, which were developed by Kellogg and colleagues

[58–61], have been useful for quantitative study on the lifetime and transport of photo-electrons and positive holes in AgX grains, and are now being used to study the carrier transport in organic semiconductors [62]. Dielectric loss method was developed by van Biesen [63] and was used to measure the dark conductivity of AgX grains with interstitial silver ions as carriers. Electron spin resonance (ESR) spectroscopy was applied to the study of the behavior of positive holes in sensitizing dye layers on AgX grains by the present author in 1975 [64–66] and is now being used to study the electronic carriers in organic semiconductors [67,68]. Electrochemical methods for measuring reduction (E_R) and oxidation potentials (E_{OX}) of dyes on the basis of molecular orbital theory by the present author in 1967 [30] have been used to estimate the heights of the lowest unoccupied molecular orbital (LUMO) and highest occupied molecular orbital (HOMO) of dyes with remarkable improvement in the precision of E_R and E_{OX} of dyes [69].

1.3 Community for Photographic Science

The results of research on photographic science and technology have been presented at scientific meetings, and subjected to vivid discussions among the participants. It is obvious that such scientific meetings have contributed to the progress in photographic science and technology for many years.

Table 1.1 Chronology of the International Congresses of Photographic Science

No.	Year	Place	No.	Year	Place
1	1889	Paris, France	17	1961	Paris, France
2	1891	Brussels, Belgium	18	1963	Turin, Italy
3	1900	Paris, France	19	1965	Paris, France
4	1905	Liege, Belgium	20	1967	Tokyo, Japan
5	1910	Brussels, Belgium	21	1970	Moscow, Russia (USSR)
6	1925	Paris, France	22	1874	Dresden, Germany (GDR)
7	1928	London, Great Britain	23	1978	Rochester, USA
8	1931	Dresden, Germany	24	1982	Cambridge, Great Britain
9	1935	Paris, France	25	1986	Cologne, Germany
10	1948	Liege, Belgium	26	1990	Beijing, China
11	1949	Zurich, Switzerland	27	1994	Rochester, USA
12	1950	Brussels, Belgium	28	1998	Antwerp, Belgium
13	1951	Paris, France	29	2002	Tokyo, Japan
14	1953	London, Great Britain	30	2006	Rochester, New York
15	1956	Cologne, Germany	31	2010	Beijing, China
16	1959	Liege, Belgium	32		

The most important meeting in this field is the International Congress of Photographic Science (ICPS), which was initiated in 1889 in Paris, and has continued since then nearly every four years [70,71]. Table 1.1 lists the years and places for ICPS. The Congress was renamed the International Congress of Imaging Science (ICIS) in 2000 owing to the change in the situation in this field. The international symposium solely devoted to the activity of photographic science and technology has been the International East–West Symposium [72,73], initiated by P. B. Gilman and held in 1984, 1988, 1992, 1997, 2000, and 2004. The present author has been involved in the organization of all of them as the representative of the participants from the East.

The International Committee on the Science of Photography (ICSP) was formed in 1962 to arrange such international meetings in this field [74], and was also renamed the International Committee for Imaging Science (ICIS) in 2000 due to changes in the field. The successive presidents have been J. Eggert, W. F. Berg, H. Pietch, F. Granzer, T. Tani, G. House, and now M. Kriss.

1.4 Organization for Research on Photographic Science and Technology

Research and development of photographic films and papers have been carried out in universities and photographic companies for many years. The present author was involved in basic research on photographic science in Kikuchi's laboratory of the University of Tokyo from 1963 to 1968, and in Ashigara Research Laboratories of Fuji Photo Film Company, Ltd, since then. A comparison of the two, from the viewpoint of the difference between basic research activities for exploratory results and that for the development of new products, is described in detail in Chapter 11.

Progress in photographic science has been brought about not only by scientists in universities, but also by those in photographic companies. In the latter case, the activity of basic research laboratories has played a central role, shifting the center of photographic science research away from universities as the photographic film industry has grown and created new knowledge more frequently than universities.

A basic research laboratory was founded in Ashigara Research Laboratories in 1970 by Y. Koseki, and has been contributing to photographic science ever since. It enhance scientific activities at Ashigara Research Laboratories, and made it possible to develop new products on the basis of photographic science. The laboratory's presence enhanced the recognition of the importance of specialists and their training. One of the roles of the basic research laboratory was to predict the future prospect of AgX photographic films. For example, the appearance of MAVICA, which was developed for the first time in 1981 with a solid-state device as a photosensitive element, caused many investigators in the field to conclude that AgX photographic films would disappear soon. The present author predicted that AgX photographic films would continue to grow for many years [75]. In fact, AgX photographic films did continue to grow and enjoyed their best period in the 1990s, although they suddenly declined a quarter of a century following the appearance of MAVICA. The evaluation of the present state with respect to the ultimate limit and prediction of room and route for improvement were very important for the development of new products, and have been carried out by the fundamental research laboratory in Ashigara Research Laboratories, as described in detail in Chapters 6 and 11.

As stated in the previous section, various scientific meetings have contributed to the progress of photographic science and technology, and have provided photographic companies new knowledge useful for the development of new photographic film products. It is noted that not only universities, but also photographic companies have contributed to the progress in photographic science through the above-stated scientific meetings, and that the stimulation and interactions among participants in the meetings have raised the quality of new knowledge, and have made it possible for the participants to gain more knowledge in exchange.

References

1. J. M. Eder, *History of Photography,* trans. E. Epstein (Columbia Univ. Press, New York, 1945).
2. H. Gernsheim, A. Gernsheim, *The History of Photography* (Oxford Univ. Press, London, 1955).
3. E. Ostroff (ed.), *Pioneers of Photography: Their Achievements in Science and Technology* Springfield, VA: SPSE, Society for Imaging Science and Technology, 1987).
4. J. L. Marignier, *Nature*, 346, 115 (1990).
5. E. Kaprelian, *Photographic Engineering*, 1(2), 42 (1950).
6. N. Sasaki, K. Takahashi, H. Ikoma, *J. Soc. Photogr. Sci. Tech. Jpn.*, 52, 41 (1989).
7. N. Sasaki, *J. Soc. Photogr. Sci. Tech. Jpn.*, 67, 504 (2004).
8. T. Tani, *J. Imaging Sci. Technol.*, 51(2), 110 (2007).
9. J. Sturge, V. Walworth, A. Shepp (eds), *Imaging Processes and Materials: Neblette's Eight Edition* (New York: Van Nostrand Reinhold, 1989).
10. J. R. Thirtle, L. K. Tong, L. J. Fleckenstein, in T. H. James (ed.), *The Theory of the Photographic Process,* 4th edn (New York: Macmillan, 1977), Chapter 12.
11. P. Klause, in Sturge, Walworth, Shepp (eds), *Imaging Processes and Materials*, Chapter 4.
12. T. Tani, *Physics Today*, 36 (September 1989).
13. T. Tani, *Photographic Sensitivity: Theory and Mechanisms* (Oxford/New York: Oxford University Press, 1995), Chapter 1.
14. Tani, *Photographic Sensitivity*, Chapter 4.
15. T. Tani, *Imaging Sci. J.*, 55, 65 (2007).
16. S. E. Sheppard, A. P. H. Trivelli, R. P. Loveland, *J. Franklin Inst.*, 200, 51 (1925).
17. R. W. Gurney, N. F. Mott, *Proc. R. Soc. London, Ser. A* 164, 151 (1938).
18. J. W. Mitchell, *Recent Prog. Phys.*, 20, 433 (1957).
19. J. W. Mitchell, *J. Phys Chem.*, 66, 2359 (1962).
20. J. W. Mitchell, *Photogr. Sci. Eng.*, 22, 249 (1078).
21. J. W. Mitchell, *Photogr. Sci. Eng.*, 25, 170 (1981).
22. J. F. Hamilton, in James (ed.), *The Theory of the Photographic Process*, 105–32.
23. J. F. Hamilton, *Adv. Phys.*, 37, 359 (1988).
24. N. F. Mott, *Photogr. J.*, 88B, 119 (1948).
25. A. Terenin, I. Akimov, *J. Phys. Chem.*, 69, 730 (1965).
26. H. Meier, *Spectral Sensitization* (London/New York: Focal Press, 1968).
27. Th. Foerster, *Ann, Phys. VI*, 2, 55 (1948).
28. Th. Foerster, Discuss. Faraday Soc., No. 27, 7 (1954).
29. Th. Foerster, in O. Sinanoglue (ed.), *Modern Quantum Chemistry*, Part III (New York: Academic Press, 1965), 93.
30. T. Tani, S. Kikuchi, *Photogr. Sci. Eng.*, 11, 129 (1967).
31. H. Buecher, H. Kuhn, B. Mann, D. Moebius, L. von Szentpaly, P. Tillman, *Photogr. Sci. Eng.*, 11, 233 (1967).
32. L. von Szentpaly, D. Moebius, H. Kuhn, *J. Chem. Phys.*, 52, 4618 (1970).

33. H. Kuhn, in W. F. Berg, U. Mazzucato, H. Meier, G. Semerano (eds), *Dye Sensitization: Proceedings Symposium, Bresanone, 1967* (London/New York: Focal Press, 1970), 199.
34. H. Kuhn, D. Moebius, *Angew. Chem. Intern. Edn.*, 10, 620 (1971).
35. D. Mobius, *Photogr. Sci. Eng.*, 18, 413 (1974).
36. R. Steiger, H. Hediger, P. Junod, H. Kuhn, D. Mobius, *Photogr. Sci. Eng.*, 24, 185 (1980).
37. K. Seki, T. Tani, H. Ishii, *Thin Solid Films*, 273, 20 (1996).
38. K. Seki, H. Yanagi, Y. Kobayashi, T. Ohta, T. Tani, *Phys. Rev.*, B49, 2760 (1994).
39. G. Scheibe, D. Brueck, *Z. Elektrochem.*, 54, 403 (1950).
40. G. Scheibe, D. Brueck, F. Doerr, *Ber. Deutsch. Chem. Ges.*, 85, 867 (1952).
41. R. W. Berriman, P. B. Gilman, Jr., *Photogr. Sci. Eng.*, 17, 235 (1973).
42. T. H. James, *Photogr. Sci. Eng., 16, 120* (1972).
43. B. O'Regan, M. Graetzel, *Nature*, 353, 737 (1991).
44 A. Hagfeldt, M. Graetzel, *Chem. Rev.*, 95, 49 (1995).
45. C. R. Berry, D. C. Skillman, *Photogr. Sci. Eng.*, 6, 159 (1962).
46. E. Klein, E. Moisar, *Photogr. und Wiss.*, 11, 3 (1962).
47. E. Klein, E. Moisar, *Ber. Bunsenges. phys. Chem.*, 67, 349 (1963).
48. E. Moisar, E. Klein, *Ber. Bunsenges. phys. Chem.*, 67, 949 (1963).
49. E. Moisar, *J. Soc. Photogr. Sci. Technol. Jpn.*, 54, 273 (1991).
50. Anonymous, *Res. Discl.* (Item 22534) 225, 20 (January 1983).
51. J. T. Kofron, R. E. Booms, *J. Soc. Photogr. Sci. Technol. Jpn.*, 49, 15 (1986).
52. A. F. Sowinski, P. J. Wightman, *J. Imaging Sci.*, 31, 162 (1987).
53. R. E. Jacobson, N. R. Axford, A. Ford, *J. Photogr. Sci.*, 38, 140 (1990).
54. D. M. Sturmer, D. W. Heseltine, in James (ed.), *The Theory of the Photographic Process*, Chapter 8.
55. E. E. Jelly, *Nature*, 138, 1009 (1936).
56. G. Scheibe, *Angew. Chem.*, 49, 563 (1936).
57. G. Scheibe, A. Mareis, R. Schiffmann, *Z. Phys. Chem.*, B., 49, 324 (1941).
58. L. M. Kellogg, N. B. Liebert, T. H. James, *Photogr. Sci. Eng.,* 16, 115 (1972).
59. L. M. Kellogg, *Photogr. Sci. Eng., 18, 378* (1974).
60. L. M. Kellogg, J. Hodes, J. Muenter, 'Measurement of Photocarrier Decay Times in Silver Halide Emulsion Grains Using Radiofrequency Techniques', in Proc. of *the Intern. East-West Symposium* (Springfield, VA: IS&T, 1988), C11–15.
61. K. Hosoi, A. Hirano, T. Tani, *J. Imaging Sci. Technol.*, 46, 76 (2002).
62. J. M. Warman, and A. M. van de Croats, *Mol. Cryst. Liq. Cryst.*, 396, 41 (2005).
63. J. van Biesen, *J. Appl. Phys.*, 41, 1910 (1970).
64. T. Tani, *Photogr. Sci. Eng.*, 19, 356 (1975).
65. T. Tani, *J. Appl. Phys.*, 62, 2456 (1987).
66. T. Tani, Y. Sano, *J. Appl. Phys.*, 69, 4391 (1991).
67. K. Marumoto, Y. Muramatsu, S. Ukai, H. Ito, and S. Kuroda, *J. Phys. Soc. Jpn.*, 73, 1673 (2004).
68. K. Marumoto, S. Kuroda, T. Takenobu, Y. Iwasa, *Phys. Rev. Lett.*, 97, 256603 (2006).
69. J. Lenhard, *J. Imaging Sci.*, 30, 27 (1986).
70. T. Tani, A. Muenter, *J. Imaging Sci. Tech.*, 46, iv (2002).
71. T. Tani, A. Muenter, *Imaging Sci. J.*, 54, 244 (2006).
72. T. Tani, *J. Soc. Photogr. Sci. Technol. Jpn.*, 48, 116 (1985).
73. T. Tani, *J. Soc. Photogr. Sci. Technol. Jpn.*, 69, 279 (2006).
74. C. Roberts, *Chemistry International*, 9 (1), 6 (1987).
75. T. Tani, J. Imaging Sci., Z9, 93 (1985).

2

Structure and Preparation of Silver Halide Grains

The preparation of silver halide (AgX) grains is one of the most important technologies for AgX photographic films and papers with high capability, and significant progress was achieved when the AgX grains were made controllable with respect to size, shape, halide distribution, and crystal defects. Mono-dispersed AgX grains with well-designed structures were thus being realized.

Silver halide grains are composed of AgBr, AgCl, and mixed crystals of AgBr with AgI and/or AgCl and of AgCl with AgBr and/or AgI. Although AgI was used at the beginning of photography as a photosensitive material for daguerreotype, it has not been used since. AgX grains are preferred because of their suspensions in aqueous gelatin solutions, which are called photographic emulsions. In this chapter, extensive descriptions of the structure and preparation of emulsion grains with AgBr and AgCl as their main bodies are provided, along with brief ones on AgI emulsion grains at the end of each section.

2.1 Structures of Silver Halide Grains

2.1.1 Structure of Regular Silver Halide Grains

The crystal structure of AgX grains used for photographic films is face-centered cubic, belonging to the space group *Fm*3*m*. Although the grains have the same crystal structure in common, they are widely varied in shape, being covered with specific faces on the grain surfaces, and exhibiting a variety of crystal habits. The average size of the AgX grains usually used for highly sensitive photographic films is about 1 μm in equivalent diameter, ranging from less than 0.1 μm to several micrometers, depending on their purpose.

From the viewpoints of shape and structure, AgX grains in photographic emulsions are classified as regular or irregular. Regular grains have homogeneous crystal forms, are free from crystal defects, and are covered with a specific face. They can have one of seven shapes: cubic, octahedral, rhombic dodecahedral, icositetrahedral, trisoctahedral, tetrahexahedral, or hexaoctahedral, depending on the face, which appears on the surface of each grain as illustrated in Figure 2.1 [1]. The difference in the shape of these grains is called crystal habit [1, 2]. In photographic emulsions, only {100} and {111} faces are stable on the surfaces of AgBr grains, thus giving a cubic or octahedral shape, respectively, to stable regular grains, as shown in Figure 2.2. On the surfaces of AgCl grains, only {100} faces are stable and give a cubic shape, while {111} can be stable in the presence of appropriate crystal habit modifiers [3–6]. Regular grains with shapes other than cubic and octahedral can be prepared in the presence of appropriate crystal habit modifiers, which can stabilize the corresponding faces on the grain surfaces. Regular grains with shapes other than cubic, octahedral, or dodecahedral are not stable when the crystal habit modifiers are desorbed from their surfaces, and therefore are not practical for photographic

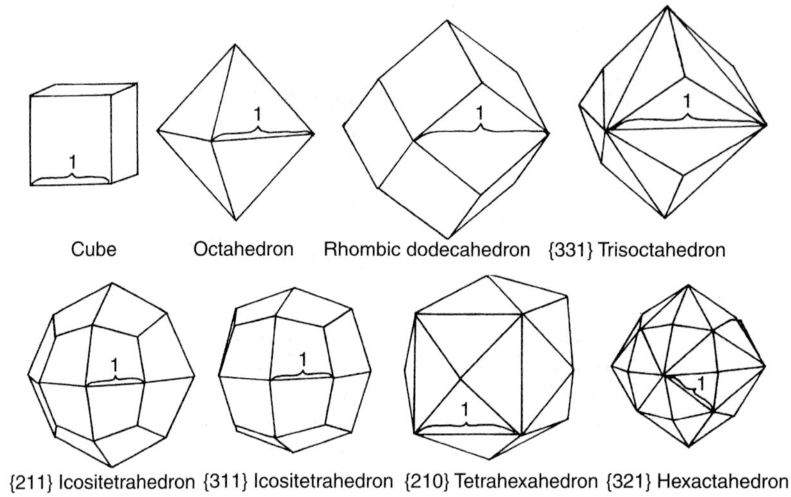

Figure 2.1 Regular AgBr grains as classified and prepared by Maskasky. Reproduced with permission from Reference [1], copyright 1986 IS&T, The Society for Imaging Science and Technology.

emulsions. Although rhombic dodecahedric crystals were observed for AgCl [7] and AgBr [8,9], they did not have any advantage in photographic capability over cubic and octahedral ones. Maskasky prepared all the regular grains for photographic emulsions by use of various crystal habit modifiers [1,2].

Since the surface of AgX grains is usually composed of {100} and/or {111} faces, the ratio of the area of the {100} face and {111} face to the total surface area (P_{100} and P_{111}, respectively, and $P_{100} + P_{111} = 1$) is important. Determining P_{100} and P_{111} by using the electron microscope is possible for some AgX grains with simple shapes [10] but is difficult for most grains. The present author developed a method for determining P_{100} and P_{111} by taking into consideration the Kubelka–Munk equation, where $(1-R_{\infty})^2/2R_{\infty}$ is proportional to the concentration of the dye responsible for R_{∞}, which is the reflectance of a thick emulsion layer of a photographic emulsion [11], and that the absorption spectra of a sensitizing dye, in a family of 9-methyl-thiacarbocyanine dyes including 3,3″-bis(4-disulfobutyl)-9-methyl-thiacarbocyanine, on {100} and {111} faces of AgX grains and in gelatin phase are considerably different from each other. The dye is first adsorbed to the {100} face, then is adsorbed to the {111} face after the {100} face is saturated, and finally exists in the gelatin phase after the {111} face is saturated [12]. This method has been used for various purposes, including the characterization of round corners of cubic AgBr grains [13] and side-surfaces of twinned tabular AgBr grains [14].

The surface structure of silver halides, in particular the structure of the {111} surface, has been a point of argument [2]. Since a simple truncated {111} surface would consist of a full layer of either Ag or Br ions and create a large surface dipole with atomically different top and bottom crystal surfaces, reconstruction should occur, resulting in a half-filled surface layer of either Ag or Br ions. Several ideas, including the hexagonal model and the alternate-row model, were proposed on the basis of the results through various means [15–18]. For the structure of {100} faces of AgBr, it has been indicated that

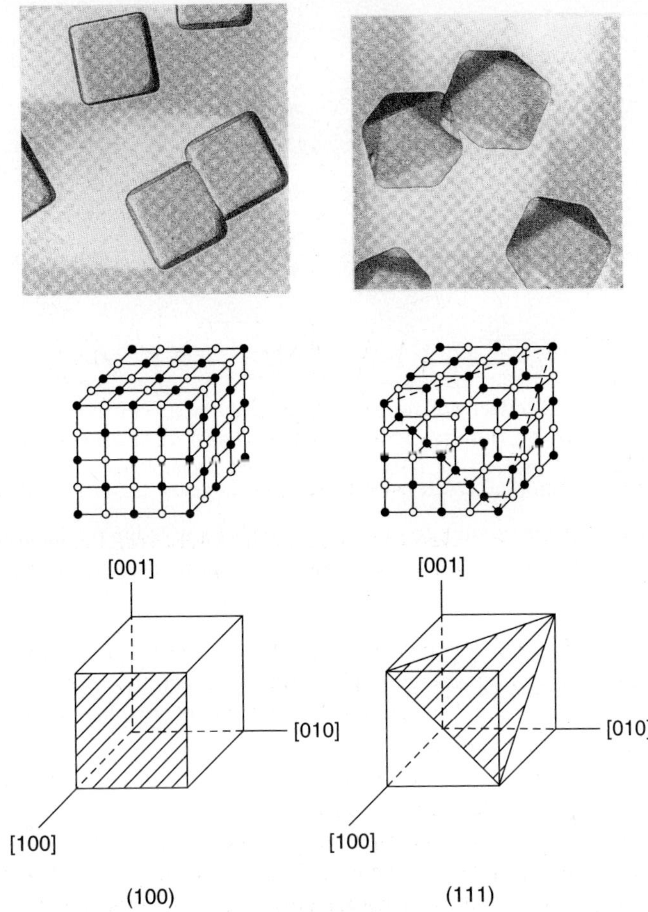

Figure 2.2 {100} and {111} faces of face-centered cubic crystal and electron micrographs of regular AgBr grains with {100} and {111} faces on their surface (i.e., cubic and octahedral grains).

relaxation occurs with the displacement of Br ions slightly above the normal surface plane and Ag ions below the surface place, thus yielding a 'rumpled' effect (Figure 10.6) [16–18]. The rumpled effect should cause the formation of aligned dipoles on the surface and influence the property of AgX, providing an interesting point of view, as described in Section 10.5.

By applying the molecular orbital method to the alternate-row model of a {111} face and {100} face of AgBr, Hoffmann et al. proposed the idea that two adjacent silver ions on the surface are likely to be subjected to the Jahn–Teller effect by bringing about pairwise distortion in order to stabilize a photoelectron, as illustrated in Figure 2.3 [19–21], giving a new point of view for the trapping of photoelectrons at the surface of AgX grains, as described in Chapter 3.

The atomic force microscope has been used to observe bromide ions on the surface of AgBr [22–29], and has revealed various structures including steps on the surface, although

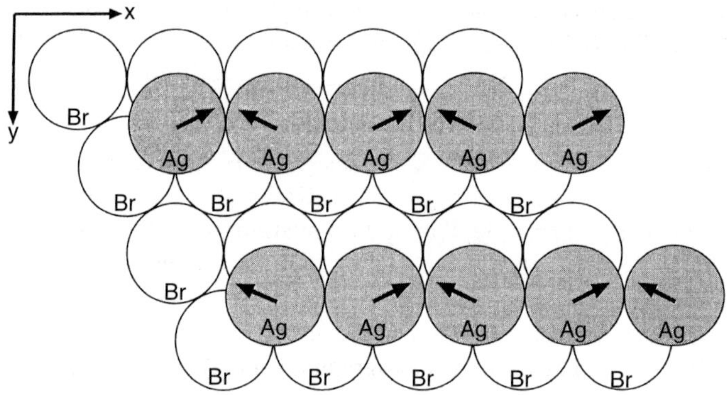

Figure 2.3 The Jahn–Teller effect with the pairwise distortion of silver ions on a {111} face of AgX as theoretically proposed by Hoffman et al. Reproduced with permission from reference [19], copyright 1998 IS&T, The Society for Imaging Science and Technology.

its resolution was not high enough to give any definite knowledge on the concentration of kink sites on the surfaces, which was most influential on the properties of AgX grains, as described in Chapter 3.

Silver iodide is different in crystal structure from AgCl and AgBr, and exists in three crystal modifications [30,31]. Above about 147°C, AgI forms a body-centered cubic phase, designated as α-AgI. Below 147°C, AgI forms a hexagonal Wurtzite-type structure designated as β-AgI and a cubic sphalerite-type structure designated as γ-AgI. Since AgI has never been used for photographic films, studies on AgI emulsion have scarcely been carried out [32]. For photothermography [33,34], a proposal to use AgI emulsions by taking advantage of these properties, namely by fixing AgI grains during thermal development,

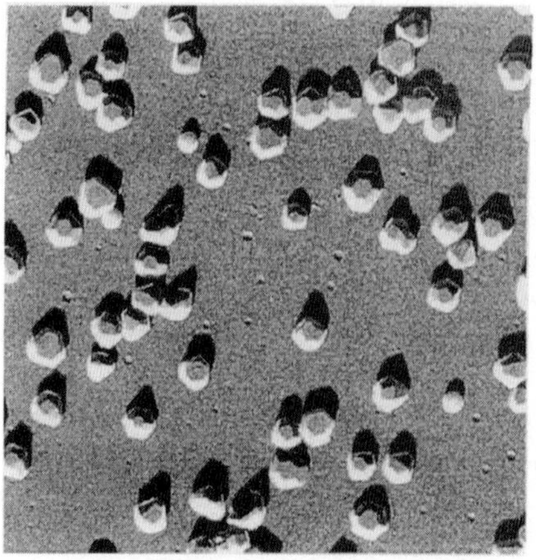

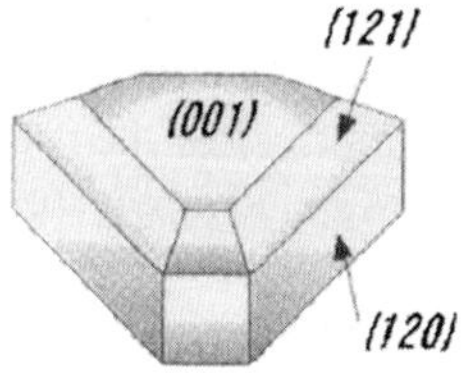

Figure 2.4 An electron micrograph of AgI grains with an average diameter of 0.14 μm (left) and a drawing of the structure of a typical AgI grain (right). Reproduced with permission from Reference [37], copyright 2007 IS&T, The Society for Imaging Science and Technology.

was recently made. [35,36]. Thus, nanoparticles of AgI, composed of mixtures of its β and γ phases as shown in Figure 2.4, were prepared for photothermographic films [35,37].

2.1.2 Structure of Irregular Silver Halide Grains

There are a wide variety of irregular AgX grains. Twin planes are frequently observed in AgX emulsion grains, resulting in various shapes. Klein, Metz, and Moisar classified those twinned grains by analyzing the number and orientation of the twin planes in each grain [38,39]. Among those grains, twinned tabular grains are popular and most important for AgX photographic films. Although studied for many years [40,41], the preparation and sensitization of twinned tabular AgX grains have improved significantly since the early 1980s due to the advantages of their photographic capabilities over that of other grains [42–45]. The main composition of twinned tabular grains should be AgBr because most twinned tabular grain surfaces are composed of {111} faces, which are stable on the surfaces of AgBr grains, but not on the surfaces of AgCl grains.

Figure 2.5 shows an electron micrograph of twinned tabular AgBr grains and an illustration of their structure. As seen in this figure, a hexagonal tabular AgX grain has at least two twin planes on {111} lattice planes in parallel to its main surfaces and uneven side surfaces, each of which is divided into three parts by two lines of intersections between the corresponding side surface and two twin planes. Figure 2.6 illustrates the arrangement of {111} planes around a twin plane, indicating that a twin plane is formed as a result of a stacking fault in the above-stated arrangement. Studies were made to prove the existence of parallel twin planes in tabular AgBr by X-ray diffraction analysis [40], to characterize the parallel twin planes [46–48], and to directly observe the twin planes on an electron micrograph of a cross section of a tabular AgBr grain [49–52].

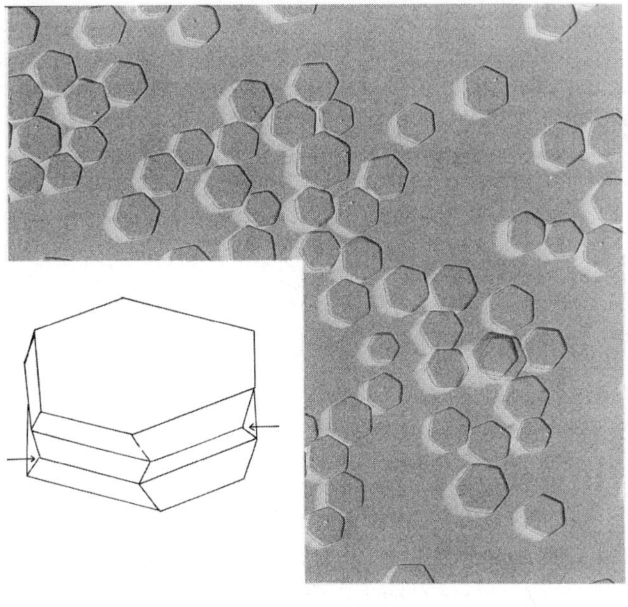

Figure 2.5 An electron micrograph of tabular AgBr grains (courtesy of M. Saito [2]) and an illustration of the structure of a tabular grain.

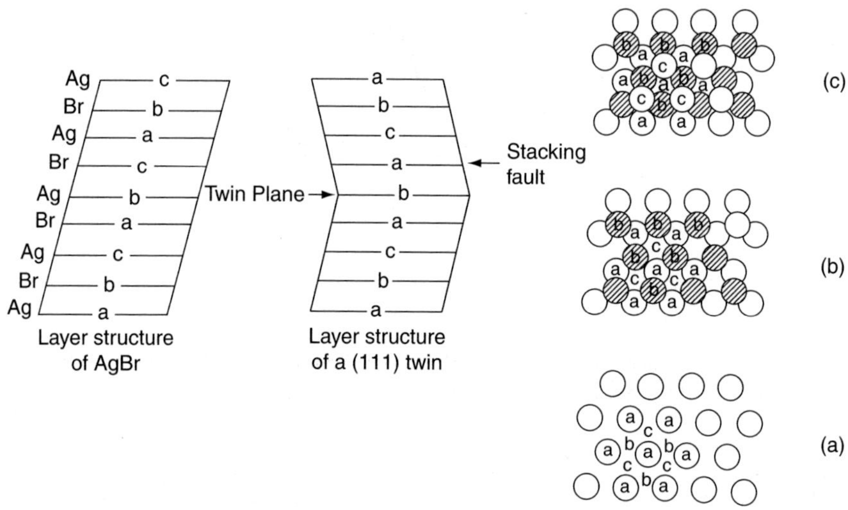

Figure 2.6 The structure of twin planes in AgBr.

Although it was believed for many years that all the side surfaces of twinned tabular AgX grains were composed of {111} faces, as shown in Figure 2.7(a), Jagannathan et al. proposed, based on their ball model simulation, that the central part of the side surfaces of a twinned tabular AgX grain is composed of a {100} face, as shown in Figure 2.7(b), which remains during growth of the grain, and that the large growth rate of {100} faces as compared to that of {111} faces under high pAg conditions is the reason for the anisotropic growth of twinned tabular AgX grains [53–55]. They pointed out that the growth of a new layer initiated on the trough could not extend over the ridge in Figure 2.7(a), since the convex angle at the ridge was much larger than 180°. The above-stated method for evaluating P_{100} of AgX grains [12] was successfully used to prove the existence of {100} planes on the side surfaces of twinned tabular AgBr grains [14].

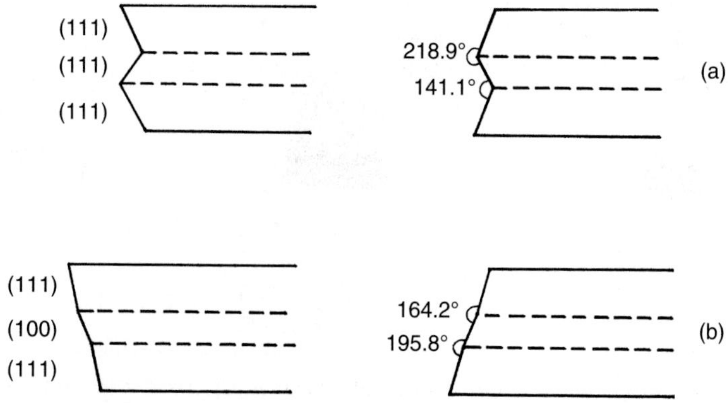

Figure 2.7 The ridge-trough (a) and rough-smooth (b) structures for the side surfaces of a twinned tabular AgX grain [53].

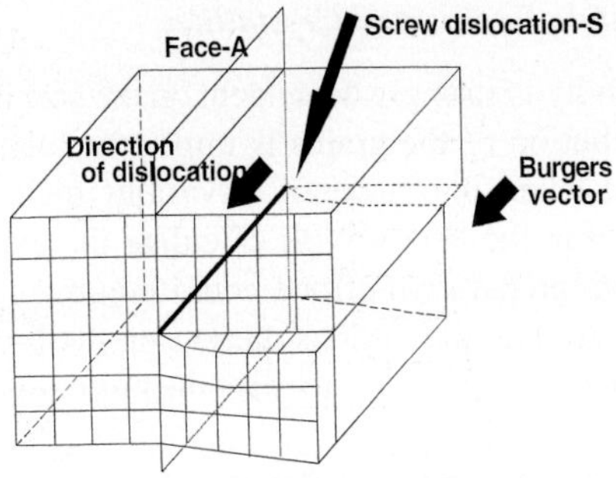

Figure 2.8 A screw dislocation formed in AgX crystal. Reproduced with permission from Reference [59], copyright 1998 IS&T, The Society for Imaging Science and technology.

In addition to twinned AgBr grains, various tabular AgX grains were developed. Twinned AgCl grains were prepared in the presence of crystal habit modifiers, which made {111} faces of AgCl grains stable [3–6]. Tabular AgBr grains, which had an {100} face on their surface and did not contain any twin plane, were prepared by Mignot et al. [56]. Tabular AgCl grains with {100} faces on the surface were also developed [57–59]. It was revealed that a small amount of AgX nuclei with varying composition brings about the formation of screw dislocations, as illustrated in Figure 2.8. Since screw dislocations enhance crystal growth, the screw dislocations thus introduced initiated the anisotropic growth of the grains. As shown in Figure 2.9, it was observed by use of TEM that two lines of screw dislocations extended from one of the corners of each tabular AgCl grain with {100} faces on the surfaces [59]. Namely, two screw dislocations, formed at a corner of a cubic nucleus enhanced the crystal growth in different directions, causing the formation of tabular rectangular grains.

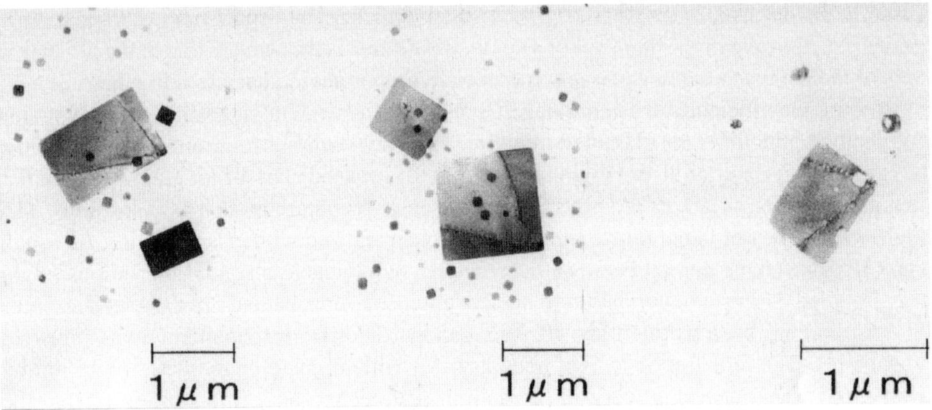

Figure 2.9 Transmission electron micrographs of tabular AgCl grains with {100} faces on their surfaces, in which two screw dislocation lines extend from one of the corners of each grain. Reproduced with permission from Reference [59], copyright 1998 IS&T, The Society for Imaging Science and Technology.

2.1.3 Sizes of Silver Halide Grains and Crystallites

Since photographic sensitivity is strongly dependent on the size of AgX grains, knowledge of the size and size distribution of the grains is important for photographic science and technology, and has been taken into account in various methods of measurement and treatment. While determining the best way to calculate the average size of AgX grains from their size distribution, preparation efforts could minimize differences in the results among the ways adopted. Among the various measuring methods, descriptions are made below on those often applied to AgX grains to meet the prerequisites of photographic films and papers.

In the case of regular grains, the edge length of a grain is measured by electron microscopy and used as its size. However, various irregular grains including tabular ones, as well as regular grains, are important for photographic films. In addition, it is difficult to determine the volume of regular grains by electron microscopy, since their corners are usually rounded. Thus, grain volume is usually accepted for the various AgX grains used for photographic films. As a result, for most AgX grains including regular ones, the diameter of a sphere, whose volume is assumed to be the average volume (i.e., equivalent diameter), is used as the average grain size in various methods.

One of the methods for measuring grain volume has been Coulter Counter, which utilizes the increase in resistance that occurs when grains suspended in an electrolyte pass through a small aperture with a simultaneous flow of electric current. The resulting voltage pulses are proportional in magnitude to the volume of the grains causing it [60]. Another method determines AgX grain volume by direct electrolytic reduction of AgX grains by a micro-electrode, following the original concept of Moller [61], and refined by Holland et al. [62,63]. This method has been able to precisely measure the volume of AgX grains and thus evaluate the efficiency of latent image formation on the grains [64].

In addition to these methods, electron microscopy has been frequently used to evaluate the volume of AgX grains. In particular, the volume of tabular AgX grains has been evaluated by measuring their projective area and thickness, which was determined by the length of its shade in the electron micrograph of their carbon replicas. The aspect ratio of tabular AgX grains (i.e., the ratio of the diameter of a circle, whose area is the same as that of the projective area of the grains, to the thickness) was especially important in characterizing them.

Because the efficiency of latent image formation is dependent on grain size, AgX grains for photographic films are classified into three categories: class I for grain sizes larger than 0.5 μm, class II for grain sizes ranging from 0.5 to 0.1 μm, and class III for grain sizes smaller than 0.1 μm. Class I grains are used for highly sensitive materials such as color films. The highest efficiency of latent image formation has been realized with class II grains. Although class III grains have not yet been well characterized, they have potential for various future use. Detailed description in relation to nuclear emulsion films is given in Chapter 7.

Attention has been paid to class III AgX nanoparticles from fundamental and practical viewpoints. The preparation of AgI nanoparticles with an average diameter of 13 nm by Berry [65] was pioneering work in the field of nanotechnology. Berry observed that the absorption edge of the exciton band of AgI was blue-shifted with decreasing grain size. AgX nanoparticles prepared for basic research have been $\leq$ 10 nm in diameter [66–71]. While the grains produced for some photographic products are around 40 nm [35,37], these grains belong to class III, and were so small that their size and shape were not clear in their electron micrograph.

Analysis of X-ray diffraction profiles of AgX emulsion grains in the light of Scherrer's equation, which make it possible to determine the average size of crystallites with an upper limit of 0.15 μm [72–78], has been improved to be applicable to AgX crystallites with an upper limit of 0.4 μm [79], and to provide a method for measuring the size of small AgX grains belonging to class III on the basis of the following examination. For cubic AgBr grains, in which dislocations were not intentionally introduced, the crystallite size was the same as the grain size. On the other hand, the former is smaller than the latter for the grains, in which crystal defects were intentionally introduced [79]. Namely, it was confirmed by this method that many regular AgX grains in emulsions were crystallographically perfect, whereas AgX emulsion grains, which contained crystal defects, were divided into small crystallites, forming mosaic structure.

2.1.4 Distribution of Halide Ions in Silver Halide Grains

The design of the distribution of halide ions in grains with mixed crystals of silver halides is important for achieving high photographic capability, and depends on the solubility between different silver halides. It is known that AgCl and AgBr can dissolve in each other with any percentage, whereas the solubilities of AgI in AgCl and AgBr are limited [80]. A grain composed of different silver halides usually incorporate various kinds of crystal defects. In particular, grains with hetero-junction of different AgX compositions have been studied extensively, since the interfaces of the hetero-junction initiate the formation of defect sites suitable for latent image formation. One of them was the so-called double-structured grain [81]. Octahedral and tabular double-structured grains with ArBrI in its core and AgBr in its shell have been useful for highly sensitive photographic films, especially for color films. Originally, the high sensitivity of double-structured grains was attributed to the facts that the incident light was absorbed more efficiently in the core than in the shell, while the electrons created in the core could migrate and contribute to the latent image formation on the shell of the grains. It turned out later that the dislocations created on the interface between the core and shell of the grain enhanced the latent image formation on the surfaces of the grains [82]. By using the enhanced lateral growth of a thin tabular AgBr-rich grain, iodide ions can be deposited exclusively on the side surfaces and thus incorporated along its edge portions during its growth. As shown in Figure 2.10, iodide ions could be incorporated into the portion near the edges of tabular AgBr grains [82,83]. As a result, edge dislocations, as illustrated in Figure 2.11, form along the edge, and can make the tabular grains highly sensitive by acting as shallow electron traps and suitable for emulsions in color films [84].

Another type of halide distribution in AgX grains provides epitaxial grains [85–88]. Maskusky studied epitaxial grains composed of AgCl and AgI [85]. Sugimoto and Miyake studied epitaxial grains composed of AgCl and AgBr [86]. Figure 2.12 shows an electron micrograph of epitaxial grains composed of cubic AgCl crystals as their hosts and AgBr crystals at their corners. Similarly, epitaxial grains with cubic AgCl crystals as their hosts and small AgI crystals at their corners could be formed. These grains can provide a model for emulsion grains in color papers. Other typical epitaxial grains are composed of tabular AgBr or AgBrI crystals as hosts and various insoluble silver salts at the corners

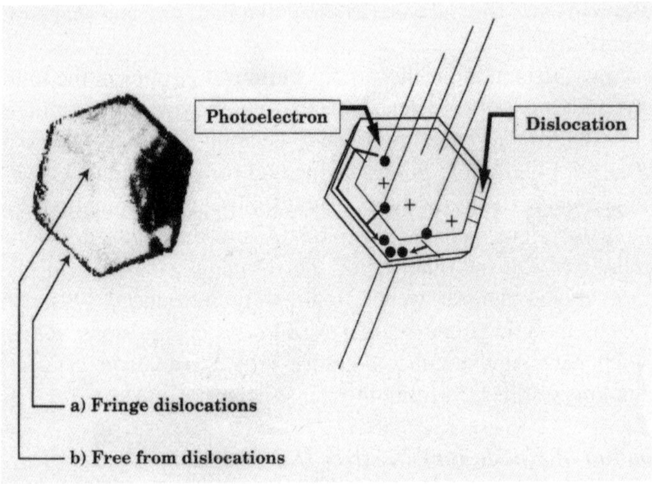

Figure 2.10 (Left) Transmission electron micrograph of a thin tabular grain of AgBrI with rich fringe dislocations, which were introduced along the edge of the grain by the incorporation of iodide ions. (Right) Illustration showing the function of fringe dislocations, which localize photoelectrons along the edge of the grain by acting as shallow electron traps.

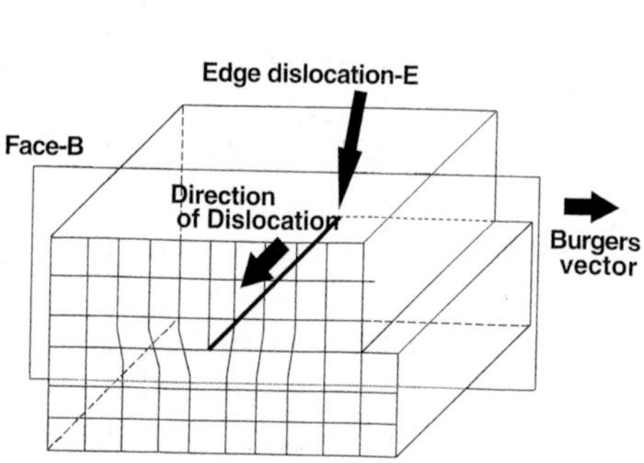

Figure 2.11 An edge dislocation formed in AgX crystal. Reproduced with permission from Reference [59], copyright 1998 IS&T, The Society for Imaging Science and Technology.

[87]. It was demonstrated that crystals of various insoluble silver salts including AgCl, AgBr, AgIBr, and AgSCN were epitaxially formed at the corners of tabular AgX grains in the presence of appropriate site directors, which retarded the growth of (111) faces of the tabular grains relative to the growth of the faces at the corners [87]. These grains can be sensitive enough for highly sensitive layers in color films. As one type of elaborately prepared epitaxial AgX grains, tabular silver iodobromide crystals were deposited on an AgI seed grain having four truncated hexagonal bipyramids, and were then forced to grow to form a hollow AgX grain [88].

Knowledge of the distribution of halide ions in AgX grains has therefore been important and obtained by use of various experimental techniques including X-ray photoelectron spectroscopy (XPS) [89], ion scattering spectroscopy (ISS) [90,91], secondary ion mass

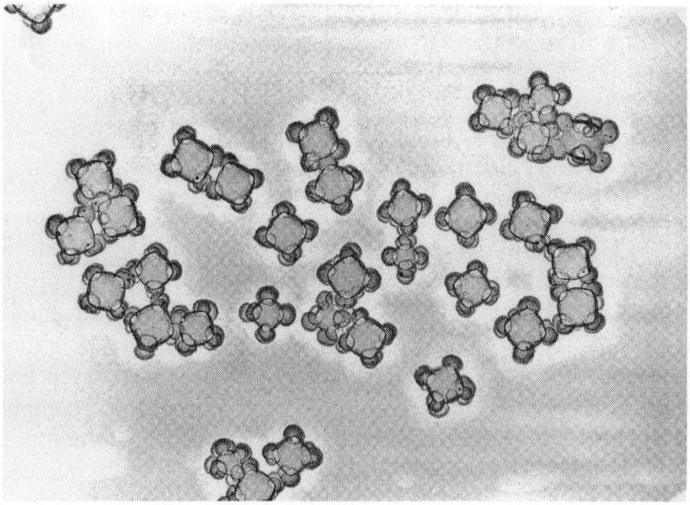

Figure 2.12 An electron micrograph of epitaxial grains, each of which is composed of a cubic AgCl grain as a host and eight AgBr(Cl) grains as guests at the corners (courtesy of T. Shiozawa).

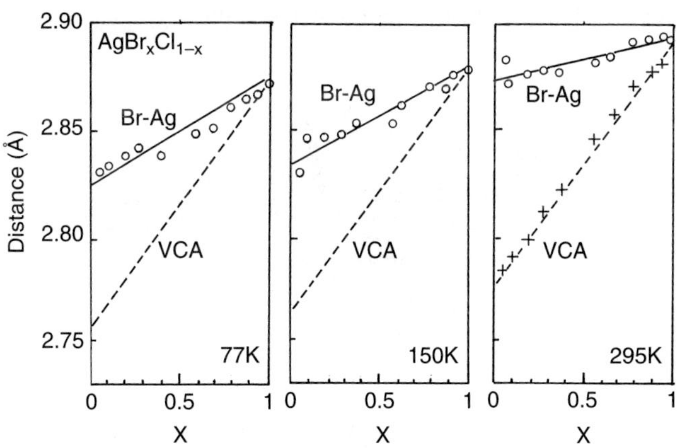

Figure 2.13 Temperature dependence of inter-atomic distances of Br–Ag and Cl–Ag in AgBrCl as determined by the measurement of EXAFS of AgBrCl emulsion grains after annealing [101]. The broken straight lines indicate the average X(Br, Cl)–Ag distance predicted with Vegard's law.

spectroscopy (SIMS) [92, [93], backscattered electron imaging [94,95], and energy dispersive spectroscopy [95] in addition to X-ray diffraction analysis.

The average distance between a silver ion and its nearest neighbor halide ion in pure AgX has been determined by measuring its X-ray diffraction. In the case of grains composed of a mixture of different silver halides, the average distance between a silver ion and its nearest neighbor halide ion changes linearly with its halide composition [96–99], forming Vegard's law. However, extended X-ray absorption fine structure (EXAFS) measurements have revealed that the real distance between a silver ion and its nearest neighbor halide ion in a mixed crystal does not obey Vegard's law [100,101]. As shown in Figure 2.13 for the case of AgBrCl emulsion grains [100,101], the distance between a bromide ion and its nearest neighbor silver ion is larger than the average distance between a silver ion and its nearest neighbor halide ion, and the discrepancy decreases with increasing temperature.

2.1.5 *Defects in Silver Halide Grains*

Various kinds of crystal defects are introduced into AgX emulsion grains and play important roles in photographic processes. Those defects include twin planes as stated and illustrated in Figures 2.5–2.7, edge and screw dislocations, as illustrated in Figures 2.11 and 2.8, respectively, and stacking faults. Mitchell and co-workers observed dislocations for the first time when using AgX crystals [102,103]. The observations were made with crystals of AgBr, and the dislocations were made visible by decorating them with photolytic silver clusters. Farnell and co-workers observed dislocation lines in AgBrI emulsion grains by decorating them with photolytic silver [104]. Hamilton [105] and Shiozawa [106] observed edge dislocation lines in transmission electron micrographs of twinned tabular AgBr grains, and Oyamada et al. observed screw dislocation lines in {100} tabular AgCl grains, as shown in Figure 2.9 [59]. Figure 2.10 is a transmission electron micrograph showing the absence of any dislocation in an AgBrI tabular grain except its edge region and the generation of dislocations along with the edge of the grain owing to the incorporation of iodide ions [83]. Mitchell reviewed the formation and observation of dislocations in silver halides [107], indicating that only the mismatch in lattice constant between silver halides with different halide composition could introduce stable dislocations into silver halides.

2.2 Preparation of Silver Halide Grains

An AgX photographic emulsion is prepared as a result of the double decomposition reaction between water-soluble silver salt and water-soluble halide salt to form insoluble AgX grains in an aqueous gelatin solution. According to Berry and Skillman [108], the four basic processes of AgX precipitation are nucleation, growth, Ostwald ripening, and recrystallization. This section is devoted to the description of the preparation of AgX grains with emphasis on the above-stated processes and on the practical conditions for preparing AgX grains for photographic films.

2.2.1 *Reaction for Formation of Silver Halide Grains*

In an aqueous solution, which is saturated with AgX salt, the activities of silver ions $[Ag^+]$ and halide ions $[X^-]$ are related to each other by the activity solubility product K_{sp}:

$$[Ag^+][X^-] = K_{sp}, \tag{2.1}$$

where X^- is either Cl^-, Br^-, or I^-. The value of K_{sp} of an AgX is independent of the nature and the concentration of the other electrolytes present, and is approximately obtained by considering the concentration of the ions, instead of their activities, when the ionic strength of the medium is low. The solubility product is frequently employed in this approximate form. When it is defined that $-\log[Ag^+] = pAg$, $-\log[X^-] = pX$, and $-\log K_{sp} = pK_{sp}$,

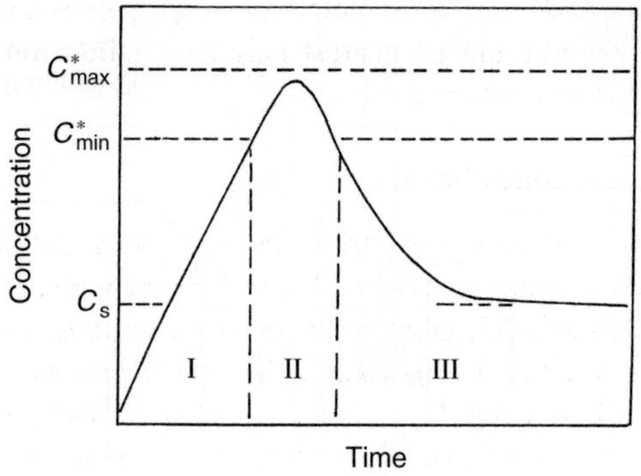

Figure 2.14 Nucleation and growth of AgX grains during precipitation, where C_s is solubility, C_{min}*, and C_{max}* are the minimum and maximum concentration for nucleation, respectively. The regions I, II, and III indicate the period before nucleation, nucleation and growth periods, respectively [109].

$$pAg + pX = pK_{sp}. \tag{2.2}$$

The value of K_{sp} could be the measure for the solubility of AgX.

Figure 2.14 illustrates a typical pattern of the precipitation with concentration of reactants as a function of time (i.e., Lamer diagram) [109]. In the case of AgX, the product of the concentrations of silver and halide ions can be applied to the ordinate of Figure 2.14. There are three important concentrations of reactants, i.e., the concentrations corresponding to solubility (C_s), critical supersaturation ($C_{min}{}^*$), above which the rapid spontaneous nucleation of AgX takes place, and critical limiting supersaturation ($C_{max}{}^*$), respectively.

While the concentration increases with time after adding silver and halide ions by a double-jet precipitation, the spontaneous nucleation of AgX begins to take place rapidly when the concentration exceeds $C_{min}{}^*$, and then shortly ceases, because the concentration begins to decrease and become lower than $C_{min}{}^*$. In order to control the uniformity of the size of emulsion grains, it is important to keep the concentration of the reactants lower than $C_{min}{}^*$ after the initial nucleation in order to prevent further nucleation. Thus, all the reactants added after the initial nucleation are used to grow the nuclei formed in the initial nucleation stage. The growth rate at $C_{min}{}^*$ is called the critical growth rate.

2.2.2 *Apparatuses and Methods for Preparation of Silver Halide Grains*

Typical methods for forming AgX photographic emulsions are single- and double-jet precipitations. In single-jet precipitations, an aqueous solution of water-soluble silver salt is usually added to an agitated aqueous solution of gelatin and water-soluble halide salt(s) in a reaction vessel. In double-jet precipitation, aqueous solutions of a water-soluble silver salt and a water-soluble halide salt(s) are usually simultaneously added to an agitated aqueous gelatin solution in a reaction vessel. A double-jet precipitation in which the concentration of free silver ions in the reaction vessel (C_{Ag}) is controlled is called 'controlled double-jet precipitation', and is useful for controlling the size and shape

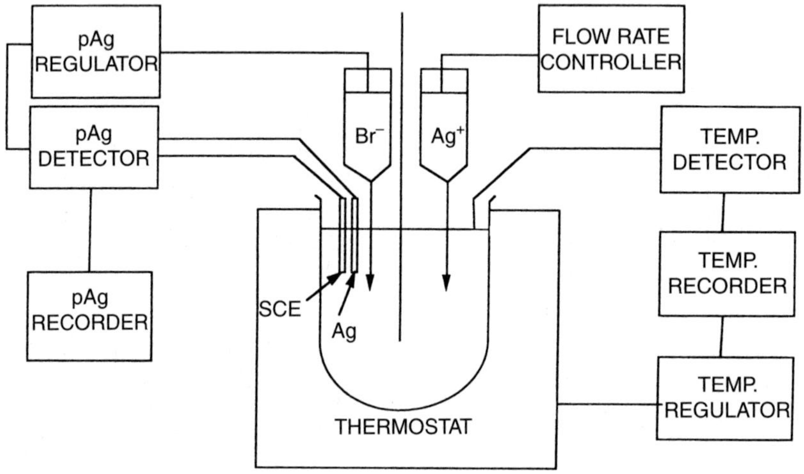

Figure 2.15 A block diagram of an apparatus for controlled double-jet precipitation of AgX grains in photographic emulsions.

of emulsion grains to form mono-dispersed AgX grains [110–114]. The mono-dispersed grains were of great importance for understanding the physical properties and photographic behavior of AgX grains and for the development of new emulsion technologies.

Figure 2.15 illustrates an apparatus for controlled double-jet precipitation, in which the temperature and C_{Ag} in a reaction solution and the addition rate of aqueous solutions of reactants can be controlled in the preparation of desired AgX grains. In the case of the precipitation of AgBr grains, the addition rate of an aqueous $AgNO_3$ solution determines the growth rate of the grains. The value of C_{Ag} in a reaction solution is measured by a silver electrode by detecting the silver potential, which is in proportion to the logarithm of C_{Ag}, and simultaneously used to control C_{Ag} by adjusting the addition rate of a KBr solution according to the measured value of C_{Ag} and Equation (2.1). It is also possible to control the pH value of a reaction solution by use of a glass electrode for pH measurement.

For the preparation of AgX emulsions, it is ideal for the addition of reactants to be uniform with practical rate by raising their concentration homogeneously within a reaction solution. However, it is not possible to realize the above-stated condition, since aqueous solutions of reactants with high concentration (e.g., 1 N) are added to a reaction solution for a short time (e.g., 1 hr) to prepare AgX emulsions for the production of photographic films and papers. In a reaction vessel, adding aqueous solutions of silver ions and halide ions in high concentrations at separate positions creates portions where the concentrations of silver ions and halide ions are deviated from those in other portions for a relatively long time before they react with each other to form AgX grains.

It has been therefore important to develop appropriate agitators that could enhance the homogeneous mixing and reaction of reactants. Some examples of proposed agitators are illustrated in Figures 2.16, 2.17, and 2.18 [115]. Figure 2.16 illustrates an apparatus designed to instantaneously dilute the added highly concentrated solutions of silver ions and halide ions in a reaction vessel prior to their reaction. Aqueous solutions of silver ions and halide ions are introduced into vertically separated chambers with strong agitators and

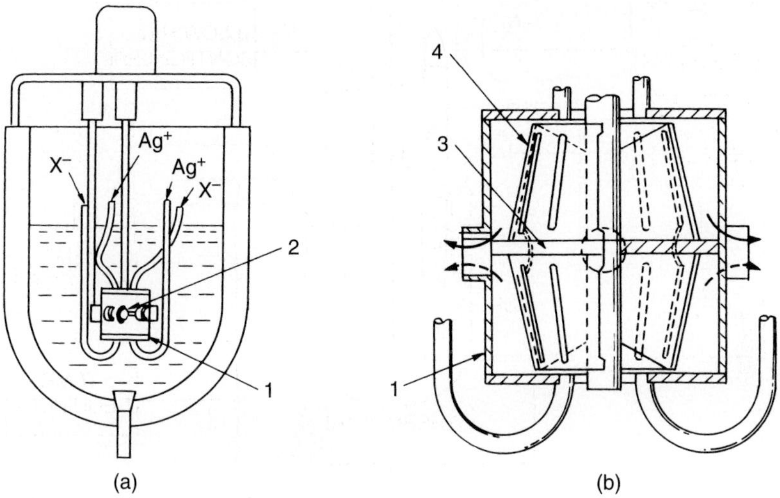

Figure 2.16 A proposed apparatus and a cylindrical casing (1) consisting of two chambers, which have openings (2) and are vertically separated by a wall (3). Cylindrical rotators (4) are installed in the chambers as strong stirrers, Aqueous solutions of silver ions and halide ions in high concentrations are introduced into the different chambers in a casing filled with a reaction solution, and are gushed out of the casing through the openings into the reaction vessel. Reproduced with permission from Reference [115], copyright 1992 H. Saitou.

a reaction solution in a casing, and are then gushed out of the chambers into the reaction vessel. The apparatus illustrated in Figure 2.17 has been proposed for solutions of silver ions and halide ions to be reacted with each other instantaneously with a static stirrer, and guided into a reaction vessel. Figure 2.18 illustrates an apparatus designed for adding solutions of silver ions and halide ions to instantaneously react with each other. Solutions of silver ions and halide ions in high concentrations are added to the positions, which are close to each other, guided with a reaction solution into a casing with a strong agitator, and gushed out of the casing into a reaction vessel.

Instead of the addition of aqueous solutions of reactants to a reaction solution, an aqueous gelatin suspension of AgX nanoparticles was added to a reaction solution. Since the solubility of AgX nanoparticles was much larger than that of ordinary AgX grains, the nanoparticles dissolved to provide reactants for the growth of the grains owing to Ostwald ripening [116]. This method is particularly effective for homogeneously adding reactants to a reaction solution and has been useful for growth with a homogeneous mixture of different halide composition and for grain growth under very low supersaturation conditions.

2.2.3 Nucleation for Silver Halide Grains

Although the nucleation process for the precipitation of AgX grains is very important, it has hardly been characterized. Experiments to examine the process of nucleation have been carried out by increasing the concentration of reactants at very slow rates,

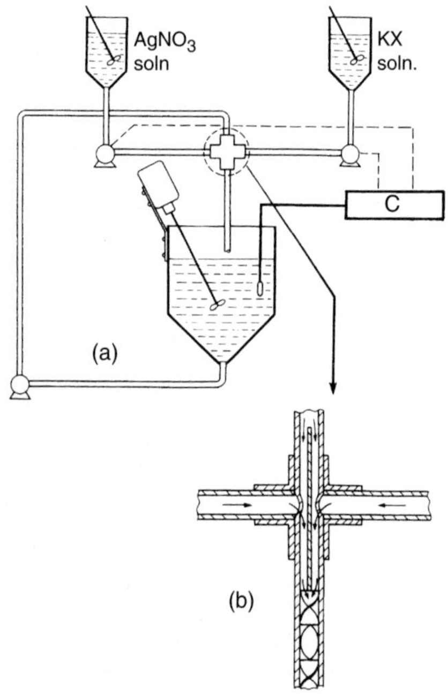

Figure 2.17 A proposed apparatus (a) where aqueous solutions of silver ions and halide ions are guided into a casing (b) where they are forced to instantaneously react with each other via a static stirrer and then gushed into a reaction vessel. Reproduced with permission from Reference [115], copyright 1992 H. Saitou.

and the addition rate of reactants in an emulsion precipitation is much larger (e.g., by a million times), resulting in the formation of much larger number of crystals per unit volume of a reaction solution, than that in the above-stated experiments [108, 118]. Random fluctuations in the concentration of reactants, leading to sufficient supersaturation, have caused nucleation to occur, whereby homogeneous nucleation has taken place in the presence of a larger number of crystals per unit volume of a reaction solution than when heterogeneous nucleation has occurred [119]. Berry and Skillman studied the nucleation process of AgX grains under typical conditions for emulsion precipitation, and found that the nucleation was homogeneous, and not dependent on the presence of impurities [108,120].

It has been proposed that small and unstable nuclei of AgX are formed first when silver and halide ions come in contact, and stable nuclei then grow at the expense of the unstable nuclei [120,12]. Several investigators have reported the C_{min}/C_S above which the spontaneous nucleation of AgX takes place [122–126]. These values range from 1.075 to 1.67, depending upon the experimental conditions. The size of the critical stable nuclei, which were reported by several investigators [127–129], ranged from 14 to 60 nm. Analyses have also been made on the number of stable nuclei formed in the nucleation process [130–135], which determined the size of the grains thus formed.

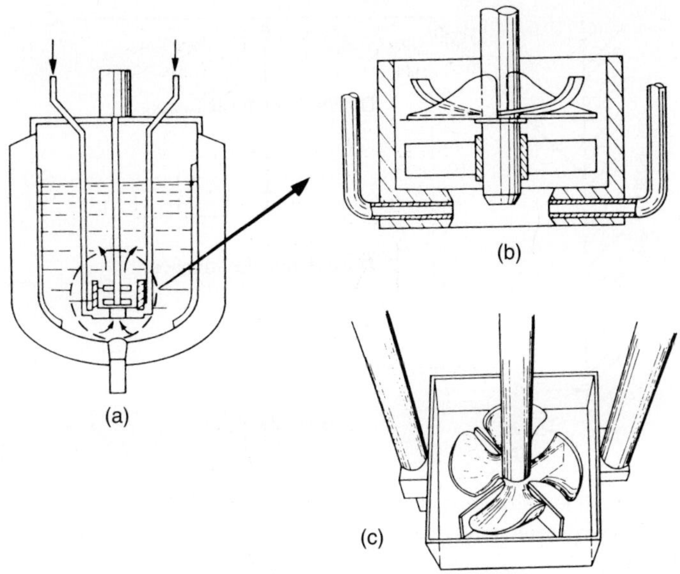

Figure 2.18 A proposed apparatus (a) and a casing (b and c) with a strong stirrer. Because of the stirrer, a reaction solution flows from an opening at the bottom to an opening at the top of the casing. Aqueous solution of silver ions and halide ions are guided by tubes, and added to the positions, which are close to each other and in front of the opening at the bottom of a casing. Reproduced with permission from Reference [115], copyright 1992 H. Saitou.

Several attempts have been made to obtain information on the unstable nuclei formed during the precipitation process [136–138].

2.2.4 Diffusion- and Reaction-Controlled Grain Growth of Silver Halide

Growth is the addition of new layers to grains that are already present. When the number of AgX grains quickly reaches a constant value and further addition of reactants causes only growth of the grains already present, it is possible to determine their growth rate based on the shape and size of the grains and to study the rate-limiting process for growth. Under typical conditions for emulsion precipitation, AgX grains have been grown according to a diffusion- or reaction-limited scheme, depending on the experimental conditions employed. While fruitful knowledge has been accumulated for many years [118], descriptions in this section on the growth of AgX grains are made useful for practical emulsion precipitation.

Figure 2.19 illustrates the distribution of the concentration of reactants near the surface of growing grains, where C and C_i are the concentrations in the bulk and at the grain surface, respectively [139]. The difference between C and C_i is the driving force for diffusion, and the difference between C_i and C_s is the driving force for reaction. Typical growth processes are diffusion- and reaction-controlled.

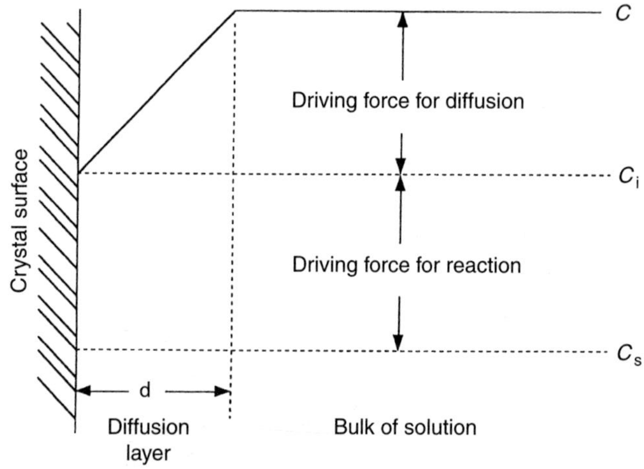

Figure 2.19 Concentration distribution of reactant in the vicinity of the surface of a growing crystal, where C, C_i, and C_S are concentrations of reactants in the bulk and at the surface, and solubility of the crystal, respectively.

The diffusion-controlled grain growth is important for preparing AgX grains with uniform size distribution (i.e., mono-dispersed grains). By applying Fick's diffusion law to a growing spherical grain (Figure 2.19), the one-dimensional growth rate dr/dt is given as [123, 125, 118]

$$\mathrm{d}r/\mathrm{d}t = DV_{\mathrm{m}}(1/r + 1/\delta)(C - C_{\mathrm{i}}), \tag{2.3}$$

where D is the diffusion constant of reactants, V_m is the volume of 1 mole of AgBr, and δ is the width of the diffusion layer. Equation (2.3) shows that the growth rate increases with decreasing r and δ. The agitation of a reaction solution and the overlapping of the diffusion layer of the grains with each other bring about the decrease in δ, thus increasing dr/dt. When r/δ is neglected, it is regarded that dr/dt is proportional to reciprocal of r. Namely, dr/dt of a small grain is larger than dr/dt of a large grain under diffusion-limited grain growth. This tendency was experimentally confirmed by several investigators [125,126, 140,141], as shown in Figure 2.20. During grain growth under a diffusion-controlled condition, a small grain grows faster than a large grain, and the difference in r among initially formed grains decreases, finally bringing about the formation of mono-dispersed emulsion grains. This was experimentally confirmed by Wey and Strong [124]. They mixed two kinds of AgBr emulsion grains with differences in edge length, and observed the change in the difference during growth under different supersaturation conditions. The difference decreased under a high degree of supersaturation, whereas it increased under a low degree of supersaturation.

The following types of reaction-controlled grain growth mechanisms have been proposed [142]: polynuclear layer and mononuclear layer. The rate-determining step of the former is the surface nucleation, which is required many times for each layer to grow, and the rate-determining step of the latter is the surface nucleation, which is followed by instantaneous spreading of the layer. It is believed that dr/dt is proportional to r^2 for

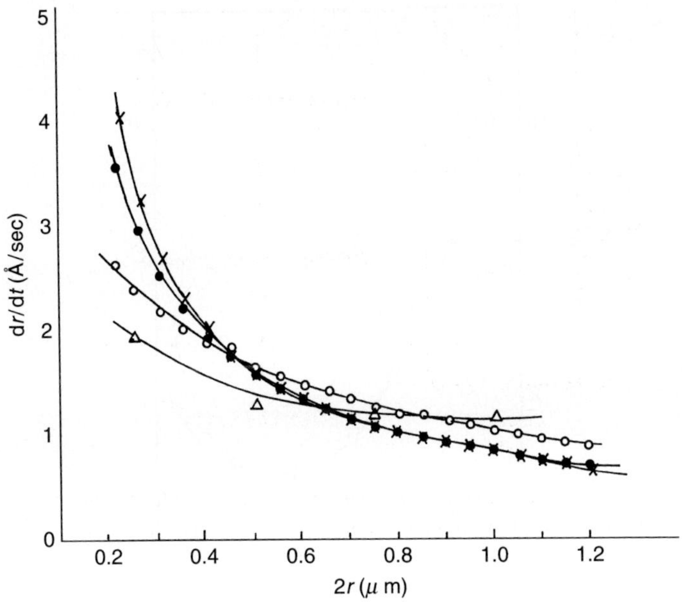

Figure 2.20 Size dependence of critical growth rates of cubic AgBr (●, Wei and Strong [140]; △, Lewis [141]), cubic AgCl (×, Strong and Wei [125]), and octahedral AgBr (○, Saitou [126]) grains.

mononuclear-layer mechanism, and dr/dt is proportional to r^0 for polynuclear-layer mechanism [143]. It is therefore hard to prepare mono-dispersed AgX grains by the reaction-controlled grain growth, since a small grain cannot grow faster than a larger grain, and even slower than that. Thus, the difference in r among initially formed grains cannot decrease during their growth.

Analyses were made on the dependence of the grain growth upon various factors including the crystal habit of AgX, pAg in a reaction solution, solubility of AgX, and population of growing grains [140, 144–146]. Jagannathan and Wey observed the increase in growth rate caused by the increase in the population of growing grains under diffusion-controlled conditions. This phenomenon was ascribed to the decrease in δ in Eq. (2.3) due to the overlap of the diffusion layers of the grains [143, 144]. Figure 2.21 shows the solubility of AgBr and the critical growth rates of {100} and {111} faces as a function of pBr [146]. As seen in this figure, both solubility and critical growth rates are steeply dependent upon pBr, and the larger the solubility is, the higher is the critical growth rate. In the range where the solubility is small, the growth rates of {100} and {111} faces are small, and equal to each other, indicating that the growth of AgBr in this region is diffusion-controlled. In the range where pBr is small and the solubility is high, the critical growth rates of {100} and {111} faces are high and different from each other. It was reported that the growth of cubic grains was diffusion-controlled, whereas that of octahedral ones was reaction-controlled plus diffusion-controlled in this region [145, 140].

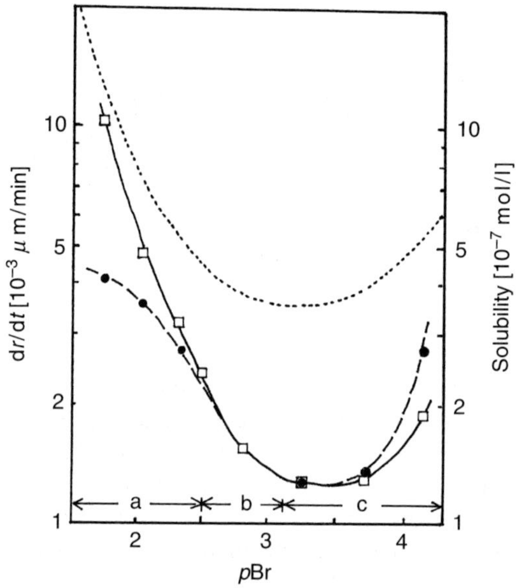

Figure 2.21 Critical growth rates of {100} and {111} faces (a solid curve with □ and a broken curve with ●, respectively) and solubility (a dotted curve) of AgBr as a function of pBr in a reaction vessel [146].

The fact that the growth rate was steeply dependent upon the solubility of AgX could indicate that the reaction rate at the grain surface was high enough for the supply rate of reactants to contribute to the growth rate. It is important to increase the growth rate and the supply rate to form large AgX grains for the development of highly sensitive photographic films, since intrinsic sensitivity is, in principle, proportional to the average volume of AgX grains and the spectral sensitivity of dyed grains is proportional to their average surface area [147,148]. The increase in the supply rate of reactants can be achieved by increasing the addition rate of the water-soluble silver salt and halide salt in concert with an increase in the area of the grain surface during the precipitation [2] and by increasing the solubility of AgX in the presence of some silver-complexing agent [149–152]. Although ammonia has been the most popular among those agents, and has been used for the preparation of large AgX grains for a long time [118], it has forced AgX emulsions to be prepared under alkaline conditions. As a result, many new silver-complexing agents have been developed to prepare large AgX grains under controlled conditions.

2.2.5 Grain Growth for Control of Crystal Habit of Regular Silver Halide Grains

Many discussions have been made on the factors controlling the shape of AgX emulsion grains. In order to control the shape of emulsion grains, it is essentially important to control the concentration of halide ions during the precipitation of AgX grains in the presence of a protective colloid (i.e., gelatin in most cases) in a reaction solution. Figure 2.22 schematically illustrates one of the most convincing models for the effect of pX on the shape of AgX emulsion grains [153,154,10] by showing the structure and growth of

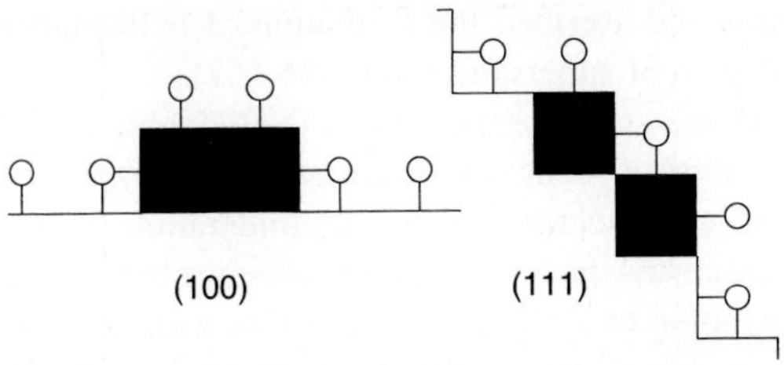

Figure 2.22 Two-dimensional nuclei (shaded areas) on {100} and {111} faces and adsorbed halide ions (○).

{100} and {111} face of AgX with nuclei (shaded squares) and adsorbed halide ions(open circles). As schematically shown in Figure 2.2 as well as in Figure 2.22, the {100} face is regarded as flat, while the {111} face is rough in atomic scale. It is considered that a nucleus on the {100} face is unstable and stabilized by adsorbed halide ions and that the stabilization of the nuclei enhances the growth of the {100} face. On the other hand, nuclei are considered to be able to find a stable site on the {111} face but cannot be stabilized by adsorbed halide ions, since the {111} face is rough. Therefore, the increase in halide ion concentration in a reaction solution enhances the growth of the {100} face more strongly than that of the {111} face.

This model explains why the precipitation of AgBr grains under low pBr conditions results in octahedral grains with a {111} face on their surfaces, the precipitation under high pBr conditions in cubic grains with a {100} face, and the precipitation under intermediate pBr conditions in tetradecahedral (i.e., cuboctahedral) grains with both (100) and (111) faces with variation of P100 depending on pBr in a reaction solution. Thus, mono-dispersed cubic, octahedral, and tetradecahedral AgX emulsion grains were prepared [110–114], and have been refined for many years. As stated earlier, a Br ion can be regarded as crystal habit modifier for AgBr. Various kinds of crystal habit modifiers have been developed, and made available for the preparation of all the possible regular grains of AgBr, as illustrated in Figure 2.1 [1].

2.2.6 Anisotropic Growth of Tabular Silver Halide Grains

As stated earlier, twinned tabular AgX grains have become most important for highly sensitive emulsions in color films. Using the X-ray diffraction method, Berriman and Herz found that twinning took place on {111} faces of AgBr grains as a result of a stacking fault [40]. Since it is commonly observed that twinning takes places at high pAg, Berry et al. have proposed that the complex ion $AgBr_3^{2-}$ fails to occupy the normal site owing to the distance between two bromide ions in it, which is larger than that on the {111} surface of AgBr, and thus brings about the formation of a twin plane [155]. Saito has experimented on the formation of tabular grains, and concluded that twining takes place at random among grains, and can be enhanced by the conditions with a high degree of supersaturation [2], including a high concentration of halide ion, a high addition rate of reactants, low temperature, a high concentration of iodide ions, a low concentration of gelatin, and so on. It is considered that a stacking fault is a beginning, and is easily followed by the formation of a twin plane under conditions such as a high degree of supersaturation. Jagannathan described similar results and ascribed the formation of twin planes to instability of the {111} face at a high degree of supersaturation [156,157].

On the other hand, Mumau et al. proposed that the twinning in AgBr grains takes place as a result of the coagulation of seeds (the coagulation model) [158]. Studying the relation between the concentration of gelatin and the agglomeration of AgBr grains [159–162], Antoniades and Wey supported the coagulation model by observing that the frequencies of the coagulation and twinning of AgBr grains increases with decreasing the concentration of gelatin, which acts as a protective colloid for the grains [161,162]. Urabe et al. also supported the coagulation model on the basis of the model experiments to observe the coagulation of octahedral AgBr grains for twinning [52]. They observed the correspondence between the size of the octahedral grains and the distance between two twin planes in twinned AgBr grains thus formed. The coagulation model was further supported by their observation that the frequency of twinning reached maximum at the pAg value, which made the surfaces of the grains electrically neutral, since it is considered that the frequency of the coagulation should be largest when repulsive interaction among grains due to the presence of electric charges on the surface is absent.

Hamilton et al. proposed that side surfaces of twinned tabular AgX grains are composed of the ridge-trough structure, having reentrant grooves as illustrated in Figure 2.7(a), and that the reentrant grooves bring about the anisotropic growth of twinned grains owing to the fact that the nucleation on the trough takes place much faster than that on the main surfaces of a twinned grain [46]. However, Jagannathan et al. proposed that the side surfaces of twinned tabular grains are composed of the rough-smooth structure with both {111} and {100} faces as shown in Figure 2.7(b), and that the large growth rate of {100} faces over that of {111} faces at low pBr is responsible for the anisotropic growth of twinned tabular AgX grains [53–55].

By analyzing the change in the thickness and projection area of tabular AgBr grains during their growth at nearly critical supersaturation, Hosoya and Tani could obtain their lateral and vertical critical growth rates as a function of the equivalent circular diameter (D_C, the diameter of a circle whose area is equal to the projection area) and pBr of the reaction solution during precipitation (Figure 2.23) [14]. At pBr 3.3, the lateral and vertical critical growth rates were small, equal to each other, and decreased with increasing D_C. The growth of tabular grains at pBr 3.3 was thus in a similar fashion with those of octahedral and cubic grains at the same pBr as seen in Figure 2.20 and were diffusion-controlled. The lateral critical growth rate became much larger than the vertical critical growth rate as the pBr decreased, indicating the peculiar growth behavior of tabular grains. The lateral critical growth rate at pBr 2.63 still decreased with increasing D_C. It is therefore possible to prepare mono-dispersed tabular AgBr grains at the pBr region around this value, although the aspect ratio of the obtained tabular grains is limited.

Although the aspect ratio of tabular AgBr grains formed at pBr 2.0 is much larger than that of the tabular grains formed at 2.63, the lateral critical growth rate at pBr 2.0 steeply increased with increasing D_C. Sugimoto, and Karpinski and Wei studied the rate and mechanism of the growth of tabular AgX emulsion grains under the condition that the degree of supersaturation was reduced to enhance the anisotropic growth of tabular grains [163,164]. The growth rate of tabular grains increased as their size increased, indicating that the dispersity of their grain size increased as the grains grew under that condition.

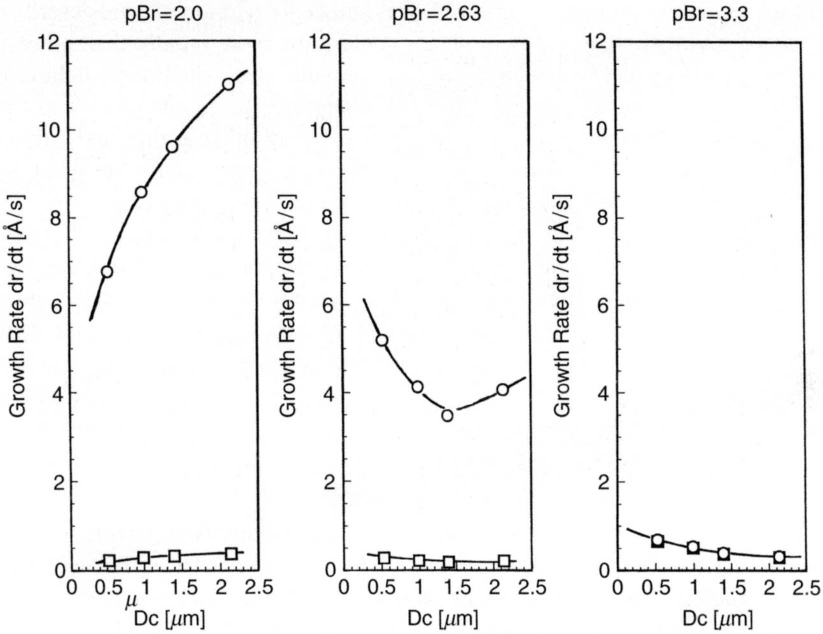

Figure 2.23 Lateral (○) and vertical (□) growth rates of tabular AgBr grains with thickness of 0.1 μm as functions of equivalent circular diameter (D_C; diameter of a circle whose area is equal to the projection area) and pBr of the reaction solution during the growth of the grains. Reproduced with permission from Reference [14], copyright 1996 IS&T, The Society for Imaging science and technology.

These results indicate that it is in general difficult to prepare mono-dispersed tabular AgBr grains with large aspect ratio.

The number of twin planes per grain formed at nucleation could be evaluated by analyzing the shape of AgX grains, which were able to be grown without loss of characteristics by maintaining the number of nuclei, neither decreasing owing to their dissolution nor increasing owing to the formation of new nuclei during the growth process [165,166]. The dependence of the number of twin planes per grain on the probability of twin plane formation was in accordance with Poisson distribution, and indicated the random formation of twin planes on {111} surfaces.

2.2.7 Ostwald Ripening for Selective Formation of Tabular Silver Halide Grains

In Ostwald ripening, a strong tendency exists for the smaller crystals to dissolve and the larger crystals to grow due to the fact that the logarithm of the solubility of spherical particles is proportional to the reciprocal of the size of the particles. Ostwald ripening occurs predominantly at higher temperatures, in the presence of solvents for silver halides, and when there is a wide distribution of grain sizes.

Since twin planes are formed by chance and randomly distributed among grains, it is difficult to introduce only two or more parallel twin planes into each grain at the nucleation stage of the formation of emulsion grains. It was known for a long time and described later that a tabular grain with parallel twin planes could grow more rapidly than other grains, since its side surfaces had such a structure as to provide sites, which were believed to be suitable for incorporation of reactants [46,118]. Therefore, during the process of Ostwald ripening, the pK_{sp} of tabular grains was larger than that of other grains, and only tabular grains could grow with the dissolution of other grains, and finally survived to give emulsions composed of only tabular grains with parallel twin planes [2].

Some grains with more than two twin planes which are not parallel with each other might grow as rapidly as tabular grains. It is therefore important to control the degree of supersaturation at the nucleation stage to avoid the formation of a grain that has more than two twin planes in it. Since twining takes place by chance, depending upon the degree of supersaturation, it is possible, by controlling the degree of supersaturation, to prepare, in an emulsion, a considerable number of AgX grains with two parallel twin planes together with grains containing less than two twin planes along with only a small number of grains containing more than two twin planes. This emulsion could then be subjected to the above-stated Ostwald ripening to give an emulsion composed of almost only tabular grains with paralell twin planes [2].

2.2.8 Recrystallization for Grains with Mixed Crystal of Silver Halides

Recrystallization is the process by which a crystal's composition changes. This occurs when crystals of different composition are present, since the state of the minimum energy is a solid solution of intermediate composition. It has been shown that when a mixture of silver halides is present, solid solutions can form owing to the increase in entropy. Recrystallization has been related to the formation of AgX grains, which were composed of silver halides of different halide composition and developed for practical photographic films and papers. Among these grains, double-structured [81,82] and epitaxial grains [85–88] played important roles.

A double-structured grain, composed of AgBrI with high concentrations of AgI as its core and AgBr as its shell, was designed for highly sensitive materials such as color films. However, it proved difficult to prepare, since the recrystallization took place during the deposition of the shell on the core of the grains. This difficulty could be overcome by increasing the rate of the deposition of the shell in concert with the increase in surface area of the growing grains so that the rate of the deposition of the shell could exceed the rate of the recrystallization without forming new nuclei.

There are two important epitaxial AgX grains. One of them is a model for the AgX emulsion grains of color papers, and is composed of cubic AgCl grains with epitaxially formed AgBr or AgI at their corners. The other is a good candidate for highly sensitive AgX grains in color films, and is composed of tabular AgX grains with epitaxially formed precipitates such as AgCl, AgBr, and AgSCN at the corners. The formation and preservation of epitaxial grains depends on the control of the recrystallization taking place between host and guest crystals in the grains by newly developed technologies. One was the reduction of growth of the host crystal surfaces by placing so-called site director compounds on the crystal surfaces [87].

2.2.9 Control of Size of Silver Halide Grains

The size of AgX grains is essentially important, since it is one of the most influential factors of the sensitivity and image quality of photographic films and papers, and it is influenced by several factors. For example, the size increases by prolonged growth process with increasing degree of super-saturation according to Figure 2.14 and by solubility, which depends on temperature, halide composition, and silver ion concentration in a reaction solution according to Figure 2.21. In addition, chemical species named ripeners such as ammonia and various solvents for AgX are useful for increasing the size of AgX grains [118,167], whereas chemical species known as restrainers, which are strongly adsorbed to AgX grains, are useful for decreasing the size of grains [118,168]. In Section 2.1.3, AgX grains are classified into three groups with respect to their size. The formation of large AgX grains belonging to class I (i.e., ≥ 0.5 μm) was important for the development of highly sensitive photographic films, and was realized by solvents for AgX including ammonia, anisotropic growth of tabular grains, and the controlled double-jet method. Although very small AgX grains, which are nanoparticles of AgX and belong to class III (i.e., ≤ 0.1 μm), are also important as described below and in Chapter 7, they are still considered difficult from various viewpoints.

AgX nanoparticles have attracted the attention of many scientists, since they provide the possibilities for studying the physical properties of quantum-confined AgX crystals [169–171], and for developing new photographic films for nuclear physics [172–174], holography, and photothermography [36]. Various methods for preparing AgX nanoparticles have been used. AgBr nanoparticles can be prepared by using inverse micelles, in which nanoparticles grow when small water pools containing salts collide [175,176]. AgX nanoparticles have been prepared in a non-aqueous solution containing a restraining polymer [177]. AgCl nanoparticles have been found to form in Bridgman grown KCl single crystals heavily doped with ultrapure AgCl powder [178]. AgBr nanoparticles in glass have been formed by adding silver and halide salts to the normal constituents of an alkali boroaluminosilicate glass, which was then annealed to form the AgBr nanoparticles [179].

From the viewpoint of photographic science and technology, it is important to prepare AgX nanoparticles in aqueous gelatin solutions. Berry prepared AgI nanoparticles with an average diameter of 13 nm in an aqueous solution as early as 1967 [60]. Although AgBr nanoparticles prepared in aqueous gelatin solutions were in the 50-nm diameter size range [169,67,180], it was significant that AgBr nanoparticles less than 10 nm in diameter could be prepared in aqueous gelatin solutions [181, 70].

2.2.10 Preparation of Silver Iodide Grains

As stated in Section 2.1.1, attention was recently paid to AgI grains for photothermographic materials [35–37], which could be developed at elevated temperature without any developing solution [33,34]. Among the various phases of AgI, α phase is formed at elevated temperature. Maskasky found that AgI emulsion grains containing α-phase were more sensitive than those with β-phase [32]. As a result of the trial to apply AgI emulsions to photothermography, it was found that AgI grains were fixed during development for photothermography. AgI nanoparticles were prepared by a controlled double-jet method in a gelatin solution with the pAg value kept constant. Tabular AgI emulsion grains were prepared by means of a controlled double-jet method with the pAg value kept constant during the precipitation of the grains in the presence of an AgX solvent (3, 6-dithia-1, 8-octanediol) in an aqueous gelatin solution [36].

2.2.11 Protective Colloid for Silver Halide Grains

In order to control the shape and size of AgX grains, it is inevitable to precipitate them in the presence of some protective colloids, among which gelatin is used in most cases, although extensive efforts have been made for the development of synthetic polymers available for the precipitation of controlled AgX grains [182]. In the absence of a protective colloid, mono-dispersed AgX grains could be prepared only in a extremely diluted solution [183]. The modification and treatment of gelatins for improving the photographic performance of AgX emulsions were carried out for many years [184–186]. After extensive studies on protective colloids for AgX grains it was found that the development of modified gelatin was more promising for the progress of the preparation of AgX emulsion grains than that of synthetic polymers, and especially useful for improving the precipitation of twinned tabular AgX grains for color films. A proposal was made by Maskasky to use oxidized gelatin to improve the structure of tabular AgX grains. The adsorptivity of oxidized gelatin to AgBr grains is known to be smaller than that of ordinary gelatin owing to the oxidation of methionine, which has strong adsorptivity to AgX. This proposal initiated the development of various modified gelatins [187,188].

References

1. J. E. Maskasky, *J. Imaging Sci.*, 30, 247 (1986).
2. T. Tani, *Photographic Sensitivity: Theory and Mechanisms* (New York/Oxford: Oxford University Press, 1995), Chapter 2.
3. K. Endhou, M. Okazi, *J. Photogr. Sci.*, 36, 182 (1988).
4. T. P. Tufano, D. M. T. Chan, *J. Imaging Sci.*, 34, 44 (1990).
5. W. van Renterghen, D. Schryvers, J. van Lenduyt, D. Bollen, C. van Roost, R. de Keyzer, *J. Imaging Sci. Technol.*, 45, 83 (2001).
6. K. Ohzeki, *J. Soc. Photogr. Sci. Technol. Jpn.*, 68, 332 (2005).
7. F. H. Claes, J. Libeer, W. Vanassche, *J. Photogr. Sci.*, 21, 39 (1973).
8. C. R. Berry, *Photogr. Sci. Eng.*, 19, 17 (1975).
9. J. E. Maskasky, *J. Imaging Sci.*, 32, 95 (1986).
10. E. Moisar, E. Klein, *Bunsenges. Phys. Chem.*, 67, 949 (1963).
11. A. H. Herz, R. Danner, G. Janusonis, in *Adsorption from Aqueous Solution* (Washington, DC: American Chemical Society, 1968), 173.
12. T. Tani, *J. Imaging Sci.*, 29, 165 (1985).
13. H. Coll, R. E. Button, *J. Imaging Sci. Technol.*, 43, 85 (1999).
14. Y. Hosoya, T. Tani, *J. Imaging Sci. Technol.*, 40, 202 (1996).
15. J. F. Hamilton, L. E. Brady, *Surface Sci.*, 23, 389 (1970).
16. R. C. Baetzold, Y. T. Tan, P. W. Tasker, *Surface Sci.*, 195, 579 (1988).
17. P. Tangyunyong, T. N. Rhodin, Y. T. Tan, K. J. Lushington, *Surface Sci.*, 255, 259 (1991).
18. Y. T. Tan, K. J. Lushington, P. Tangyunyong, T. N. Rhodin, *J. Imaging Sci. Tech.*, 36, 118 (1992).

19. A. S. Malik, F. J. DiSalvo, R. Hoffman, J. T. Blair, *J. Imaging Sci. Technol.*, 42, 210 (1998).
20. A. S. Malik, J. T. Blair, W. A. Bernett, F. J. DiSalvo, R. Hoffmann, *J. Solid State Chem.*, 146, 516 (2000).
21. P. Gutta, R. Hoffmann, *J. Phys. Chem. A*, 107, 8184 (2003).
22. G. Hegenbart, Th. Muessig, *Surf. Sci. Lett.*, 275, L655 (1992).
23. E. Meyer, H. Heinzelmann, P. Grutter, H. R. Hidber, H. J. Guntherodt, R. Steiger, *J. Appl. Phys.*, 66, 4243 (1987).
24. H. Haefke, E. Meyer, H. J. Guentherodt, G. Gerth, M. Krohn, *J. Imaging Sci.*, 35, 290 (1991).
25. M. P. Keyes, E. C. Phillips, W. L. Gladfelter, *J. Imaging Sci. Technol.*, 36, 268 (1992).
26. U. D. Schwarz, H. Haefke, H. J. Guentherodt, J. Bohonek, R. Steiger, *J. Imaging Sci. Technol.*, 37, 344 (1993).
27. H. Haefke, U. D. Schwarz, H. J. Guntherodt, H. Froeb, G. Gerth, R. Steiger, *J. Imaging Sci. Technol.*, 37, 545 (1993).
28. G. Hegenbart, T. Muessig, *J. Imaging Sci. Technol.*, 37, 551 (1993).
29. M. Plomp, J. G. Buijnsters, G. Boegels, W. J. P. van Enckevort, D. Bollen, *J. Crystal Growth*, 209, 911 (2000).
30. M. D. Donohue, A. H. Herz, *Photogr. Sci. Eng.*, 25, 92 (1981).
31. G. Burley, *J. Phys. Chem.*, 68, 1111 (1964).
32. J. E. Maskasky, *J. Imaging Sci.*, 35, 89 (1991).
33. D. R. Whitcomb, *Photogr. Sci. Photochem.*, 21, 1 (2003).
34. M. R. V. Sahyun, *J. Imaging Sci. Technol.*, 49, 337 (2005).
35. H. Mifune, T. Ohzeki, K. Yamane, F. Nariyuki, K. Watanabe, M. Yoshikane, M. Nakanishi, T. Maekawa, T. Funakubo, *J. Imaging Sci. Technol.*, 51(3), 207–11 (2007).
36. T. Funakubo, H. Mifune, T. Ohzeki, K. Ohzeki, K. Yamane, *J. Imaging Sci. Technol.*, 51(3), 2012–16 (2007).
37. T. Tani, H. Mifune, S. Yamashita, S. Aiba, T. Ohzeki, K. Yamane, *J. Imaging Sci. Tech.*, 51, 202 (2007).
38. E. Klein, H. J. Metz, E. Moisar, *Photogr. Korr.*, 99, 99 (1963).
39. E. Klein, H. J. Metz, E. Moisar, *Photogr. Korr.*, 100, 57 (1964).
40. R. W. Berriman, R. H. Herz, *Nature*, 180, 293 (1957).
41. S. Urabe, *J. Soc. Photogr. Sci. Technol. Jpn.*, 66(2), 168 (2003).
42. Anonymous, *Res. Discl.* (Item 22534) 225, 20 (January 1983).
43. J. T. Kofron, R. E. Booms, *J. Soc. Photogr. Sci. Technol. Jpn.*, 49, 15 (1986).
44. A. F. Sowinski, P. J. Wightman, *J. Imaging Sci.*, 31, 162 (1987).
45. R. E. Jacobson, N. R. Axford, A. Ford, *J. Photogr. Sci.*, 38, 140 (1990).
46. J. F. Hamilton, L. E. Brady, *J. Appl. Phys.*, 35, 414 (1964).
47. J. E. Maskasky, *J. Imaging Sci.*, 31, 15 (1987).
48. J. E. Maskasky, *J. Imaging Sci. Technol.*, 40, 79 (1996).
49. R. V. Mehta, R. Jagannathan, W. K. Lam, D. L. Black, J. A. Timmons, *J. Imaging Sci. Technol.*, 39, 67 (1995).
50. S. Jagannathan, S. Chen, R. V. Mehta, R. Jagannathan, *Phys. Rev. B*, 53, 9 (1996).
51. S. Chen, S. Jagannthan, R. V. Mehta, R. Jagannathan, A. E. Taddei, *J. Imaging Sci. Technol.*, 42, 399 (1998).
52. Y. Hosoya, S. Urabe, *J. Imaging Sci. Technol.*, 42, 487 (1998).
53. R. Jagannathan, R. V. Mehta, J. A. Timmson, D. L. Black, *Phys. Rev. B*, 48(18), 13261 (1993).
54. R. V. Mehta, R. Jagannathan, J. A. Timmons, *J. Imaging Sci. Technol.*, 37, 107 (1993).
55. R. V. Mehta, R. Jagannathan, J. A. Timmons, *J. Imaging Sci. Technol.*, 40, 77 (1996).
56. A. Mignot, E. Francois, M. Catinat, *J. Crystal Growth*, 23, 207 (1974).
57. T. B. Brust, G. L. House, *J. Imaging Sci. Technol.*, 41, 413 (1997).

58. W. van Rentergen, C. Goessens, D. Schryvers, J. van Landuyt, P. Verrept, D. Bollen, Van Roost, R. de Keyzer, *J. Crystal Growth*, 187, 410 (1998).
59. T. Oyamada, T. Shiozawa, S. Yamashita, *J. Imaging Sci. Technol.*, 42, 483 (1998).
60. G. H. Klinger, F. C. Forsgard, M. V. Cwikla, in G. Semerano, U. Mazzucato (eds) *Photographic Science* (Symposium: Torino, 1963) (London/New York: Forcal Press, 1965), 47–8.
61. G. Moller, in *Abasracts of the International Congress on Photographic Science* (Moscow, 1970), 125.
62. A. B. Holland, J. R. Sawers, *Photogr. Sci. Eng.*, 17, 2951 (1973).
63. A. Holland, A. Feinerman, *J. Appl. Photogr. Eng.*, 8, 165 (1982).
64. R. K. Hailstone, N. B. Liebert, M. Levy, R. T. McCleary, S. R. Gilolmo, D. L. Jeanmaire, C. R. Boda, *J. Imaging Sci.*, 32, 113 (1988).
65. C. R. Berry, *Phys. Rev.*, 161, 848 (1967).
66. T. Tanaka, H. Saijo, T. Matsubara, *J. Photogr. Sci.*, 27, 60 (1979).
67. H. Kanzaki, Y. Tadakuma, *Solid State Communications*, 80, 33 (1991).
68. K. P. Johansson, G. McLendon, A. P. Marchetti, *Chem. Phys. Lett.*, 179, 321 (1991).
69. K. P. Johansson, A. P. Marchetti, G. L. Mclendon, *J. Phys. Chem.*, 96, 2873 (1992).
70. M. Masumoto, T. Kawamura, T. Ohzeki, S. Urabe, *Phys. Rev. B*, 46, 1827 (1992).
71. A. P. Marchetti, K. P. Johansson, G. L. McLendon, *Phys. Rev. B*, 47, 4268 (1993).
72. C. R. Berry, *J. Appl. Phys.*, 27, 636 (1956).
73. F. C. Boardman, R. H. Herz, G. C. Terry, *Nature,* 182, 1726 (1958).
74. W. Waidelich, *Z. Anges. Phys.*, 10, 525 (1958).
75. E. Pitts, F. W. Willet, *Acta Cryst.*, 14, 1302 (1961).
76. F. W. Willets, Brit. *J. Appl. Phys.,* 16, 323 (1965).
77. F. W. Willets, *J. Photogr. Sci.*, 13, 240 (1965).
78. F. W. Willets, *J. Photogr. Sci.*, 14, 433 (1966).
79. K. Miyake, T. Tani, O. Nittono, *J. Soc. Photogr. Sci. Technol. Jpn.*, 60, 4 (2000).
80. J. Pouradier, A. Pailliotet, C. R. Berry, in T. H. James (ed.), *The Theory of the Photographic Process*, 4th edn (New York: Macmillan, 1977), 3–4.
81. S. Bando, Y. Shibahara, S. Ishimaru, *J. Imaging Sci.*, 29, 193 (1985).
82. S. Urabe, T. Ohzeki, *J. Soc. Photogr. Sci. Tech.*, 69, 105 (2006).
83. K. Ohzeki, S. Urabe, T. Tani, *J. Imaging Sci.*, 34, 136 (1990).
84. T. Tani, *J. Imaging Sci. Technol.*, 51(2), 110 (2007).
85. J. E. Maskasky, *Photogr. Sci. Eng.*, 25, 96 (1981).
86. T. Sugimoto, K. Miyake, *J. Colloid Interface Sci.*, 140, 335 (1990).
87. J. E. Maskasky, *J. Imaging, Sci.*, 32, 160 (1988).
88. M. E. Irving, in the preprint of *The International East-West Symposium III: New Frontiers in Silver Halide Imaging, November 1992, Kaanapali, Hawaii*, A6.
89. T. M. Kelly, M. G. Mason, *J. Appl. Phys.*, 47, 4721 (1976).
90. Y. T. Tan, *Surface Sci.*, 61, 1 (1976).
91. L. M. Niedzwiecki, Y. T. Tan, *J. Photogr. Sci.*, 35, 155 (1987).
92. B. K. Furman, G. H. Morrison, V. I. Saunders, Y. T. Tan, *Photogr. Sci. Eng.*, 25, 121 (1981).
93. T. J. Maternaghan, C. J. Falder, R. Levi-Setti, J. M. Chabala, *J. Imaging Sci.*, 34, 58 (1990).
94. W. K. Lam, Y. T. Tan, C. Palmstrom, J. W. Mayer, *J. Photogr. Sci.*, 33, 219 (1985).
95. X. L. Gao, R. Gijbeis, B. Nlys, W. Jacob, Y. Gilliams, *J. Imaging Sci.*, 33, 87 (1989).
96. H. Chateau, *Compt. Rend.*, 248, 1950 (1959).
97. H. Chateau, M. C. Moncet, J. Pouradier, in W. Eichler, H. Frieser, O. Helwich (eds), *Wissenschaftliche Photographie* (Ergebnisse der Intern. Konf. fur Wissenschaftliche Photographie, Koln, 1956) (Damstadt: Verlag Dr O. Helwich, 1958), 16.
98. H. Chateau, *Compt. Rend.*, 249, 1887 (1959).

99. H. Chateau, M. C. Moncet, and J. Pouradier, in James (ed.), *The Theory of the Photographic Process*, 20.
100. K. J. Lushington, Y. T. Tan, in the preprint of *The International East-West Symposium II: Factors Influencing the Efficiency of Photographic Imaging, October/November 1988*, C10.
101. T. Yokoyama, F. Takamatsu, K. Seki, K. Miyake, T. Tani, T. Ohta, Japanese *J. Appl. Phys.*, 29, L1486 (1990).
102. J. M. Hedges, J. W. Mitchell, *Phil. Mag.*, 44, 223 (1953).
103. J. W. Mitchell, *J. Soc. Photogr. Sci. Technol. Jpn.*, 54, 248 (1991).
104. G. C. Farnell, R. B. Flint, J. B. Chanter, *J. Photogr. Sci.*, 13, 25 (1965).
105. J. F. Hamilton, *Photogr. Sci. Eng.*, 11, 57 (1967).
106. T. Shiozawa, *J. Soc. Photogr. Sci. Technol. Jpn.*, 34, 16 (1971).
107. J. W. Mitchell, *J. Soc. Photogr. Sci. Technol. Jpn.*, 54, 258 (1991).
108. C. R. Berry, D. C. Skillman, *J. Photogr. Sci.*, 16, 137 (1968).
109. V. V. LaMer, R. Dinegar, *J. Am. Chem. Soc.*, 72, 4847 (1950).
110. C. R. Berry, D. C. Skillman, *Photogr. Sci. Eng.*, 6, 159 (1962).
111. E. Klein, E. Moisar, *Photogr. Wiss.*, 11, 3 (1962).
112. E. Klein, E. Moisar, *Ber. Bunsenges. Phys. Chem.*, 67, 349 (1963).
113. E. Moisar, E. Klein, *Ber. Bunsenges. Phys. Chem.*, 67, 949 (1963).
114. E. Moisar, *J. Soc. Photogr. Sci. Technol. Jpn.*, 54, 273 (1991).
115. H. Saito, *Kagaku Souchi*, 34, 61 (1992).
116. S. Urabe, M. Satou, *J. Soc. Photogr. Sci. Tech. Jpn.*, 69, 340 (2006).
117. D. H. Klein, L. Gordon, T. H. Walnut, *Talanta*, 3, 177 (1959).
118. C. R. Berry, in James (ed.) *The Theory of the Photographic Process*, Chapter 3.
119. A. E. Nielsen, *Acta Chem. Scand.*, 11, 1512 (1957).
120. C. R. Berry, D. C. Skillman, *J. Phys. Chem.*, 68, 1138 (1964).
121. C. R. Berry, *Photogr. Sci. Eng.*, 20, 1 (1976).
122. R. W. Berriman, *J. Photogr. Sci.*, 12, 121 (1964).
123. A. Hirata, S. Hohnishi, *Bull. Soc. Sci. Photogr. Jpn.*, 16, 1 (1966).
124. J. S. Wey, R. W. Strong, *Photogr. Sci. Eng.*, 21, 248 (1977).
125. R. W. Strong, J. S. Wey, *Photogr. Sci. Eng.*, 6, 344 (1979).
126. M. Saito, in *The Annual Conference of Soc. Photogr. Sci. Technol. Japan, May, 1980, Tokyo*.
127. C. R. Berry, *PSA Tech. Quarterly*, 149 (1955).
128. F. Claes, R. Berendsen, *Photogr. Korr.*, 101, 37 (1965).
129. J. S. Wey, in W. R. Wilcox (ed.), *Preparation and Properties of Solid State Materials*, Vol. 5 (New York: Marcell Dekker, 1980).
130. C. W. Wagner, private communication to E. Klein and E. Moisar, cited in E. Klein, E. Moisar, in W. Eichler, H. Frieser, O. Helwich (eds), *Wissenschaftliche Photographie* (Ergebnisse der Intern. Konf. fur Wissenschaftliche Photographie, Koln, 1956) (Damstadt: Verlag Dr O. Helwich, 1958).
131. V. G. Losinov, N. B. Denisova, *Zh. Nauchn. Prikl. Photogr.*, 20, 231 (1975).
132. C. R. Berry, D. C. Skillman, *J. Phys. Chem.*, 68, 1138 (1964).
133. I. H. Leubner, R. Jagannathan, J. S. Wey, *Photogr. Sci. Eng.*, 24, 268 (1980).
134. I. H. Leubner, *J. Imaging Sci.*, 29, 219 (1985).
135. I. H. Leubner, *J. Imaging Sci.*, 31, 145 (1987).
136. C. W. Davies, A. L. Jones, *Diss. Faraday Soc.*, 5, 103 (1949).
137. D. H. Klein, L. Gordon, T. H. Walnut, *Talanta*, 3, 177 (1959).
138. T. Tanaka, *J. Inform. Record. Mater.*, 18, 249 (1990).
139. J. W. Mullin, *Crystallization* (London: Butterworths, 1961), 117.
140. J. S. Wey, R. W. Strong, *Photogr. Sci. Eng.*, 21, 14 (1977).

141. J. D. Lewis, *J. Photogr. Sci.*, 27, 24 (1979).
142. A. E. Nielsen, *Kinetics of Precipitation* (New York: Pergamon Press, 1964).
143. C. R. Berry, *Photogr. Sci. Eng.*, 18, 4 (1974).
144. R. Jagannathan, J. S. Wey, *J. Crystal Growth*, 51, 601 (1981).
145. R. Jagannathan, *J. Imaging Sci.*, 32, 100 (1988).
146. T. Sugimoto, *J. Colloid Interface Sci.*, 93, 461 (1983).
147. H. Zwicky, *Z. Wiss. Photogr.*, 53, 68 (1959).
148. T. Tani, *J. Photogr. Sci. Eng.*, 29, 93 (1985).
149. M. J. Harding, *J. Photogr. Sci.*, 27, 1 (1079).
150. T. Tani, *J. Imaging Sci.*, 32, 227 (1988).
151. A. H. Herz, in the preprint of *The International East-West Symposium III: New Frontiers in Silver Halide Imaging, November 1992, Kaanapali, Hawaii.*
152. J. Pouradier, A. Pailliotet, C. R. Berry, in James (ed.), *The Theory of the Photographic Process*, 1–12.
153. O. Knack, I. N. Stranski, *Z. Elektrochem.*, 60, 816 (1956).
154. I. N. Stranski, *VDI-Berichte*, 20, 5 (1957).
155. C. R. Berry, S. J. Marino, C. F. Oster, Jr, *Photogr. Sci. Eng.*, 5, 332 (1961).
156. R. Jagannathan, *J. Imaging Sci.*, 35, 104 (1991).
157. R. Jagannathan, V. V. Gokhale, *J. Imaging Sci.*, 35, 113 (1991).
158. C. T. Mumau, E. F. Haugh, *J. Imaging Sci.*, 30, 198 (1986).
159. M. G. Antoniades, J. S. Wey, *J. Imaging Sci.*, 36, 517 (1992).
160. M. G. Antoniades, J. S. Wey, *J. Imaging Sci. Technol.*, 37, 272 (1993).
161. M. G. Antoniades, J. S. Wey, *J. Imaging Sci. Technol.*, 39, 323 (1995).
162. M. G. Antoniades, J. S. Wey, *J. Imaging Sci. Technol.*, 42, 393 (1998).
163. T. Sugimoto, *Photogr. Sci. Eng.*, 28, 137 (1984).
164. P. H. Karpinski, J. S. Wey, *J. Imaging Sci.*, 32, 34 (1987).
165. K. Ohzeki, T. Tani, *J. Soc. Photogr. Sci. Technol. Jpn.*, 68, 332 (2005).
166. K. Ohzeki, H. Hosoya, *J. Crystal Growth*, 305, 192 (2007).
167. I. H. Leubner, *Precision Crystallization Theory and Practice of Controlling Crystal Size* (Boca Raton, FL: CRC Press, 2010), Chapter 6.
168. I. H. Leubner, *Precision Crystallization Theory and Practice of Controlling Crystal Size*, Chapter 7.
169. M. I. Freedhoff, A. P. Marchetti, in R. E. Hummel, P. Wissmann (eds), *Handbook of Optical Properties*, Vol. II: *Optics of Small Particles, Interfaces, and Surfaces* (Boca Raton, FL: CRC Press, 1997), Chapter 1.
170. H. Stolz, H. Vogelsang, W. von der Osten, in in Hummel, Wissmann (eds), *Handbook of Optical Properties*, Vol. II: *Optics of Small Particles, Interfaces, and Surfaces*, Chapter 2.
171. R. S. Eachus, A. P. Marchetti, A. A. Muenter, *Annu. Rev. Phys. Chem.*, 50, 117 (1999).
172. K. Niwa, *J. Phys. Soc. Jpn.*, 59, 871 (2004).
173. K. Niwa, *J. Soc. Photogr. Sci. Technol. Jpn.*, 67, 561 (2004).
174. T. Tani, *Radiation Measurements*, 44, 733 (2009).
175. A. P. Marchetti, K. P. Johnsson, G. L. McLendon, *Phys. Rev. B*, 42, 4268 (1994).
176. W. Chen, G. McLendon, A. P. Marchetti, J. M. Rhem, M. I. Freedhoff, C. Myers, *J. Am. Chem. Soc.*, 116, 1585 (1994).
177. M. I. Freedhoff, G. McLendon, A. P. Marchetti, *Mater. Res. Soc. Symp. Proc.*, 358, 301 (1995).
178. H. Vogelsang, O. Husberg, U. Koehler, W. von der Osten, *Phys. Rev. B*, 61, 1847 (2000).
179. A. P. Marchetti, P. J. Rodney, W. von der Osten, *Phys. Rev. B*, 64, 132201 (2001).
180. M. I. Freedhoff, A. P. Marchetti, G. L. McLendon, *J. Luminescense*, 70, 400 (1996).
181. H. Kanzaki, Y. Tadakuma, *Solid State Commun.*, 80, 33 (1991).

182. For example, T. Yano, *Photogr. Korr.*, 104, 121 (1968).
183. R. H. Ottewill, R. F. Woodridge, *J. Colloid Sci.*, 16, 581 (1961).
184. W. D. Kelly, Jr, *J. Photogr. Sci.*, 6, 16 (1958).
185. P. Lanza, M. T. Lippolis, *J. Photogr. Sci.*, 17, 155 (1969).
186. F. Moll, *J. Photogr. Sci.*, 34, 47 (1986).
187. J. E. Maskasky, *J. Imaging Sci.*, 33, 10 (1989).
188. J. E. Maskasky, *J. Imaging Sci.*, 33, 13 (1989).

3

Physical Properties of Silver Halides

Since AgBr and AgCl are the main bodies of silver halide (AgX) grains in photographic films, descriptions of the physical properties of AgBr and AgCl emulsion grains are made extensively in this chapter. Brief descriptions are also made of the physical properties of AgI grains, which were recently studied for photothermographic materials. These descriptions are followed by an analytical comparison in physical properties among AgCl, AgBr, and AgI nemulsion grains.

As described in detail in Section 4.1, photographic sensitivity is based on the formation of a latent image center on an AgX emulsion grain as a result of its light absorption. A latent image center is a cluster of Ag atoms and is formed by the electronic and ionic processes that take place repeatedly one after another in the grain when photoelectrons are created as a result of the light absorption of an emulsion grain or sensitizing dye molecules on the grain. The electronic process is the migration and capture of a photoelectron by an electron trap, which is usually provided by chemical sensitization as described in Section 4.5. The ionic process is the migration and reaction of an interstitial silver ion with the trapped electron to form a silver atom in an emulsion grain.

Therefore, the knowledge of the light absorption, and ionic and electronic properties of AgX emulsion grains are essential for understanding the mechanism of photographic sensitivity. Extensive studies has been successfully carried out by solid-state physicists on the physical properties of large AgX crystals to obtain their fundamental knowledge, which includes the characteristics of light absorption and electronic structure of AgX, formation and transport of interstitial silver ions, transport and traps of photoelectrons and positive holes. Large differences in ionic and electronic properties between large crystals and emulsion grains of silver halides have turned out to be important for understanding the mechanism of photographic sensitivity.

3.1 Light Absorption and Electronic Structure of Silver Halides

3.1.1 Light Absorption and Band Structure of Silver Halides

Figure 3.1 shows absorption spectra of AgCl and AgBr at various temperatures [1]. Each of these absorption spectra is composed of a low energy, weak and highly temperature-dependent light absorption band with an extended tail in the visible wavelength region, and an intense exciton band in the UV region. In contrast to silver halides, each of the absorption spectra of alkali halides is composed of only an intense exciton band in the UV region [2,3].

These absorption properties reflect the electronic structure of AgX, as has been studied from various aspects [4–21]. The band structures are schematically shown in Figure 3.2 for AgBr as a representative of silver halides used for photographic films and for KBr as a representative of alkali halides [22]. For convenience, Figure 3.2 also shows the simplified

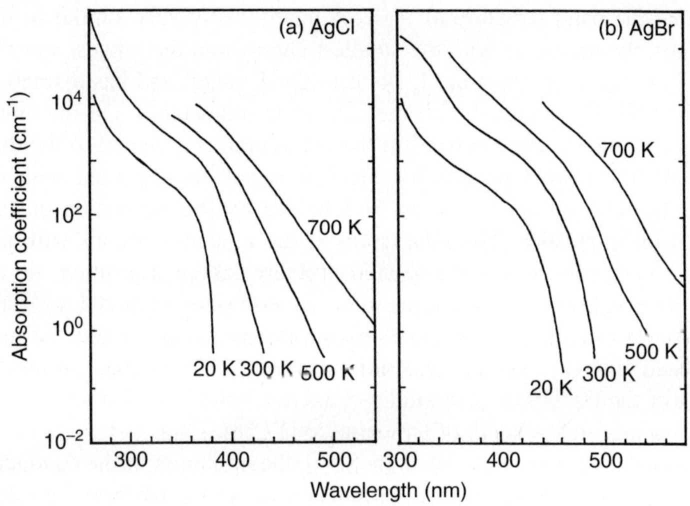

Figure 3.1 Temperature dependence of absorption spectra of AgCl (a) and AgBr (b) [1].

band structure of AgBr as a function of distance. The band with an extended tail has been assigned experimentally [4] and theoretically [5,6] to the indirect exciton transition, whereas the intense exciton band in the absorption spectrum of AgBr in the UV region and the first exciton band in KBr are assigned to the direct exciton transition. The energies of the indirect exciton transitions of AgCl and AgBr are 3.245 eV [23,24] and 2.684 eV [23], respectively.

The studies on the transport of electrons in silver halides by Brown and his collaborators [10,11] revealed that the minimum of the conduction band in AgBr and AgCl are located at the Γ point in momentum space. It has been established that the maximum of the valence band of AgBr is located along the [111] direction and at the L point [12]. It is

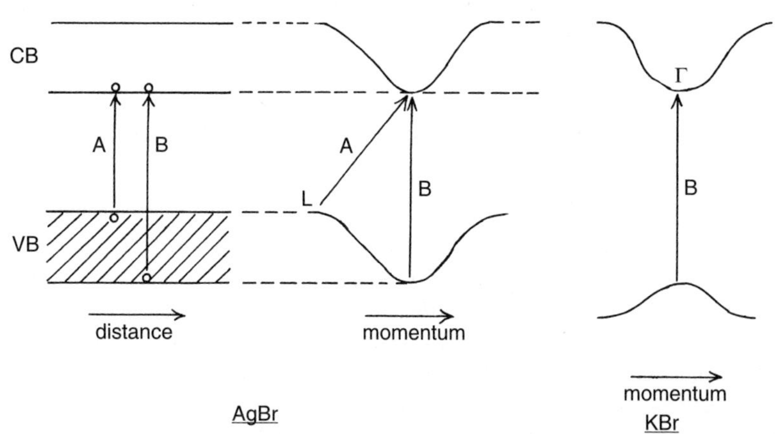

Figure 3.2 Schematic band structures of AgBr and KBr, where CB and VB are the conduction band and valence band, respectively. Arrows A and B are the indirect and direct transitions, respectively.

confirmed that the band structure of AgCl is qualitatively very similar to that of AgBr [5–8,16]. Thus, the transition with the smallest energy and the longest wavelength is the indirect exciton transition from the L point to the Γ point, and has a small probability, since it should be accompanied with the change in momentum. On the other hand, the transition by the intense exciton band in the UV region is assigned to the direct exciton transition, and has a large probability, since it is not accompanied with a change in momentum. The change in momentum is achieved by the interaction of excitons with phonons or with impurities. The interaction of the indirect excitons with phonons and impurities is clearly observed in the phonon structures of the absorption and fluorescence spectra of silver halides at low temperature, as extensively studied by Kanzaki et al. [17,18]. In accord with these two schemes for the change in momentum, the strength of the phonon-assisted indirect-exciton transition in pure AgBr was temperature-dependent, whereas that of the transition of the indirect exciton band bound at an isoelectric iodine impurity was nearly independent of temperature [17,18].

According to theoretical calculations [5–8,16], the minimum of the conduction band of silver halides is mostly composed of 5*s* orbitals of silver ions, whereas the valence band is composed of *np* orbitals of halide ions and 4*d* orbitals of silver ions. The maximum of the valence band at the L point exhibits strong *p*–*d* mixing, whereas the valence band at the Γ point exhibits little *p*–*d* mixing. The repulsive interaction between halide *np* and silver 4*d* states raises the valence band in the region away from the Γ point, bringing about the low-energy indirect exciton transition. On the other hand, the valence bands of alkali halides are essentially composed of the *np* orbitals of halogen ions [22].

The wave functions of an electron in the lowest state of the conduction band in AgCl at low temperature were studied by means of pulsed electron nuclear magnetic resonance (ENDOR). It has been revealed that this state is not only composed of Ag 5*s* orbitals, but also includes a substantial amount of Cl 4*s* orbitals [25]. This result has been supported by electronic structure calculations [26].

In contrast to AgBr and AgCl, AgI exhibits an intense and sharp absorption band at around 420 nm with steep edge beginning at about 440 nm [27,28]. This band corresponds to the direct transition of an electron from the valence band to the conduction band of AgI to form an exciton, and is intense enough to provide one of the reasons to use AgI grains for photothermographic materials [29]. The electronic structure of AgI has been theoretically calculated [30–33], and experimentally studied [33–35].

3.1.2 Absorption and Emission Spectra of Nanoparticles of Silver Halides

The absorption and emission spectra of AgX nanocrystals have attracted the attention of many scientists from the viewpoint of the quantum confinement effects on the electronic structure of silver halides. As stated above, AgCl and AgBr have indirect band-to-band lowest energy transitions, which are forbidden by the momentum selection rule, while AgI has a direct band-to-band lowest energy transition. As described in the previous chapter, nanocrystals of AgCl, AgBr, and AgI, which are small enough to exhibit the quantum confinement effects, can be prepared by various methods [36–46,28]. The best ones for preparing nanocrystals in an aqueous gelatin solution, and for reducing size and dispersity among them, have been determined not only for AgI [36], but also for AgBr [37,39].

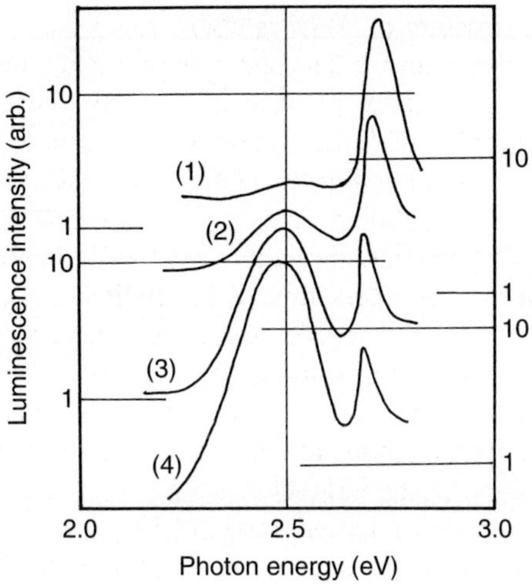

Figure 3.3 Photoluminescence of AgBr nanoparticles at 2 K. The ordinate, in log scale, is shifted in each plot. Diameters of particles in nm are: (1) 7.93, (2) 11.0, (3) 16.3, (4) 21.0 [37].

Silver bromide exhibits three luminescence spectra at low temperature. They are free exciton luminescence (FEL) at 463 nm, bound exciton luminescence (BEL) at 500 nm arising from the recombination of excitons trapped at impurity iodide ions, and donor–acceptor (D–A) luminescence at 595 nm arising from the pair recombination of charge carriers trapped by donors and acceptors [44]. When the size of nanocrystals is reduced to be comparable to that of the exciton radius, three quantum confinement effects are experimentally observed [37–41]. In the first place, the intensity of FEL with respect to that of BEL dramatically increases with decreasing size of AgBr nanocrystals, as shown in Figure 3.3. Secondly, the peak of FEL shifts to higher energy as the size decreases. Finally, the decay of indirect exciton luminescence indicates impurity exclusion. Namely, the decay curves were well reproduced by a model with the discrete number of iodide ions in a AgBr nanocrystal, indicating that many of the nanocrystals were free of impurity iodide ions.

Similarly, AgI nanocrystals exhibit increasing blue shifts of their absorption and luminescence with decreasing size. However, the luminescence arising from the recombination of the excitons is absent in very small nanocrystals of AgI [35,44]. Quantum confinement effect was observed on the self-trapping of excitons in AgCl nanocrystals at low temperature, in which only the recombination of the self-trapped exciton was observed, whereas the recombination of the excitons trapped at the bromide-containing silver ions in the first ligand shells and that between trapped electron–hole pairs were not, owing to the effects of the impurity exclusion and the confinement of an electron in a nanocrystal, respectively [42].

3.1.3 Heights of Valence and Conduction Bands of Silver Halides

The positions of the top of the valence band and the bottom of the conduction band of silver halides with respect to the vacuum level are important knowledge for the mechanism of photographic sensitivity. The height of the top of the AgX valence band is given

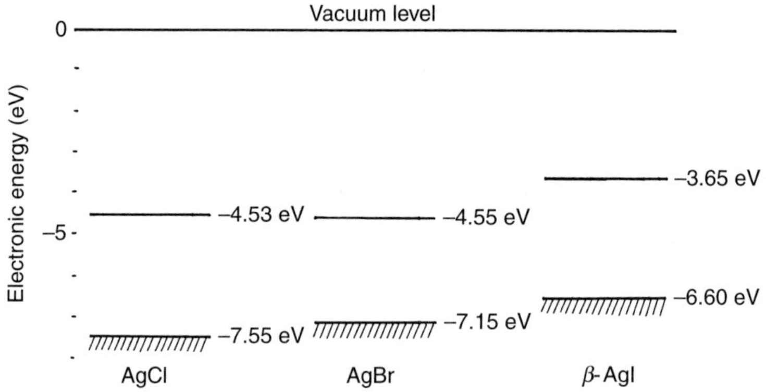

Figure 3.4 The heights of the top of the valence band (lower thick line with thin slanted lines) and the bottom of the conduction band (upper thick line) of AgBr, AgCl, and β-AgI, which were evaluated from the result of the measurement of their electronic structures in high vacuum by means of ultraviolet photoelectron spectroscopy [49].

by ultraviolet photoelectron spectroscopy (UPS) [47–49], while that of the bottom of the conduction band is evaluated based on the height of the top of the valence band and the band gap. According to Bauer and Spicer [49], the positions of the top of the AgCl, AgBr, and AgBr valence bands are situated 7.55, 7.15, and 6.6 eV, respectively, below the vacuum level. Based on this result, the electronic structures of AgCl, AgBr, and AgI are drawn in Figure 3.4.

Since each AgX grain is covered with gelatin in photographic materials, the electronic structures of the AgX layers are important in discussions on the mechanism of photographic sensitivity, and have been measured by UPS as a function of the pAg of an aqueous gelatin solution, which had been coated and dried on the layer's surface as a membrane, thin enough to allow measurement by UPS [50]. The heights of the top of the valence band of silver halides in the presence of thin gelatin membranes at varying pAg are shown as in Figure 3.5 [51]. It was noted that in the presence of the thin gelatin membranes, the heights of the top of the valence band were much higher than those of the corresponding genuine silver halides [50,51]. These results indicated the formation of dipoles aligned on the interface between the silver halides and gelatin membranes, which lowers the vacuum level of the latter with respect to the former as indicated by Seki, Tani, and others for the interfaces between substrates and organic layers [52–55]. It is also noted that the positions of the top of the valence band and the bottom of the conduction band of silver halides covered with gelatin become higher as the aqueous solution's pAg increases. This tendency was in accord with the pAg dependence of photographic phenomena. For example, the electron transfer from excited dye molecules to silver halides could be enhanced by the decrease in pAg of the emulsion examined [56], especially for the dyes, whose lowest unoccupied molecular orbital (LUMO) levels were lower than the bottom of the conduction band of the AgX examined.

The electronic structure of silver halides as a function of the distance from the surface of their crystals and emulsion grains is described in the next section as their space charge layers arising from the displacement of silver ions from the surface to the bulk. The

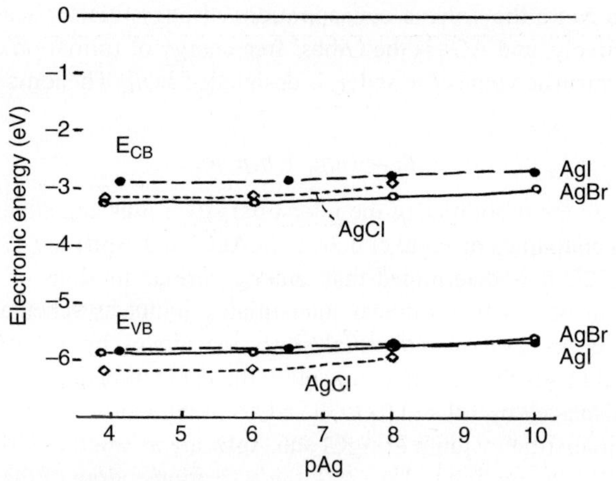

Figure 3.5 The heights of the top of the valence band (E_{VB}) and the bottom of the conduction band (E_{CB}) of thin layers of AgCl, AgBr, and AgI, which were evaporated on silver substrates, coated by aqueous gelatin solutions with pAg values on the abscissa, and dried. Reproduced with permission from Reference [51], copyright 2009 The Society of Photographic Science and Technology of Japan.

comparison in electronic structure among different silver halides has been studied with respect to the electron transfer between silver halides and compounds adsorbed to them [57–59]. The comparison in electronic structure of the interface between AgBr and sensitizing dyes was made between octahedral and cubic AgBr grains from the viewpoint of the difference in space charge between them [60]. The electronic structures of the interfaces between different silver halides were studied [61–65] as models of double-structured AgX emulsion grains and epitaxial grains as described in Sections 2.1.4 and 2.2.8. The electronic structures of the interfaces between AgX thin layers and sensitizing dyes have been studied by means of UPS by Seki and Tani [52–54] in relation to the electron transfer process in dye sensitization of AgX grains in photographic films.

3.2 Ionic Properties of Silver Halides

3.2.1 Ionic Conductivity of Silver Halides

There are two types of intrinsic point defects in ionic crystals: Frenkel and Schottky, as schematically shown in Figure 3.6. It has been determined that Frenkel defects composed of interstitial silver ions and silver ion vacancies are predominant in silver halide [66–70], and are the charge carriers for the dark conductivity (i.e., ionic conductivity) of silver halides [71]. In a large crystal, a silver ion at a lattice position jumps into an interstitial position to form an interstitial silver ion and a silver ion vacancy. The thermodynamic equilibrium is expressed by

$$n_F^2 = n_i n_v = 2N^2 \exp[-\Delta G_F/kT], \tag{3.1}$$

where n_i, n_v, and N are the volume concentrations of interstitials, vacancies, and lattice positions, respectively, and ΔG_F is the Gibbs' free energy of formation of a Frenkel pair. The equilibrium intrinsic value of n_i and n_v is designated as n_F. The ionic conductivity σ_i is expressed as

$$\sigma_i = (n_i\mu_i + n_v\mu_v)e, \tag{3.2}$$

where μ_i and μ_v are the mobilities of the interstitial silver ions and silver ion vacancies.

The formation enthalpies of Frenkel defects in AgCl and AgBr are 1.49 and 1.163 eV, respectively [71,72]. It is determined that, among various motions of interstitial silver ions, the dominant one is the collinear interstitialcy jump, as schematically shown in Figure 3.7. In this process, an interstitial silver ion moves by 1/4 <111> to a lattice position, where <111> is the size of a unit cell in the [111] direction, and the lattice silver ion is similarly displaced by 1/4 <111> to the adjacent interstitial position. The enthalpies for the collinear interstitialcy jump in AgCl and AgBr are as small as 0.018 and 0.042 eV, respectively [70,71]. According to Eq. (3.2), the activation energy of the ionic conductivity is the sum of half of the formation enthalpy of the Frenkel defects and the enthalpy for the collinear interstitialcy jump when interstitial silver ions are predominant as carriers for the ionic conductivity.

Since an electron is bound by a shallow center with an interstitial silver ion, the wave function of the electron reflects the geometric structure of the center, and has been studied experimentally [73,74,25] and theoretically [75]. Pulsed ENDOR experiments conducted on AgCl single crystals at 1.2 K gave evidence of a shallowly trapped electron that had the split interstitial silver ion as its core, as illustrated in Figure 3.7 [73,74]. This structure could explain such a low activation energy of the mobility of interstitial silver ions in AgCl, and has been supported by Hartree-Fock calculations [25].

Impurity ions can influence the concentrations of interstitial silver ions and silver ion vacancies [76–81]. For example, divalent metal ions, which substitute for silver ions at lattice positions, are associated with silver ion vacancies for neutralizing excess charges on the divalent metal ions in the lattice. With increasing concentration of such divalent metal ions, the ionic conductivity of AgX grains with interstitial silver ions as carriers decreases,

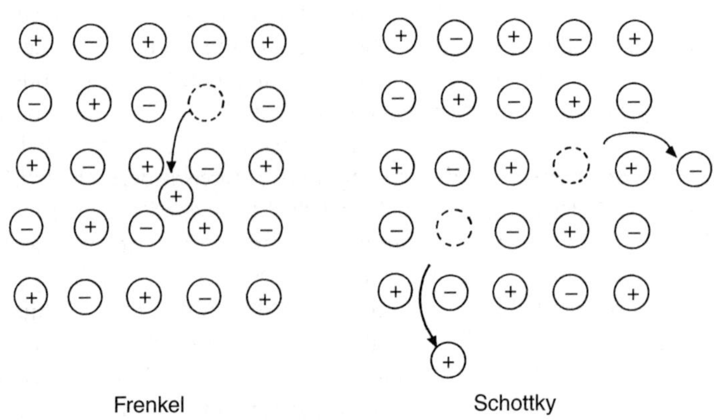

Figure 3.6 Pathways for formation of Frenkel and Schottky defects in an ionic crystal.

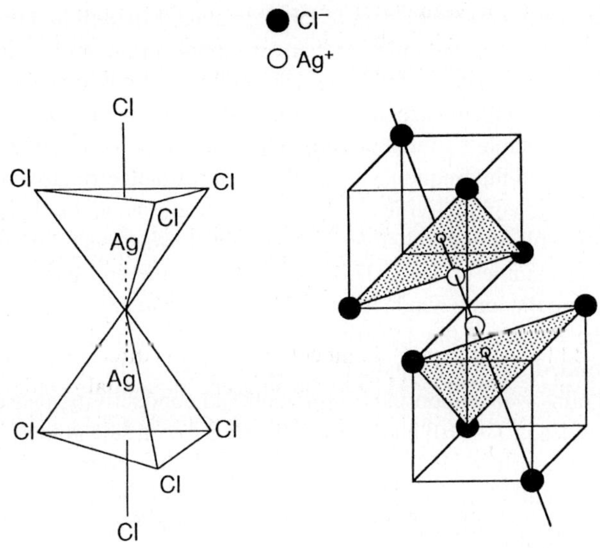

Figure 3.7 Sketches of a model for silver ion split interstitial. On the left-hand side, the {111} crystal axis is rotated to lie along the vertical axis of this page. Reproduced is cited with permission from Reference [75], copyright 1995 IOP Publishing.

and then the ionic conductivity of AgX grains with silver ion vacancies as carriers increases.

3.2.2 Ionic Conductivity of Silver Halide Emulsion Grains

The ionic conductivity of AgX grains in a photographic emulsion has been measured by various methods and found to be significantly larger than that of the corresponding large crystal. Following the experiments of Hayes and Schockley with large AgCl crystals [82,83], Hamilton and Brady applied synchronized pulses of electric field and light to a AgBr emulsion layer to observe the deviation of the location of latent image centers on a grain. The delay of the light pulse after the beginning of the field pulse made it possible to determine the time for the internal field relaxation owing to ionic conduction. The relaxation time, τ_i, is related to the ionic conductivity, σ_i, by the equation $\tau_i = \epsilon/4\pi\sigma_i$, where ϵ is the static dielectric constant of the medium. They found that the ionic conductivity of AgBr emulsion grains was by two to three orders of magnitude larger than that of a large AgBr crystal [84,85], and that the activation energy of the ionic conductivity of the emulsion grains was much smaller than that of the large crystal [86]. Observing the decrease in the emulsion grains' ionic conductivity by adding a strong silver complexing agent, 1-phenyl-5-mercaptotetrazole [85], and divalent cadmium ion to the emulsion [87], they indicated that interstitial silver ions were formed by the displacement of silver ions at surface kink sites into interstitial positions in AgBr grains, and that the concentration was large enough to account for the formation of latent image centers on emulsion grains according to the current mechanism described in the next chapter (see Figure 4.1).

Based on the ionic conductivity data of large crystals, Mitchell had previously pointed out that there was less than one interstitial silver ion per grain, and proposed the mechanism of latent image formation [88]. The above-stated results revealed that there were many interstitial silver ions in each AgBr grain, and led to the mechanism of latent image formation being acceptable by most photographic scientists, as described in Section 4.1.

Van Biesen [89] used the frequency dependence of the dielectric loss of emulsion layers to determine the ionic conductivities of the emulsion grains, and could confirm that the ionic conductivity with interstitial silver ions as carriers was much larger than that of the corresponding large single crystals. This dielectric loss method has been used by many scientists to measure the ionic conductivity of AgX grains in photographic films [90–100]. By analyzing the decay of the Dember photovoltage in emulsion grains, Saito et al. [101,102] determined the ionic conductivity of the grains. The above-stated results with AgX emulsion grains were supported by conventional conductivity measurements on thin films of silver halides [103–110] and by investigating the surface potential or the potential distribution near the surface [111–117]. In accord with the above-stated ionic conductivities, the activation enthalpies of the ionic conductivity of cubic AgCl grains, and cubic and octahedral AgBr grains, which are the sums of the formation enthalpies and the activation energies for the collinear interstitialcy jumps of interstitial silver ions in the corresponding grains, were 0.66, 0.40, and 0.36 eV [93], respectively, and much smaller than those of the ionic conductivities of the corresponding large crystals.

As stated earlier, the ionic conductivities of AgX emulsion grains are usually measured by the dielectric loss method, according to which a maximum appears in the curve of dielectric loss as a function of frequency (Figure 3.8). The observed frequency at which a maximum occurs (ω_{max}) can be related to the conductivity of AgX emulsion grains as

$$\omega_{max} = 1/\tau, \tag{3.3}$$

where

$$\tau = [\varepsilon_1'(n-1) + \varepsilon_2']/4\pi\sigma_2,$$

where ϵ_1' is the dielectric constant of gelatin, ϵ_2' and σ_2 are the dielectric constant and conductivity of AgX emulsion grains, respectively, and n is the shape factor of the grains. Therefore, the conductivity of the grains is proportional to ω_{max}. When an alternating electric field is applied to an emulsion layer, charge carriers in emulsion grains (i.e., interstitial silver ions in most cases) are displaced to give dielectric polarization (i.e., the sum of the concentration (c) and distance of displacement ($d = \mu Et$) of charge carriers, where μ is mobility, E is electric field, and t is time), following the applied field. When the frequency of the alternating field is increased, the delay of the dielectric polarization begins to take place owing to the limited concentration and mobility of the carriers, giving rise to the loss of the electric energy by way of its conversion to thermal energy (i.e., dielectric loss). With increasing the frequency of the alternating field, the dielectric loss increases, then decreases after passing its maximum at ω_{max}, and finally disappears when the frequency becomes to be too high for charge carriers to follow. Thus, ω_{max} is proportional to the product of the concentration and mobility of charge carriers (i.e., the conductivity) in emulsion grains.

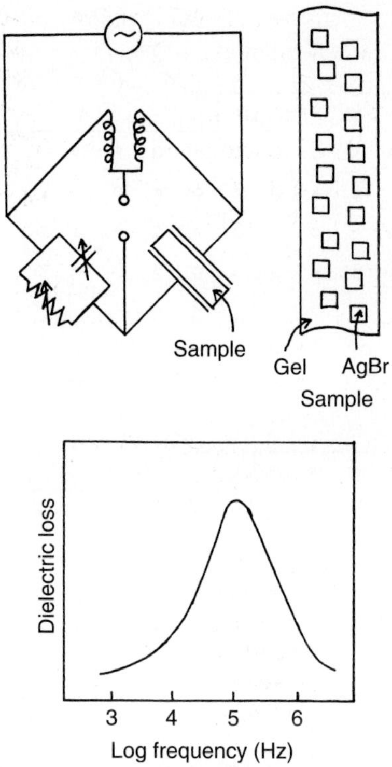

Figure 3.8 Schematic representation of dielectric loss measurement of an emulsion layer and a dielectric loss curve of an emulsion layer containing cubic AgBr grains with edge length of 0.89 μm at room temperature.

The conductivity of cubic AgBr emulsion grains as measured by the dielectric loss method was proportional to the surface-to-volume ratio of the grains, indicating that the charge carriers were created from the grain surface [90]. The identification of interstitial silver ions as charge carriers was also confirmed by the analyses of several factors influencing ionic conductivity of emulsion grains. The ionic conductivity was decreased by the incorporation into an AgX grain polyvalent metal ions, which are associated with silver ion vacancies, and the adsorption of a silver complexing agent such as 4-hydroxy-6-methyl-1,3,3a,7-tetraazaindene to the grain [89–102]. As illustrated in Figure 3.9, the ionic conductivity was increased by the adsorption of a cyanine dye with a positive charge on its chromophore, which repels a silver ion at a surface kink site into an interstitial position of the grain, and decreased by the adsorption to the grain of a dye with a negative charge on its chromoiphore [118,119]. From the correlation between the ionic conductivity of cubic AgBr grains and the adsorbed amount of a cyanine dye to the grains, it was found that cyanine dye molecules repelled silver ions at kink sites into interstitial positions and increased the ionic conductivity when the amount of adsorbed dye neared its saturated amount [120].

Figure 3.10 shows the dependence of ionic conductivity upon halide composition and crystal habit of AgX emulsion grains with an equivalent circular diameter of 0.2 μm [121].

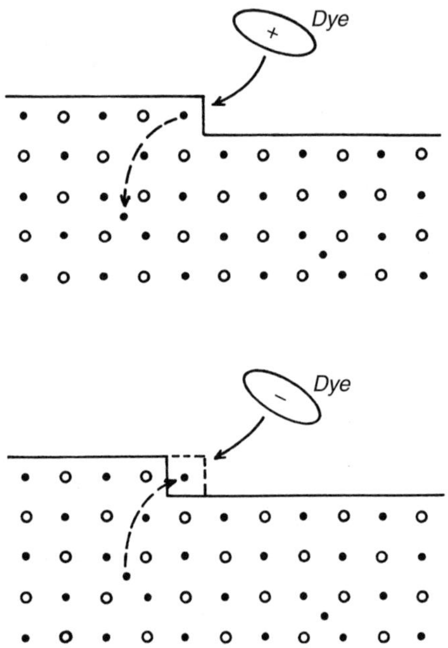

Figure 3.9 The effect of sensitizing dyes on the formation process of interstitial silver ions through the surface of AgBr grains, where closed and open circles are Ag^+ and Br^-, respectively.

It is noted that the variation of ionic conductivity was extended by more than three orders of magnitude, and mainly attributed to the variation of the energy level of silver ions at surface kink sites with respect to that of interstitial silver ions. By use of these grains, Takada reported that the ionic conductivity played an important role in the latent image formation on exposure to high-intensity light. As seen in Figure 3.10, the ionic conductivity of AgCl grains was very small. However, the ionic conductivity of AgCl grains in emulsions could be increased by two orders of magnitude when a small amount of iodide ions was added to the emulsions as shown in Figure 3.11 [122], and met the requirement for efficient latent image formation.

Figure 3.12 shows the result of the dielectric loss measurement of AgBr emulsion layers, by Ohzeki, Urabe, and Tani, as indication of the influence of the surface-to-volume ratio of the grains upon their conductivity with interstitial silver ions as charge carriers [123]. Contrary to the fact that the conductivity of octahedral grains with {111} faces on their surfaces was proportional to the surface-to-volume ratio and was much larger than that of cubic AgBr grains with {100} faces on the surfaces, the conductivity of tabular grains with mainly {111} faces on the surfaces and constant thickness was independent of the surface-to-volume ratio and even smaller than that of cubic grains. In accord with the above-stated result, the activation enthalpy of the ionic conductivity of tabular AgBr grains was 0.58 eV and larger than those of cubic and octahedral AgBr grains. Hamilton proposed that the stacking sequence at a twin boundary should not be the minimum in energy and that the layer spacing perpendicular to the boundary should be greater in the immediate vicinity of twin boundaries than that in the bulk [124,125]. Supporting Hamilton's hypothesis, the above result could indicate that a twin plane could trap

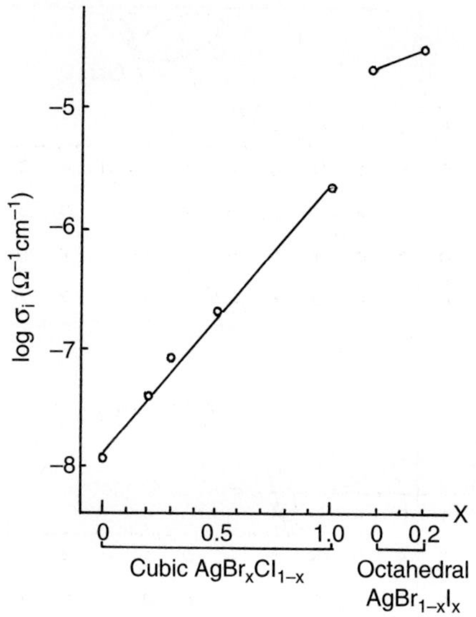

Figure 3.10 Dependence of the ionic conductivity (σ_i) upon halide compositions and crystal habit of AgX emulsion grains. The diameter of the circle, whose area was equal to the average projection area of the grains studied, was 0.2 μm. Reproduced with permission from Reference [121], copyright 1979 The Society of Photographic Science and Technology of Japan.

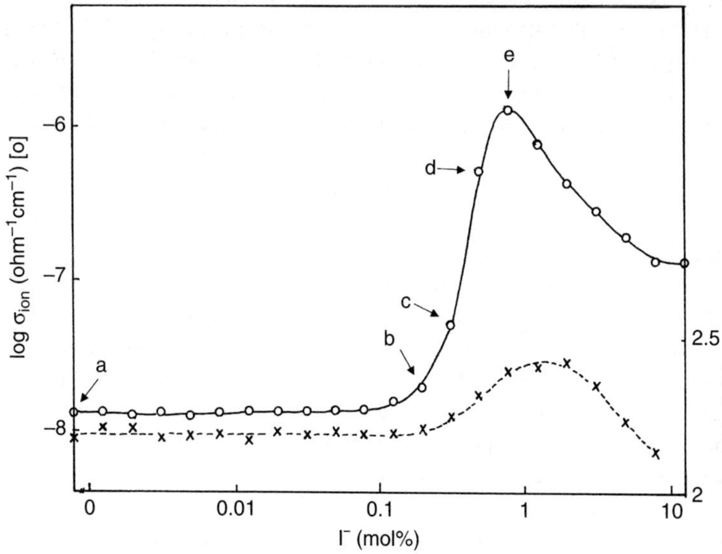

Figure 3.11 Ionic conductivity (solid line) and photographic sensitivity (broken line) of cubic AgCl grains, into which AgI, with amount indicated on the abscissa, was introduced. Reproduced with permission from Reference [122], copyright 2007 IS&T, The Society for Imaging science and Technology.

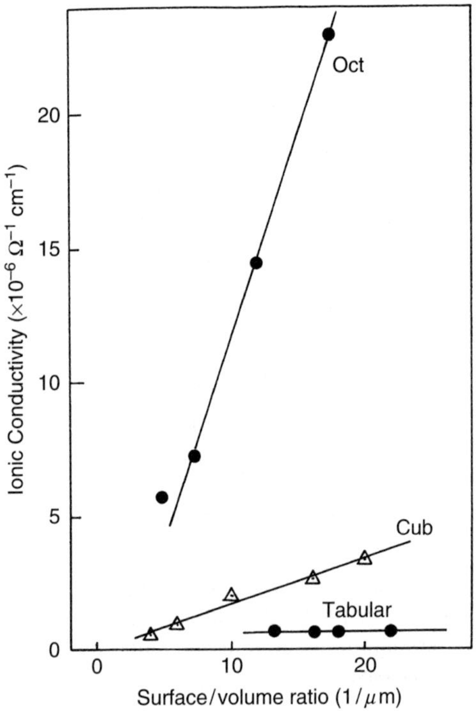

Figure 3.12 Ionic conductivities of octahedral, cubic, and tabular AgBr grains as a function of their surface/volume ratio. Reproduced with permission from Reference [123], copyright 1990 IS&T, The Society for Imaging Science and Technology.

interstitial silver ions, since they should be stable at the immediate vicinity of the twin plane than in the bulk. It is therefore considered that twin planes decrease the concentration of interstitial silver ions, and formed the electric double layers by enhancing the displacement of silver ions from surfaces to twin planes in tabular grains [123], as schematically shown in Figure 3.13.

For the application of AgI emulsion grains to photothermographic materials [29,126,127], a study was carried out on the ionic conductivity of AgI emulsion grains (Figure 2.4) with dielectric loss method [127,122], by taking into account the similarity in crystal structure [128] and dielectric loss curve between AgI grains and octahedral AgBr ones. It was found that the ionic conductivity of AgI grains was comparable to that of AgBr grains, while the activation energy for the former was larger than that for the latter due to large activation energy for the migration of interstitial silver ions in AgI [129].

3.2.3 Energy Levels of Silver Ions in Silver Halides

As shown above, the energy level of a silver ion at a surface kink site is higher than that of a silver ion at a lattice position, being responsible for the formation of interstitial silver ions with high concentration in emulsion grains. Figure 3.14 illustrates the structure of the {100} face with a kink site. Although each silver ion is surrounded by six halide ions in the

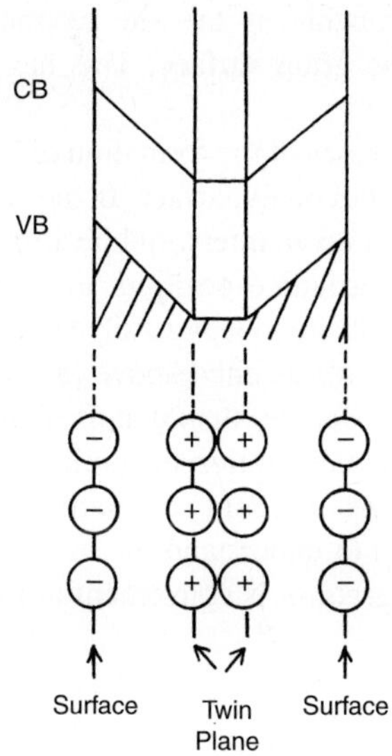

Figure 3.13. Electric double-layer model showing the electronic structure of a tabular AgBr grain, where CB and VB indicate the bottom of the conduction band and the top of the valence band, respectively. Reproduced with permission from Reference [123], copyright 1990 IS&T, The Society for Imaging Science and Technology.

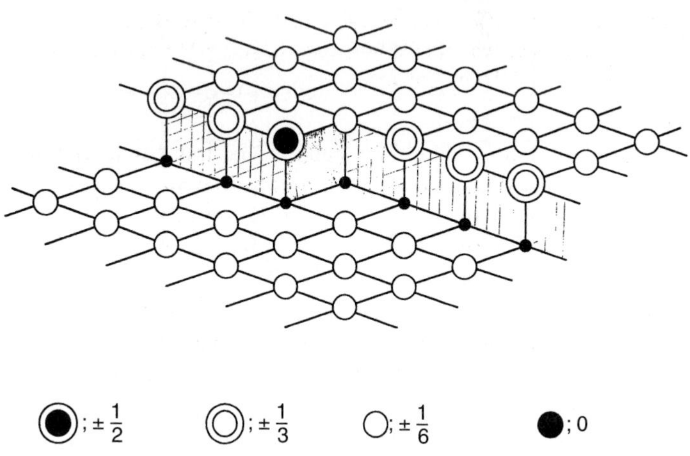

Figure 3.14 A surface kink site with electric charge of 1/2 on (100) surface of AgX crystal, where electric charges at various sites are indicated.

bulk, a silver ion at a kink site is surrounded by only three halide ions, and has therefore a half of a positive charge. It is noted that a silver ion at the corner of a cubic grain is at the same situation as a silver ion at a kink site from this viewpoint, although the number of the latter is overwhelmingly larger than that of the former. It is therefore considered that a silver ion at a kink site has the highest energy among silver ions at probable sites on the

{100} surface, and is most probable as the site responsible for the formation of an interstitial silver ion through the grain surface. This has been supported by theoretical calculation [130].

Figure 3.15 illustrates the processes of the formation of Frenkel defects in a large crystal, and in a small grain with {100} faces on its surface. In the latter case, a silver ion at a surface kink site on the surface can jump into an interstitial position to give an interstitial silver ion more easily than a silver ion at a lattice position. Furthermore, a silver ion at a lattice position can jump to a surface kink site to leave a silver ion vacancy more easily than to an interstitial position. The experiments as cited above [89–102] indicate that the formation energy of interstitial silver ions is smaller than that of silver ion vacancies, bringing about interstitial silver ions with high concentration in emulsion grains.

Figure 3.16 schematically shows the energy diagram of silver ions at various sites on the surface and in the bulk. In order to understand the movement of silver ions between the surface and interior of an AgX grain, it is convenient to regard the middle point between

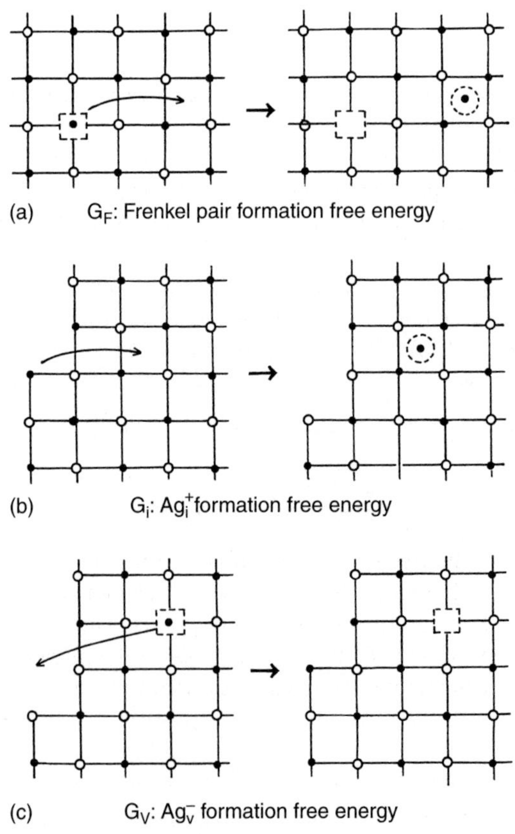

Figure 3.15 The formation of a Frenkel pair in the bulk (a), an interstitial silver ion (b), and a silver ion vacancy (c) through the surface of AgX, where closed circles and open circles are silver ions and halide ions, respectively.

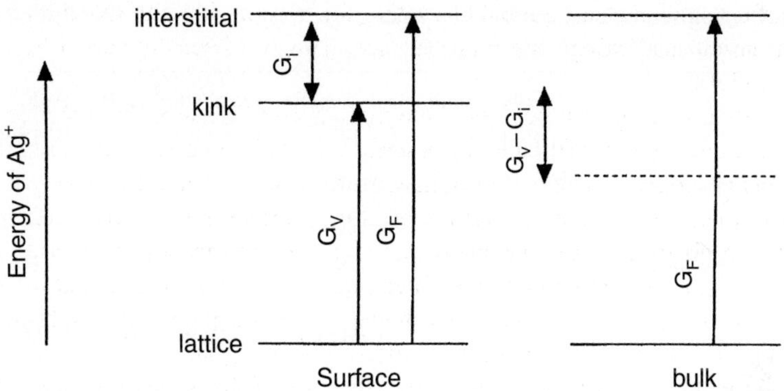

Figure 3.16 A schematic representation of the energy levels of silver ions at various sites indicating the free energies of the formation of a Frenkel pair in the bulk (G_F), and of an interstitial silver ion (G_i) and a silver ion vacancy (G_v) through surface kink sites on the grain surface, as expressed by Eqs. (3.5)–(3.8).

the free energies of silver ions at interstitial and lattice positions as the Fermi level of silver ions. Namely, the free energy of silver ions at kink sites with respect to the middle point could be the driving force for them to move to interstitial positions, and the movement comes to an end to form a space charge layer, in which the potential difference across the layer is equal to the difference of the free energy of silver ions at kink sites from the middle point.

According to Frenkel's concept [131,132], the potential profile in the space charge layer was described by Kliewer and Koehler [133], and extended by Poeppel and Blakely [134], and Hoyen and Tan [135–138]. Namely, the potential at the surface with respect to that in the interior Φ_s is given as

$$2e\Phi_s = \Delta G_i - \Delta G_v + kT\ln(2) \tag{3.5}$$

$$\Delta G_i = \Delta H_i - T\Delta S_i \tag{3.6}$$

$$\Delta G_v = \Delta H_v - T\Delta S_v, \tag{3.7}$$

where ΔG_i and ΔG_v are the free energies required to form an interstitial silver ion and a silver ion vacancy through a kink site, respectively, ΔH_i and ΔH_v are the corresponding enthalpies, ΔS_i and ΔS_v are the corresponding entropies, and the term kT ln (2) comes from the fact that the number of interstitial positions is twice the number of silver ions in the bulk of a grain. From a simple thermodynamic cycle it follows that

$$\Delta G_F = \Delta G_i + \Delta G_v, \tag{3.8}$$

where ΔG_F is the free energy for formation of a Frenkel pair. The width of the space charge layer, i.e., the Debye length λ, is given as

$$\lambda^2 = 2\pi e^2 n_i / \varepsilon\varepsilon_o kT, \tag{3.9}$$

where n_i is the concentration of interstitial silver ions and ϵ and ϵ_o are the dielectric constants of AgX and the vacuum. A proposal for measuring the potential energy profile

Table 3.1 Comparison in ionic conductivity among AgCl, AgBr, and AgI emulsion grains, where Ag_i^+ is an interstitial silver ion

	Cubic AgCl (0.2 μm)	Cubic AgBr (0.2 μm)	Octahedral AgBr (0.2 μm)	AgI (0.14 μm)
Ionic conductivity ($ohm^{-1}cm^{-1}$)	$\sim 5\times10^{-8}$	$\sim 5\times10^{-6}$	$\sim 2\times10^{-5}$	$\sim 1\times10^{-5}$
Ionic relaxation	~100 μs	~200 ns	~50 ns	~100 ns
Activation energy for ionic conditions	0.53 eV	0.39 eV	0.34 eV	0.45 eV
Activation energy for Ag_i^+ migration (E)	0.018 eV	0.042 eV	0.042 eV	0.62 eV (c∥) 0.29 eV (c⊥) 0.40 eV[a]
Formation energy of Ag_i^+ via surface	0.51 eV	0.35 eV	0.31 eV	~0.05eV

[a] $(E_{\parallel} + 2E_{\perp})/3$

and the energy required to form an interstitial silver ion and a silver ion vacancy from the grain surface was made [115–117].

A comparison in ionic conductivity was made among AgCl, AgBr, and AgI emulsion grains in Table 3.1 and Figure 3.17. It is noted that the activation energy of the ionic conductivity of AgI grains was comparable to the activation energy of the migration of interstitial silver ions in AgI [127,129]. This result indicates that only small energy is necessary for the formation of an interstitial silver ion from a surface kink site on an AgI grain, since the activation energy of the ionic conductivity is the sum of the formation energy and migration energy of an interstitial silver ion.

As seen in Table 3.1, the formation energy of interstitial silver ions through surface kink sites is in the order of AgCl, AgBr, and AgI, while the activation energy of the migration of

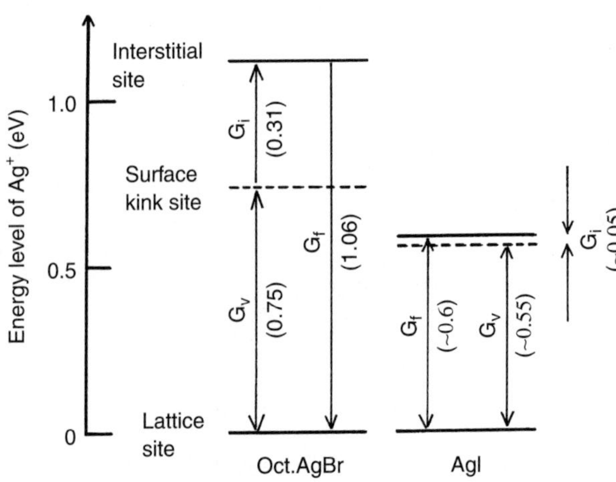

Figure 3.17 Estimated energy levels of silver ions in octahedral AgBr grains and AgI with free energies of the formation of a Frenkel pair in the bulk (G_F), and of an interstitial silver ion (G_i) and a silver ion vacancy (G_v). Reproduced with permission from Reference [127], copyright 2007 IS&T, The Society for Imaging Science and technology.

interstitial silver ions is in the reverse order [71,129]. These results should be ascribed to the ionic character of AgX, which is in the order of AgCl, AgBr, and AgI [127]. Since a silver ion at a surface kink site and an interstitial position in the crystal structure of AgX bear one half and unit positive charge, respectively, the former is more stable than the latter, and the difference in energy level between them, which is the formation energy of an interstitial silver ion, should increase with increasing ionic character from electrostatic viewpoint.

Since interstitial silver ions play an important role in latent image formation, the formation of interstitial silver ions with high concentration through surface kink sites is important for understanding the mechanism of photographic sensitivity. Figure 3.18 schematically shows the conduction and valence bands of AgX as a function of the distance from the crystal surface. It is believed that the band bending in the space charge layer have some influence on photographic sensitivity. The space charge layer might reduce the recombination between a photoelectron and a positive hole, since a photoelectron and a positive hole are forced to move in opposite directions in the space charge layer. The band bending might also influence the electron transfer processes between AgX emulsion grains and adsorbed compounds on the grain surface [60]. As stated before, interstitial silver ions migrate in AgBr and AgCl according to the collinear

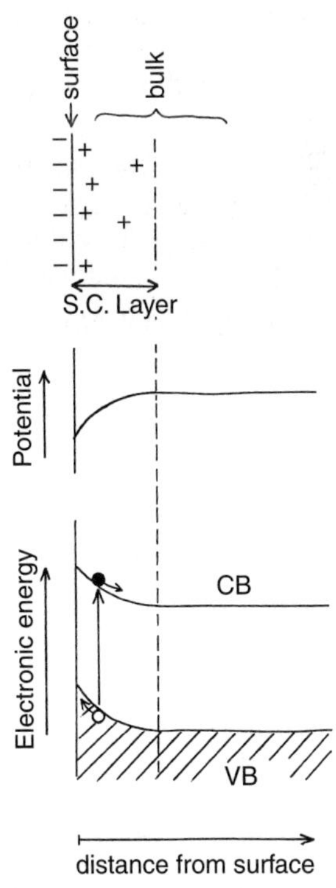

Figure 3.18 The band structure of AgX with a positive space charge layer and behavior of a photoelectron and a positive hole, where CB and VB are the conduction band and valence bands of AgX, respectively.

interstitialcy jump [70,71], whose effectiveness should increase with increasing ionic character of silver halides.

3.3 Electronic Properties of Silver Halides

3.3.1 Electronic Transport in Large Crystals of Silver Halides

As described in the previous section, excitons are first formed on exposure of AgX emulsion grains to light for taking photographs. Since the binding energies of excitons are as small as 0.022 eV for AgBr and 0.04 eV (estimated value) for AgCl [17,20], excitons in AgX emulsion grains freely dissociate into an electron and a positive hole at room temperature. The quantum yield for the photoconduction of electrons in AgBr at 300 K is therefore larger than 0.9 [139] and favorable for photographic sensitivity. It is considered that one of the reasons for those small binding energies of excitons is the large dielectric constants of silver halides. On the other hand, an exciton in an organic semiconductor hardly dissociates at room temperature, since its binding energy as given by $e^2/4\pi\epsilon r$ is as large as ~0.4 eV [140], where r is the distance between an electron and a positive hole and ϵ is the dielectric constant of its surroundings (i.e., 3.5 for an organic semiconductor).

Figure 3.19 illustrates the behavior of photoelectrons in an AgX crystal. Moving toward an anode under an applied electric field, a photoelectron is scattered many times by phonons, captured by a shallow electron trap, and thermally released from it according to the multiple trapping and release model [141]. The important properties of photoelectrons are their Hall mobility (μ_H), drift mobility (μ_D), and lifetime (τ). The Hall mobility, which is evaluated on the basis of the Hall effect and known as microscopic mobility in this case, is the velocity of photoelectrons per unit electric field when they are actually moving. The drift mobility, which is called macroscopic mobility, is their velocity per unit electric field averaged over the period in which electrons are moving and also the period in

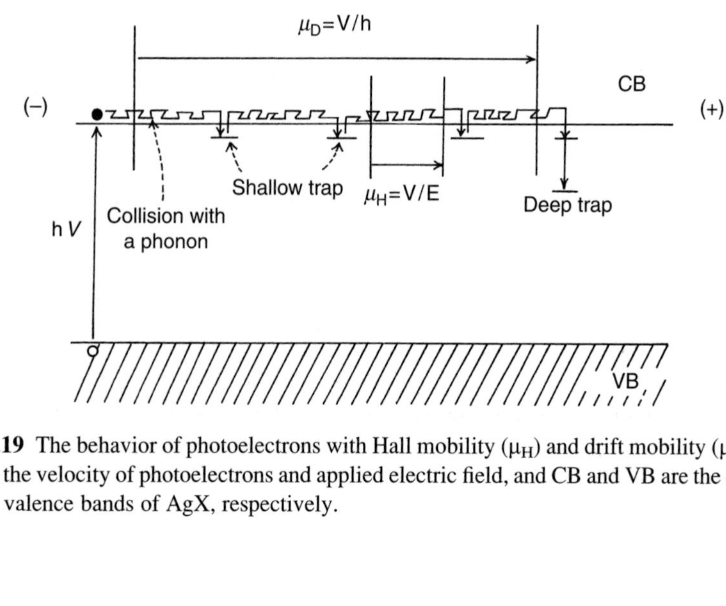

Figure 3.19 The behavior of photoelectrons with Hall mobility (μ_H) and drift mobility (μ_D). Here, v and E are the velocity of photoelectrons and applied electric field, and CB and VB are the conduction band and valence bands of AgX, respectively.

which electrons are at rest in shallow electron traps. Thus, the difference between μ_H and μ_D indicates the portion of the period in which photoelectrons are at rest in shallow traps. Similar situation and discussions are valid for the behavior of positive holes in AgX. In terms of shallow trapping, the deviation of μ_D from μ_H was interpreted by Brown and Kobayashi [142] as

$$\mu_D/\mu_H = 1/[1 + (2N_t/N_C)\exp(E/kT)], \tag{3.10}$$

where N_C is the effective density of states at the conduction edge, and N_t and E are the density and depth of the shallow traps. Equation (3.10) gives the fraction of time period during which an electron is free in the conduction band.

The electronic transport properties of AgCl and AgBr have been extensively studied and described [10,143–145]. According to those studies, a conduction electron can be represented as a charged carrier in a single isotropic and parabolic potential valley, having an effective mass of 0.399 for AgCl and 0.282 for AgBr [146].

Table 3.2 shows some electronic properties of AgX crystals. At room temperature and above, the values of electron drift mobility μ_D as given by time-of-flight measurements with high-quality single crystals of silver halides [147–149] are in agreement with values of Hall mobility μ_H obtained from the photo-Hall effect [150,151] and are 50 $cm^2V^{-1}s^{-1}$ for AgCl and 60 $cm^2V^{-1}s^{-1}$ for AgBr. The mobility is limited, not by shallow electron traps but by scattering on longitudinal optical (LO) phonons, whose probability is proportional to the LO phonon density. The mobility thus increases with decreasing temperature, as shown for AgBr in Figure 3.20 [16,149,152].

Table 3.2 Electronic properties of AgBr

Carrier	Sample	μ_D (cm^2/Vs)	τ (μs)	l (Eq. (3.11))	Reference
Electron	Large crystal	60	0.1–10	4–40 (*1)	149–151
	Emulsion grains	0.2	3	1.2[a]	153
		0.6–0.8			154,155
		2.1	0.1	2.3[a]	156
				0.4[b]	202
				0.5[c]	203
Positive hole	Large crystal	1.1	0.1–10	0.5–5[a]	171–173
				0.33[d]	206,207
				0.2[d]	208
	Emulsion grains	0.001	1.5	0.19[a]	153

[a] Evaluated according to Eq. (3.12) with observed values of μ_D and τ.

[b] Evaluated on the basis of the distance between photolytically formed adjacent silver particles on AgBr grains.

[c] Evaluated on the basis of the AgX grain size, from which the proportionality of sensitivity to the grain volume began to deviate.

[d] Evaluated on the basis of the distance to which positive holes could migrate to reach the surface of an AgBr crystal.

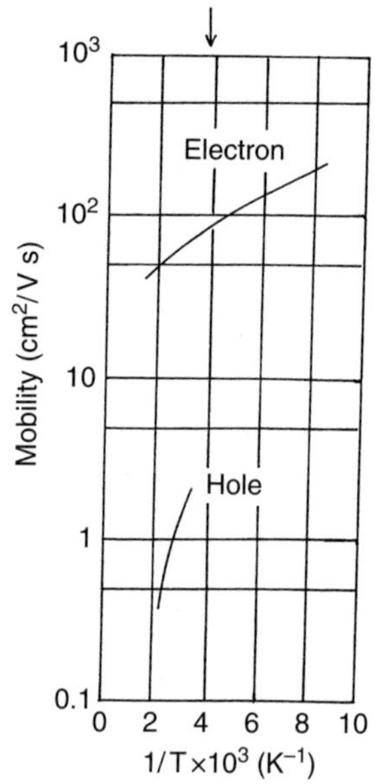

Figure 3.20 Mobilities of electrons and holes in a large crystal of AgBr in the range near room temperature. Reproduced with permission from Reference [15], copyright 1980 IS&T, The Society for Imaging Science and Technology.

The drift mobility of electrons in AgX emulsion grains was measured by several groups of workers [153–156] with values depending upon the methods and emulsion grains and included in Table 3.2. However, it is obvious that the drift mobility of electrons in emulsion grains was considerably smaller than that of electrons in a large crystal, indicating that there were many shallow traps in emulsion grains. Details of those traps are described in the next section.

The low-temperature transport properties of holes in silver halides are quite different from those of electrons [157–159]. In AgCl, positive holes are self-trapped at liquid helium temperature. Electron paramagnetic resonance studies have revealed that a positive hole is located on a silver ion with displacement of the nearest-neighbor chloride ions (Figure 3.21) [15,157,158,160–162]. The thermal ionization energy of the self-trapped hole was estimated as 0.12 eV [163,164]. Although the self-trapped hole is unstable at temperature above about 50 K, the self-trapped state severely limits the drift mobility of positive holes in AgCl at room temperature, which is about 10^{-2} $cm^2V^{-1}s^{-1}$ as given by measurements of diffusion effect [165] and much smaller than that of photoelectrons. Although it has been found from cyclotron resonance studies that positive holes are not self-trapped at liquid helium temperature in AgBr [12,166–170], the drift mobility [171–173] and Hall mobility [152,174] of positive holes in AgBr at room temperature are about 1 $cm^2V^{-1}s^{-1}$ and are much smaller than that of photoelectrons, steeply increasing with decreasing temperature (Figure 3.20) [15,174–176]. These mobility characteristics of positive holes in AgBr have been also attributed to the self-trapped state, which is regarded as metastable state in AgBr [177,178].

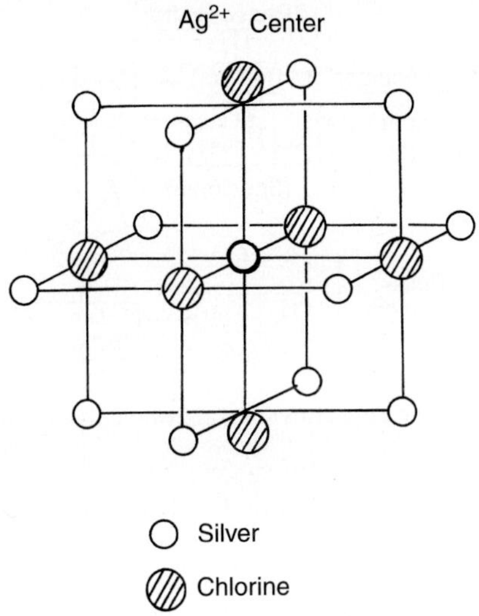

Figure 3.21 The Jahn–Teller distortion of the six chloride ions around a self-trapped hole (Ag^{2+}) in AgCl.

The drift mobility of positive holes in a large AgBr crystal is a little smaller than the Hall mobility [171–173], indicating that the movement of a positive hole in AgBr is influenced by a trapping level, whose depth is estimated as 0.44 eV according to Eq. (3.10). According to Kanzaki, a silver ion vacancy is a dominant hole center, and its binding energy was estimated as 0.4 eV from the luminescence modulation spectra [15]. Although the measured values of the drift mobility of positive holes in emulsion grains are scarce [153], it is also obvious that the drift mobility of positive holes is much smaller than the Hall mobility, indicating that there are many shallow traps. Details of those traps are described in the next section.

3.3.2 Electronic Transport in Emulsion Grains of Silver Halides

The photoconductivity and electron lifetime in AgX emulsion grains were measured by several methods. West reported the results of classical DC photoconductivity measurements on emulsion coatings, demonstrating the coincidence of the spectral range of the photoconductivity with that of the photographic sensitivity of spectrally sensitized emulsions in 1951 [176]. Extending the Haynes–Shockley experiments [82,83] with the application of synchronized pulses of light and electric field to emulsion layers as described in the previous section, Hamilton and Brady measured the electron lifetime in various emulsion grains [84,85,153] and observed a striking correlation between ionic field decay times and electron lifetimes. The results supported the absence of deep electron traps in the emulsions as proposed by Mitchell [179] and were in accord with physical properties of AgX and photographic phenomena [180]. The correlation indicated that the electron lifetime was determined by the ionic event at an otherwise shallow trapping center.

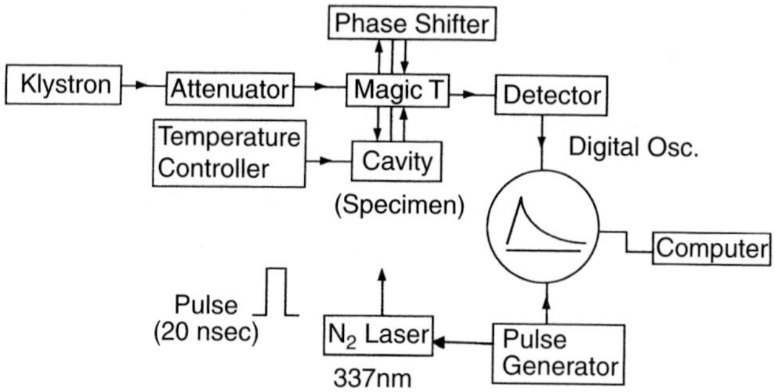

Figure 3.22 A block diagram of an apparatus for microwave photoconductivity measurements.

The electron lifetimes in emulsion grains are now frequently measured by a microwave photoconductivity technique, which was first reported by Baranov and Akimov [181], and was extended by Kellogg and co-workers [182,183], and followed by many groups of workers [154–156,184–190]. Figure 3.22 shows a diagram of an apparatus for the microwave photoconductivity measurement. The exposure of a sample (an emulsion layer) in a tuned microwave cavity to a light pulse creates electronic charge carriers, which can absorb the microwave energy and detune the microwave cavity to give the signal with intensity in proportion to the concentration of electronic charge carriers. It is considered that only free electrons are usually responsive to the signal of microwave photoconductivity, since the mobility of photoelectrons is much larger than that of positive holes in AgX as shown in Table 3.2. The time resolution of microwave photoconductivity was improved to several nanoseconds [154–156,188].

The Dember photovoltage in AgX emulsion grains was observed by Levy [191–193] and used by several workers [101,194–196]. However, it was not effectively used for the study of the electronic property of AgX emulsion grains.

By means of the extended Haynes–Shockley experiment, Hamilton and Brady measured the lifetime of positive holes in tabular AgBr emulsion grains as 15 μs [153]. Since positive holes are less mobile than photoelectrons, the detection of positive holes in AgX emulsion grains could be realized when a photoconductivity apparatus with high sensitivity was applied to the emulsion layers, in which the number and/or mobility of photoelectrons were reduced. Kellogg, Hodes, and Muenter developed such a sensitive apparatus by using radio frequency instead of microwave in the photoconductivity apparatus, and observed the lifetime of positive holes in 0.5-μm octahedral AgBr emulsion grains as 26 μs at room temperature [196]. The radio frequency photoconductivity method was followed [197] and improved [198] and effectively used for time-resolved photoconductivity measurements with high sensitivity.

In contrast to a microwave photoconductivity apparatus, in which a narrow strip of an AgX emulsion layer is put in a cavity, a wide strip is placed between two electrodes in a radio frequency photoconductivity apparatus. The sensitivity of a radio frequency photoconductivity method is therefore much higher than that of a microwave photoconductivity method. However, the former resolves only several microseconds, while the latter resolves several nanoseconds. In accord with the ionic relaxation time in AgX grains in Table 3.1, photoelectrons disappeared within a microsecond in fine AgBr grains, while they survive more than a microsecond in AgCl grains [51]. Namely, a radio frequency photoconductivity method is available for the measurements of time-resolved photoconductivity of positive holes in fine AgBr grains [196–199,51] and that of photoelectrons in AgCl grains [51,200].

The diffusion length of charge carriers l is important [201] for the analysis of size dependence of photographic sensitivity, and was estimated by

$$l = (kT\mu_d\tau/e)^{1/2}, \tag{3.11}$$

where μ_D and π are the drift mobility and lifetime of charge carriers, respectively. The values of l of electrons and positive holes in silver halides were evaluated on the basis of Eq. (3.11) with measured values of μ_D and τ, and are listed in Table 3.2. One of those methods was the evaluation of l on the basis of the distance between two adjacent photolytically formed silver particles on AgBr grains, which corresponded to the diffusion length of electrons to reach the silver particles [202]. Another method was the evaluation of l on the basis of the maximum grain size in the range with the ratio of sensitivity to grain volume, which indicated that electrons at any place in a grain could reach a growing image center [203–205]. The diffusion length of positive holes in silver halides were also evaluated on the basis of the distance from the interior to the surface of an AgBr crystal, to which positive holes could migrate for their escape to the vacuum [206–211] and listed in Table 3.2.

3.4 Electron and Positive Hole Traps in Silver Halides

3.4.1 Intrinsic Traps for Electrons and Positive Holes in Silver Halides

Shallow electron levels in large AgCl and AgBr crystals were first detected by Brandt and Brown as transient, exposure-induced absorption spectra in the far infrared at 9.3 K [212]. The spectrum had absorption peaks identified as the $1s$–$2p$ and $1s$–∞ transitions of a hydrogenic center consisting of an electron loosely bound by the Coulomb potential of a charged defect. The energies of the $1s$–$2p$ and $1s$–∞ transitions were 33.7 and 36.2 meV, respectively, for AgCl, and 20.8 and 23.8 meV, respectively, for AgBr. For a hydrogenic system, neglecting central core corrections, the $1s$–$2p$ transition energies would give binding energies of 45 meV for AgCl and 28 meV for AgBr. Determining the spectral dependence of actinic-exposure-induced absorption and conductivity, Sakuragi and Kanzaki [213,214] attributed the absorption to an electron center. Subsequent magnetic resonance studies demonstrated that those traps were paramagnetic, consisting of an electron that was shallowly bound to a charged center [215–217]. The energetics of a carrier bound to a charged center received much attention from theorists [142, 212, 218–221]. The estimated orbital radii are 15.1 Å for AgCl and 23.9 Å for AgBr.

It is noted that predominant intrinsic electron traps in large AgCl and AgBr crystals are positively charged centers with a shallow potential well with a radius of several lattice spacings and a binding energy of some 20–40 meV and have little influence upon the drift mobilities of photoelectrons in both AgCl and AgBr at room temperature or above. Silver halides with intrinsic shallow electron centers are in strong contrast to the alkali halides in which stable color centers such as the F center with a binding energy of as large as 2–3 eV are efficiently produced. It is considered that one of the reasons for the small binding energy of the intrinsic electron centers is the large dielectric constants of silver halides. Some consideration was made of similar shallow levels associated with the partially charged surface and internal kink and jog sites [180].

Intrinsic shallow electron traps in AgCl at low temperature were subjected to further studies with pulsed ENDOR [73,74,25]. It was revealed that the density of an electron bound by the trap deviated from spherical symmetry, although the electron was mainly contained in a hydrogen-like 1*s* orbital. The split interstitial geometry, as illustrated in Figure 3.7, was proposed for an intrinsic shallow electron trap in AgCl at low temperature.

It is quite probable that intrinsic electron traps on the surface of AgX emulsion grains are deeper than the above-stated positively charged center in the bulk [180, 190], as shown in detail later in this section. This is supported by the following proposal. By applying the molecular orbital method to the alternate-row model of a {111} face and {100} face of AgBr, Hoffmann et al. proposed the idea that two adjacent silver ions on the surface were likely to be subjected to the Jahn–Teller effect due to pairwise distortion to stabilize a photoelectron as illustrated in Figure 2.3 [222,223].

Descriptions are made on intrinsic traps for positive holes in AgCl and AgBr including the self-trapped states and silver ion vacancies in the previous section. The behaviors of photo-holes in AgBr and AgCl emulsion grains were studied by means of multifrequency electron paramagnetic resonance (EPR) and ENDOR [224]. In AgBr, band-gap excitation produces a neutral complex formed from hole-trapping by the four nearest neighbors of a surface silver ion vacancy (*V*) $[(Br_4)^{3-}V]$, which reacts with gelatin to produce a transient radical. In AgCl, band-gap excitation produces the self-trapped hole $(AgCl_6)^{4-}$, which is released and retrapped by a center containing four equivalent chloride ions.

The microwave photoconductivity measurement of AgX emulsion grains reflected intrinsic electron traps in AgX emulsion grains [154–156,182–190]. It is noted that mono-dispersed grains with definite shape could be prepared for AgX with varied halide composition by the controlled double-jet method as described in Chapter 2, and made it possible to get comprehensive experimental results for the electron transport in AgX emulsion grains. One of the most comprehensive observations and explanations for the decay of photoelectrons in AgCl and AgBr emulsion grains is described below [190].

The decay of photoelectrons in cubic AgBr emulsion grains with an edge length of 0.6 μm was measured at room temperature by means of a microwave photoconductivity apparatus with a high time resolution, indicating two components as shown in Figure 3.23. The decay probability (i.e., $1/\tau$, where τ is the electron lifetime) of the first component was proportional to the surface/volume ratio (i.e., $1/d$ where d is the edge length) of the grains as shown in Figure 3.24, indicating that photoelectrons were captured by intrinsic traps on the grain surface according to reaction-controlling scheme and came in equilibrium between the conduction band and the traps.

On the other hand, $1/\tau$ of the second component was proportional to the square of the surface/volume ratio of the grains (i.e., $1/d^2$) as seen in Figure 3.24. As described in the previous section, the concentration of interstitial silver ions was proportional to the surface/volume ratio (i.e., $1/d$) of the grains. In addition, $1/\tau$ was proportional to the concentration of electron traps on the grain surface (i.e., proportional to the surface/volume ratio of the grains), when the electron-trapping was reaction-controlling. It is

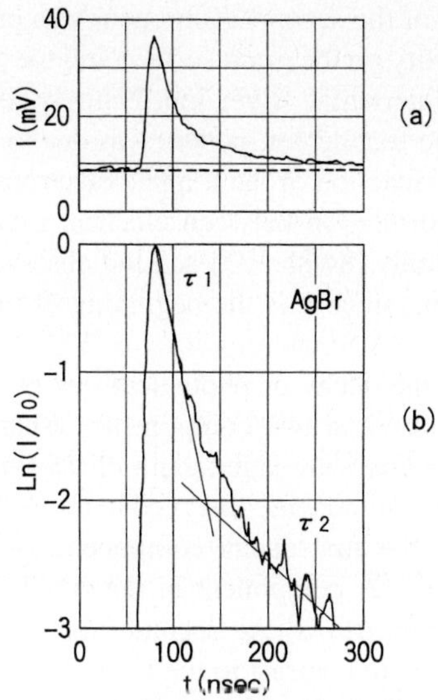

Figure 3.23 The signal of microwave photoconductivity measurement at room temperature for cubic AgBr emulsion grains with an edge length of 0.6 μm (a) and an analysis for the decay kinetics of the above signal (b), indicating two components (τ_1 and τ_2, respectively). Reproduced with permission from Reference [190], copyright 1995 IS&T, The Society for Imaging Science and Technology.

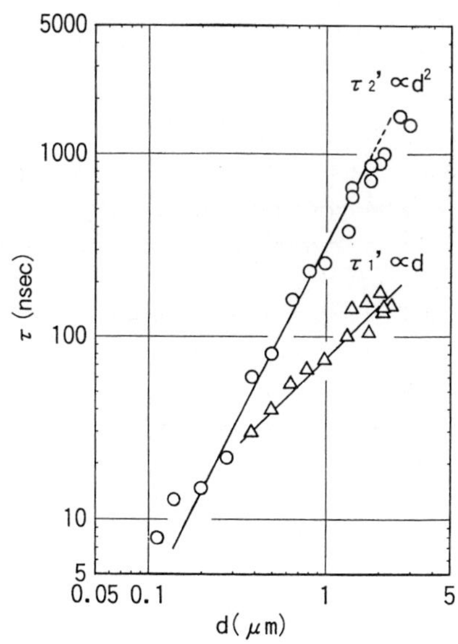

Figure 3.24 The dependence of the electron lifetimes for the first and second components in the decay of photoconductivity (τ_1 and τ_2, respectively) upon the edge length (d) of cubic AgBr emulsion grains at room temperature. Reproduced with permission from Reference [190], copyright 1995 IS&T, The Society for Imaging Science and Technology.

therefore considered that $1/\tau$ of the second component was proportional to the product of the electron-trapping probability on the grain surface and the probability of the reaction of a trapped electron with an interstitial silver ion. Namely, this result indicated that the decay of photoelectrons in the second component was due to the electron-trapping on the grain surface followed by the reaction of the trapped electrons with interstitial silver ions. This is in accord with the correlation between electron lifetimes and field-decay times observed by Hamilton and Brady [84,85,153] and the observation by Kaneda [185]. This decay component is important, since it is the beginning of the formation of latent image centers.

As shown in Figure 3.25, the decay of photoelectrons in cubic AgCl emulsion grains with edge length of 0.6 μm also had two components. When the chloride content (x) in cubic $AgCl_xBr_{1-x}$ emulsion grains with edge length of 0.6 μm was increased as shown in Figure 3.26, the second decay component, observed in the AgBr grains, disappeared when x exceeded about 50%, and the first decay component, observed in the AgBr grains, corresponded to the second decay component in the AgCl grains. However, τ (i.e., the reciprocal of the decay probability) in the second component in the AgCl grains was proportional to d^2 of the grains, in contrast to the case of the first component of the AgBr grains, in which τ was proportional to d of the grains.

According to the reaction-controlling scheme, the electron-trapping probability is expressed as [154,155],

$$1/\tau = 1/\tau_B + 6k/d, \tag{3.12}$$

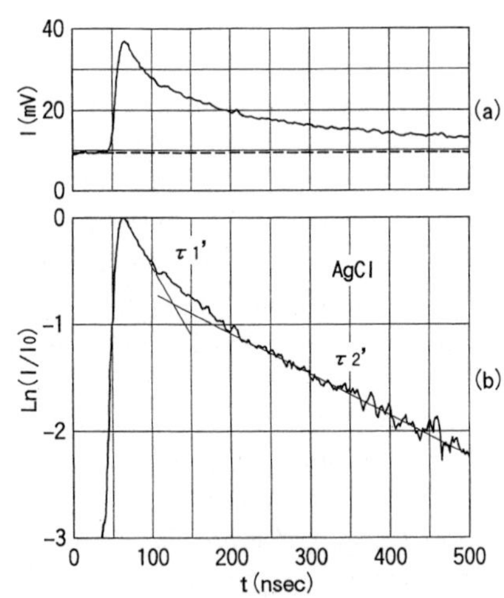

Figure 3.25 The signal of microwave photoconductivity measurement at room temperature for cubic AgCl emulsion grains with an edge length of 0.6 μm (a), and an analysis for the decay kinetics of the above signal (b), indicating two components (τ_1' and τ_2', respectively). Reproduced with permission from Reference [190], copyright 1995 IS&T, The Society for Imaging Science and Technology.

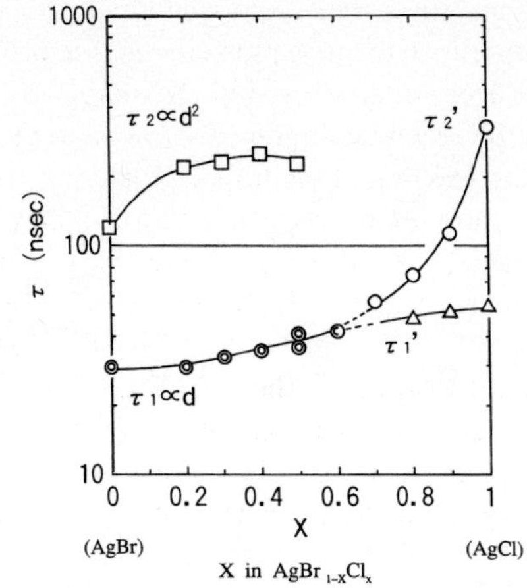

Figure 3.26 Electron lifetimes in cubic $AgCl_xBr_{1-x}$ emulsion grains with an edge length of 0.6 μm with variation of x at room temperature. Reproduced with permission from Reference [190], copyright 1995 IS&T, The Society for Imaging Science and Technology.

where τ_B is the electron lifetime in the bulk and k is the rate constant. When $1/\tau_B$ is negligibly small, τ is proportional to d of the grains. According to the diffusion-controlling scheme, the electron-trapping probability is expressed as [154,155],

$$1/\tau = 1/\tau_B + 4\pi^2 D/d^2, \tag{3.13}$$

where D is the diffusion coefficient. When $1/\tau_B$ is negligibly small, τ is proportional to d^2 of the grains.

From Eqs. (3.12) and (3.13), the proportion of τ in the second component in the AgCl grains to d^2 of the grains indicates that photoelectrons were trapped on the grain surface according to a diffusion-controlling scheme. Namely, photoelectrons were immediately and irreversibly captured by electron traps on the surface of AgCl grains when they arrived on the grain surface. This was confirmed by observing that the electron lifetime in the second decay component in the AgCl grains was independent of the presence of strong extrinsic electron traps (e.g., sulfur sensitization centers; see Section 4.5.2) in contrast to that of the first decay component in the AgBr grains, in which τ was significantly decreased by the extrinsic electron traps (Figure 3.27). The above-stated results are consistent with Keevert's observation that the electron lifetime is much shorter than the field-decay time in AgCl emulsion grains [195]. In the presence of a silver complexing agent, 4-hydroxy-6-methyl-1,3,3a,7-tetraazaindene (TAI), the electron-trapping on the AgCl grains was reaction-controlling. Therefore the above-stated electron traps on the surface of the AgCl grains are concluded to be silver ions at some surface sites, presumably at kink sites on the grain surface. Little is known about the fast component in the decay of photoelectrons in AgCl emulsion grains.

Silver halide emulsion layers on film bases were exposed to light pulses for several nanoseconds through neutral density filters in order to measure their microwave photoconductivity, and then developed in order to measure their photographic response under

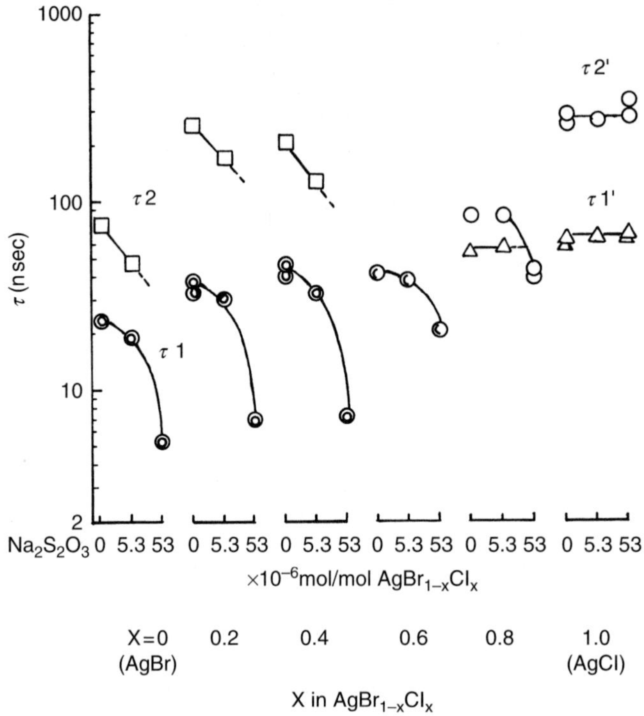

Figure 3.27 Dependence of electron lifetimes in cubic $AgCl_xBr_{1-x}$ emulsion grains with a variation of x upon extrinsic electron traps (i.e., sulfur sensitization centers provided by the digestion of the emulsions in the presence of $Na_2S_2O_3$ at 60°C for 30 min with the amount shown on the abscissa). Reproduced with permission from Reference [190], copyright 1995 IS&T, The Society for Imaging Science and Technology.

the same exposure conditions [122]. Figure 3.28 shows the photoconductivity and the photographic response as functions of exposure to light at 355 nm for several nanoseconds. In most cases, photoconductivity was in linear relationship with the square root of exposure, and therefore determined by the recombination between photoelectrons and positive holes. Although the light absorption at 355 nm was in the order of AgI, AgBr, and AgCl, the photoconductivity and photographic response were in the order of AgCl, AgBr, and AgI, being in accord with fact that the activity of positive holes is in the order of AgI, AgBr, and AgCl [127,51]. Extremely low sensitivity of AgI grains for photographic response should be at least in part due to the low developability of AgI grains [127,122].

Crystal defects in AgX provide electron traps, decreasing the lifetime of photoelectrons. Those defects include dislocations and twin planes. Hayes and Shockley [82] and Suptitz [225] found that the motion of photoelectrons was impeded by a strained region. A strain-induced reduction of the photoconductive response has also been reported [226–230]. Ozeki, Urabe, and Tani have confirmed that the electron lifetime in tabular AgBr grains, which contained twin planes, was shorter than those in cubic and octahedral AgBr grains, which did not contain such crystal defects [123].

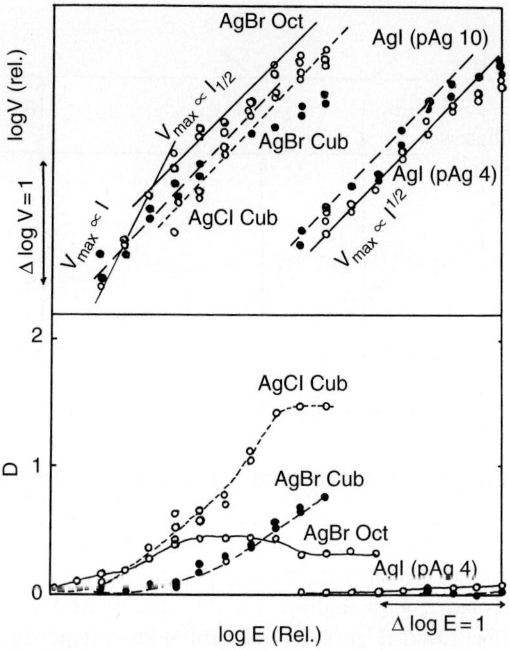

Figure 3.28 Photoconductivity (as signal intensity) with a 35-GHz microwave photoconductivity apparatus in an emulsion layer (top) and optical density of the layer, which was exposed for the microwave photoconductivity measurement and then developed, fixed, and dried (bottom) as a function of the relative exposure, for the layers containing cubic AgCl grains (0.2 μm), cubic and octahedral AgBr grains (0.2 μm), and AgI grains (0.14 μm). Reproduced with permission from Reference [122], copyright 2007 IS&T, The Society for Imaging Science and Technology.

3.4.2 Extrinsic Traps for Electrons and Positive Holes in Silver Halides

There are various kinds of extrinsic traps for electrons and holes in silver halides. They include uncharged ion centers, charged ion ones, and neutral ones. Another type of electron traps is a sulfur sensitization center, which enhances the electron-trapping by intrinsic shallow electron traps (i.e., interstitial silver ions). Typical neutral centers are dimers of silver atoms, which constitute reduction sensitization centers. Trapping of electronic carriers by sulfur sensitization and reduction sensitization centers are described later in this section.

Uncharged ion centers include shallow isoelectronic (i.e., uncharged) traps for holes such as I^- in AgBr, I^- in AgCl, and Br^- in AgCl. According to Kanzaki and Sakuragi, the trapping of a positive hole at a residual iodide impurity was dominant in nominally pure AgBr at liquid helium temperature, and the localization energy of the hole trapped at iodide ion could be estimated as 0.04 eV for an isolated iodide ion and as 0.1 to 0.15 eV for an iodide ion pair [231,17]. The deeper localization at an impurity pair compared to a single impurity was observed commonly for isoelectric traps in compound semiconductors [237].

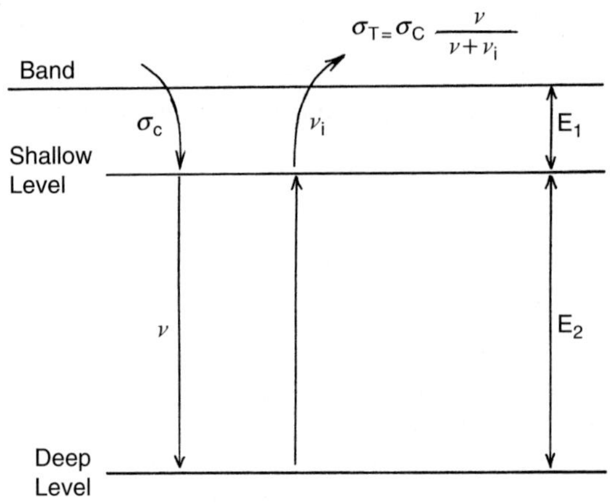

Figure 3.29 Schematic diagram of the transitions in the trapping process at deep charged centers in silver halides [232,233].

There are various charged ion centers that provide shallow Coulombic levels only or deep valence levels along with shallow ones, depending on their electronic structures. A deep valence level is provided by an atomic orbital in an impurity ion or in a molecular orbital of a ligand associated with it. As shown in Figure 3.29, Lax outlined the general scheme for trapping into any deep level [232], according to which a charge carrier was first captured into a shallow Coulombic level and then slowly relaxed into a deeper valence level associated with the trap [233]. The over-all cross-section of the deep charged trap was given by

$$\sigma = \sigma_C \eta, \tag{3.14}$$

where

$$\eta = \gamma/(\gamma + \gamma_i), \tag{3.15}$$

where γ was the rate of de-excitation from the shallow level to the deep and γ_i was the ionization rate of the carrier from the shallow level [233].

There are several examples of extrinsic shallow charged centers in silver halides. In relation to the absorption for electrons bound to charged intrinsic centers in silver halides, Sakuragi and Kanzaki [213,214] found similar absorptions for electrons bound to the Coulomb centers of doped closed-shell divalent cations such as Pb^{2+}, Cd^{2+}, and Fe^{2+}, which do not provide atomic orbitals capable of deeply capturing electrons in the conduction band of silver halides [234].

Doped transition-metal ion complexes provide electron traps with widely varying depth in silver halides. The presence of deep valence levels depends on the electronic energy levels of the *d* orbitals in the metal ions, which split owing to their interaction with ligands. The degree of splitting increases with increasing degree of interaction. An example of ligands that weakly interact with metal ions is a chloride ion, while an example of ligands that strongly interact is a cyanide ion.

This trapping mechanism by the former complex ions was observed and extended for Ir^{3+} in AgBr by Eachus and Olm [217,235]. Namely, Ir^{3+} shortened the photoelectron lifetime in AgBr to the range of pico- or nanoseconds [154,156] by shallow trapping,

$$[(IrCl_6)^{3-}2V]^0 + e^- \rightarrow [(IrCl_6)^{3-}2Ve^-]^-, \tag{3.16}$$

whereas the paramagnetic Ir^{2+} was formed to give the electron spin resonance (ESR) signal on a much longer time scale through the electronic relaxation

$$[(IrCl_6)^{3-}2Ve^-]^- \rightarrow [(IrCl_6)^{4-}2V]^-, \tag{3.17}$$

and ionic relaxation,

$$[(IrCl_6)^{4-}2V]^- \rightarrow [(IrCl_6)^{4}V]^0 + ,V^-, \tag{3.18}$$

where V is a silver ion vacancy.

The thermal trap depth for $[(IrCl_6)^{4-}V]^0$ was measured as 0.46 eV by kinetic ESR spectroscopy. In AgBr, $[(IrCl_6)^{4-}V]^0$ has a lifetime of about 20 ms at room temperature and decays by several competitive pathways, among which was the reduction of an interstitial silver ion to produce a silver atom at an adjacent site [236]. Another important pathway was the thermally assisted excitation of the trapped electron to the conduction band. It was retrapped many times at other Ir^{3+} sites, so that the time before the electron was permanently annihilated by silver formation at an image center was increased. It is therefore concluded that the addition of Ir^{3+} reduces high-intensity reciprocity law failure of emulsions when the above-stated decay mechanism predominates [217].

Although both Rh^{3+} and Ir^{3+} trap electrons to produce Rh^{2+} and Ir^{2+} and to reduce the photoelectron lifetime to about 10^{-8}–10^{-9} s [154,156], their effects are dramatically different. In the case of rhodium, the trapped electron state is so long lived that all of Rh^{3+} in the emulsion grain can be saturated. Kinetic ESR studies of the $[(RhCl_6)^{4-}]^0$ center in AgBr yield a value of about 0.8 eV for the thermal trap depth and a lifetime of $>10^5$ s at 300 K [237]. The sensitivity of the emulsion is reduced as a result of Rh^{3+} doping, but the contrast is increased, making it suitable for applications in the printing industries [217, 238,239]. The detailed structure of $[(RhCl_6)^{4-}]^0$ has been studied by means of ENDOR spectroscopy [240]. The conditions of Rh^{3+}[241] and Ir^{3+}[242] doped in AgCl were studied by means of magnetic resonance methods.

The above-stated processes as expressed by Eqs. (3.16) through (3.18) and exemplified by Ir^{3+} and Rh^{3+} could provide the general sequence of the trapping of electronic carriers by deep charged centers. A variety of trapping processes were studied on transition-metal ion complexes by varying the transition metals and ligands and the site charge and conformation of dopant-vacancy(ies) complexes [217,241–250]. Among them, some examples of unique ligands are $[Ru(NO)Cl_5]^{2-}$ [248] and [Cl5Ir(N-methylpyrazinium)]$^-$ [249,250]. The former complex deeply traps conduction band electrons and the resultant $[Ru(NO)Cl_5]^{3-}$ center has a lifetime of at least 1.6×10^9 s, and a trap depth in excess of 1.2 eV [248].

The incorporation of [Cl5Ir(N-methylpyrazinium)]$^-$ in AgCl were confirmed by means of multi-frequency EPR spectroscopy, ENDOR spectroscopy, and advanced computational techniques, and paved the way for incorporating organic moieties into silver halides. Figure 3.30 illustrates the structure of this complex and its surroundings in AgCl. On exposure to light, this dopant captures an electron in its LUMO, which is composed of a π^* orbital of the organic moiety [246,247].

Transition-metal complexes that do not introduce empty, dopant-derived levels into the band gap of AgX act as shallow electron traps, and are those with widely split d-orbitals as exemplified by metal cyanides such as $[Fe(CN)_6]^{4-}$ [251–253] and $[Ru(CN)_6]^{4-}$ [254].

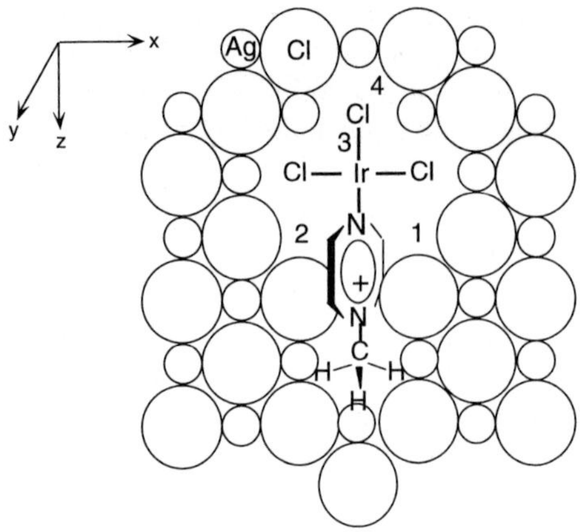

Figure 3.30 Model calculated for the geometry of the $[Cl_5Ir(NMP)]^-$ center with four bound vacancies. The dopant substitutes for $(Ag_2Cl_7)^{5-}$. The Cl^- ligands coordinated to Ir above and below the plane are omitted for clarity. The NMP ligand is rotated 45° out of the plane. Sites for the bound vacancies are marked 1–4, with position 3 above the plane and position 4 below the plane. Reproduced with permission from reference [249], copyright 2000 IOP Publishing.

It was determined that the thermal trap depth of the dopant $[Fe(CN)_6]^{4-}$ was 21 meV in AgBr. Introducing these extrinsic shallow electron traps reduces the time electrons spend free, and thus reduces the number of free electrons recombining with trapped positive holes such as those trapped by dye molecules (i.e., dye positive holes).

Among the various electron traps introduced onto AgX emulsion grains, sulfur sensitization centers and sulfur-plus-gold sensitization centers are unique and most important for efficient latent image formation [255] as described in detail in Section 4.5.2, and characterized by their electronic properties, which have been studies by many groups. The original hypothesis of Gurney and Mott was that a silver sulfide speck has a vacant electronic energy level below the bottom of the conduction band of AgX and acts as an electron trap [256,257]. The electron-trapping by sulfur sensitization centers has been confirmed by several groups [175,176,183,190,258–262].

Quantitative knowledge of sulfur sensitization and sulfur-plus-gold sensitization centers was reported afterwards. It is known that, being excited by red light, these centers can inject electrons into the conduction band of AgX with the aid of thermal energy and thus give rise to red sensitivity. Hamilton, Harbison, and Jeanmaire estimated the depth of these electron-trapping centers on octahedral AgBr emulsion grains by analyzing the temperature coefficient of their red sensitivity [260] and found that the thermal depth of sulfur sensitization centers was 0.33 eV and of sulfur-plus-gold sensitization centers was 0.19 eV. Using a thermally stimulated current technique, Kellogg and Hodes estimated the thermal depth of sulfur sensitization centers to be 0.39 eV and that of sulfur-plus-gold sensitization centers to be 0.17 eV on octahedral AgBr emulsion grains [261].

By means of an excitation spectroscopy using luminescence modulation, Kanzaki and Tadakuma observed that the electrons localized at sulfur sensitization centers give a transient absorption band, which indicated that the optical and thermal depths of sulfur sensitization centers for electron trapping are 0.5 and 0.3 eV, respectively [262] on AgBr emulsion grains. They also estimated that a sulfur sensitization center is composed of a dimer of Ag_2S by observing the relative number of centers formed as a function of the amount of Ag_2S. A similar conclusion has been also presented by Keevert and Gokhale [263]. According to Tadakuma, Yoshida, and Kanzaki's estimation, based on a quantitative analysis of the amounts of Ag_2S and sulfur sensitization centers with Smoluchouski's equation, the number of sulfur sensitization centers were as many as $2800/\mu m^2$ of (100) face of AgBr emulsion grains, when the grains were optimally sulfur-sensitized [264].

The present author proposed a model for a sulfur sensitization center, as illustrated in Figure 3.31 [265,266], consisting of a dimer of substitutional sulfide ions associated with two interstitial silver ions to compensate for the excess negative charges on the sulfide ions. An interstitial silver ion has a hydrogen-like orbital, which can loosely bind an

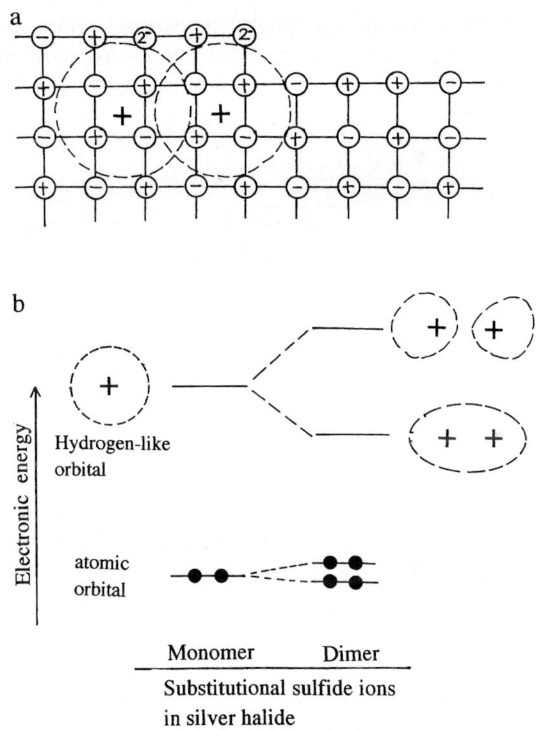

Figure 3.31 The proposed model for a sulfur sensitization center. (a) The structure of a sulfur sensitization center composed of a dimer of substitutional sulfide ions in an AgX grain, where circles with + and −, and 2− are silver ion, bromide ion, and a substitutional sulfide ion, respectively, in an AgBr grain. Two interstitial silver ions for the compensation of excess negative charges at the sulfide ions are placed at probable positions. Two broken circles centered at the interstitial silver ions represent hydrogen-like orbitals accompanying the interstitial ions. (b) The electronic energy levels of a monomer and a dimer of interstitial silver ions in an AgBr grain, where + is an interstitial silver ion. Reproduced with permission from reference [267], copyright 2000 IS&T, The Society for Imaging Science and Technology.

electron. While the two hydrogen-like orbitals interact with each other to give bonding and anti-bonding orbitals, the former acts as an electron trap for a sulfur sensitization center. A proposal was also made for a sulfur-plus-gold sensitization center formed by replacing one of the two interstitial silver ions with a gold ion [267]. Details are given in Section 4.5.3.

Examples of neutral species acting as trapping centers on AgX are reduction sensitization centers, which are composed of dimers of silver atoms [255]. There are two kinds of reduction sensitization centers. As described in 4.5.3, one of them (i.e., P center) is formed adjacent to a positively charged kink site on the grain, and acts as an electron trap. On the other hand, the other (i.e., R center) is formed adjacent to a neutral kink site, and act as a positive hole trap [268–273]. Tani et al. confirmed that there are two types of reduction sensitization centers, both of which increase photographic sensitivity. Microwave and radio frequency photoconductivity methods for measuring the photoconductivity of electrons and positive holes, respectively, in fine AgBr emulsion grains are available [274]. As shown in Figure 3.32, the photographic sensitivity of fine AgBr grains increases through two steps as the amount of a reduction sensitizer increases. The first and second arrows in this figure indicate the thresholds for the two steps. As seen here, the reduction

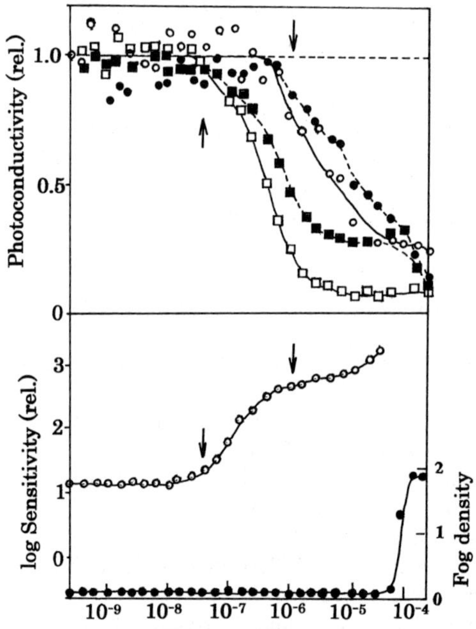

Figure 3.32. Sensitivity (○) and fog density (●) (bottom) and photoconductivity (top) of reduction-sensitized octahedral AgBr emulsion grains as a function of the amount of DMAB used for the reduction sensitization of the grains. The photoconductivity measurements were made using four methods (9-GHz microwave photoconductivity at –150°C (●), 35-GHz microwave photoconductivity at room temperature (○), radio frequency photoconductivity at room temperature with (□) and without (■) addition of an electron-trapping dye (panakryptol yellow; 1.1 m mol/mol AgBr) to emulsion layers [274]).

sensitization centers, which were ascribed to R centers, reduced the photoconductivity of positive holes, while they did not decrease that of photoelectrons. On the other hand, reduction sensitization centers, which were ascribed to P centers, reduced the photoconductivity of electrons. As described in Section 4.4.2, the capture of a positive hole by an R center is followed by the reaction, which results in the creation of a free electron available for latent image formation [275–278].

The trapping of positive holes and electrons by sensitizing dyes and desensitizing dyes is described in detail in Chapter 5.

Berry [279] outlined the depth (E) and resident time of electronic charge carriers at various traps (t) by use of Marker's equation [280]:

$$E = \ln(10^{12}t)kT. \tag{3.19}$$

According to his estimation, the values of t are 1 μs, 10 s, and 10 years, when the values of E are 0.3, 0.7, and 1.2 eV, respectively.

References

1. Y. Okamoto, *Nachr. Akad. Wiss. Goettingen, Math.-Physik. Kl. IIa, Math.-Physik.-Chem. Abt.*, 275 (No. 14, 1956).
2. W. Martienssen, *J. Phys. Chem. Solids*, 2, 257 (1957).
3. T. Tomiki, T. Miyata, H. Tsukamoto, *Z. Naturforsch.*, 29a, 145 (1974).
4. F. C. Brown, T. Masumi, H. H. Tippins, *J. Phys. Chem. Solids*, 22, 101 (1961).
5. F. Bassani, R. S. Knox, W. B. Fowler, *Phys. Rev.*, 137, A1217 (1965).
6. P. M. Scop, *Phys. Rev.*, 139, A934 (1965).
7. R. B. Aust, *Phys. Rev.*, 170, 784 (1968).
8. A. D. Brothers, D. W. Lynch, *Phys. Rev.*, 180, 911 (1969).
9. W. B. Fowler, *Phys. Stat. Sol.* (b), 52, 591 (1972).
10. F. C. Brown, in B. Hannay (ed.), *Treatise on Solid State Chemistry*, Vol. 4 (New York: Plenum Press, 1976), Chapter 7.
11. F. C. Brown, in J. H. Crawford Jr, L. M. Slifkin (eds), *Point Defects in Solids*, Vol. 1 (New York: Plenum Press, 1972), Chapter 8.
12. H. Tamura, T. Masumi, *Solid State Commun.*, 12, 1183 (1973).
13. M. G. Mason, *Phys. Rev.*, B11, 5094 (1975).
14. J. Tazela, N. J. Shevchik, W. Braun, A. Goldmann, M. Cardona, *Phys. Rev.*, B12, 1557 (1975).
15. H. Kanzaki, *Photogr. Sci. Eng.*, 24, 219 (1980).
16. A. B. Kunz, *Phys. Rev.*, B26, 2070 (1982).
17. M. Ueta, H. Kanzaki, K. Kobayashi, Y. Toyozawa, E. Hanamura, *Exitonic Processes in Solids* (New York: Springer, 1986), Chapter 6.
18. H. Kanzaki, S. Sakuragi, *J. Phys. Soc. Jpn.*, 24, 652 (1968).
19. H. Kanzaki, S. Sakuragi, *J. Phys. Soc. Jpn.*, 24, 1184 (1968).
20. H. Kanzaki, S. Sakuragi, *J. Phys. Soc. Jpn.*, 29, 924 (1970).
21. M. G. Mason, Y. T. Tan, T. J. Miller, G. N. Kwawer, F. C. Brown, A. B. Kunz, *Phys. Rev.*, B42 (5), 2996 (1990).
22. J. A. Krumhansl, in R. G. Breckenridge, B. R. Russel, E. E. Hahn (eds), *Photoconductivity Conference* (New York: Wiley, 1956), 450.
23. B. L. Joestin, F. C. Brown, *Phys. Rev.*, 148, 919 (1966).
24. K. Nakamura, W. von der Osten, *J. Phys.* C, 16, 6669 (1983).

25. M. T. Bennebroek, A. Arnold, O. G. Poluektov, *Phys. Rev.*, B54, 11276 (1996).
26. P. K. de Boer, R. A. de Groot, *J. Phys. Chem. A*, 103, 5113 (1999).
27. R. Hirsch, R. W. Pohl, *Z. Phys.*, 48, 384 (1928).
28. M. Cardona, *Phys. Rev.*, 129, 69 (1963).
29. H. Mifune, T. Ohzeki, K. Yamane, F. Nariyuki, K. Watanabe, M. Yoshikane, M. Nakanishi, T. Maekawa, T. Funakubo, *J. Imaging Sci. Technol.*, 51(3), 207 (2007).
30. P. V. Smith, D. A. Keen, *Europhys. Lett.*, 23, 129 (1993).
31. B. N. Onwuagba, *Solid State Commun.*, 97, 267 (1996).
32. A. Laref, W. Sekkal, A. Zaoui, *J. Appl. Phys.*, 86(8), 4435 (1999).
33. A. Goldmann, J. Tejeda, N. J. Shevchik, M. Cardona, *Phys. Rev.*, B10, 4388 (1974).
34. S. Ves, D. Gloetzel, M. Cardona, H. Overhof, *Phys. Rev.*, B24, 3073 (1981).
35. C. R. Berry, *Phys. Rev.*, 161, 848 (1967).
36. T. Tanaka, H. Saijo, T. Matsubara, *J. Photogr. Sci.*, 27, 60 (1979).
37. H. Kanzaki, Y. Tadakuma, *Solid State Commun.*, 80, 33 (1991).
38. K. P. Johansson, G. McLendon, A. P. Marchetti, *Chem. Phys. Lett.*, 179, 321 (1991).
39. M. Masumoto, T. Kawamura, T. Ohzeki, S. Urabe, *Phys. Rev.*, B46 1827 (1992).
40. K. P. Johansson, A. P. Marchetti, G. McLendon, *J. Phys. Chem.*, 96, 2873 (1992).
41. A. P. Marchetti, K. P. Johansson, G. L. McLendon, *Phys. Rev.*, B47 4268 (1993).
42. H. Vogelsang, O. Husberg, U. Koehler, W. von der Osten, *Phys. Rev.*, B61, 1847 (2000).
43. A. P. Marchetti, P. J. Rodney, W. von der Osten, *Phys. Rev.*, B64, 132201 (2001).
44. M. I. Freedhoff, A. P. Marchetti, G. L. McLendon, *J. Luminescense*, 70, 400 (1996).
45. M. I. Freedhoff, A. P. Marchetti, in R. E. Hummel, P. Wissmann (eds), *Handbook of Optical Properties*, vol. II: *Optics of Small Particles, Interfaces, and Surfaces* (Boca Raton, FL: CRC Press, 1997), Chapter 1, pp. 1–30.
46. H. Stolz, H. Vogelsang, W. von der Osten, in Hummel, Wissmann (eds), *Handbook of Optical Properties*, vol. II: *Optics of Small Particles, Interfaces, and Surfaces*, Chapter 2, pp. 31–53.
47. C. W. Peterson, *Phys. Rev.*, 148, 335 (1966).
48. M. Cardona, *Phys. Rev.*, B12, 1557 (1975).
49. R. S. Bauer, W. E. Spicer, *Phys. Rev.*, B14, 4539 (1976).
50. T. Tani, Y. Inami, *J. Appl. Phys.*, 88, 3601 (2000).
51. T. Tani, *J. Soc. Photogr. Sci. Technol. Jpn.*, 72, 88 (2009).
52. K. Seki, T. Tani, *Oyo Butsuri*, 60, 1018 (1991).
53. K. Seki, H. Yanagi, Y. Kobayashi, T. Ohta, T. Tani, *Phys. Rev.*, B49, 2760 (1994)
54. K. Seki, T. Tani, H. Ishii, *Thin Solid Films*, 273, 20 (1996).
55. H. Ishii, K. Sugiyama, E. Ito, K. Seki, *Adv. Mater.*, 11, 605 (1999).
56. R. W. Berriman, P. B. Gilman, Jr, *Photogr. Sci. Eng.*, 17, 235 (1973).
57. W. Vanassche, *J. Photogr. Sci.*, 21, 180 (1973).
58. P. B. Gilman, Jr, *Photogr. Sci. Eng.*, 18, 475 (1974).
59. C. R. Berry, *Photogr. Sci. Eng.*, 19, 93 (1975).
60. T. Tani, T. Suzumoto, K. Yoshihara, *J. Imaging Sci. Tech.*, 49, 526 (2005).
61. F. Granzer, *J. Imaging Sci.*, 33, 207 (1989).
62. B. Pischel, F. Granzer, *J. Imaging Sci.*, 35, 136 (1991).
63. F. Granzer, B. Pischel, *J. Soc. Photogr. Sci. Technol. Jpn.*, 53, 148 (1990).
64. F. Granzer, *J. Soc. Photogr. Sci. Technol. Jpn.*, 54, 284 (1991).
65. M. Kawasaki, H. Hada, H. Uchida, *J. Appl. Phys.*, 60, 3945 (1086).
66. F. Seitz, in S. Fujisawa (ed.), *Photographic Sensitivity* (Hakone Symposium, 1953), Vol. 1 (Tokyo: Maruzen, 1956), 5.
67. C. R. Berry, *Phys. Rev.*, 82, 422 (1951).
68. J. D. Eshelby, *J. Appl. Phys.*, 24, 1249 (1953).
69. C. R. Berry, *J. Appl. Phys.*, 24, 658 (1953).

70. R. D. Fouchaux, R. O. Simons, *Phys. Rev.*, 136A, 1664 (1964).
71. R. J. Friauf, in A. Baldereschi, W. Czaja, E. Tosatti, M. Tosi (eds), *The Physics of Latent Image Formation in Silver Halides* (Singapore: World Scientific, 1984), 79.
72. K. J. Kao, Doctoral Thesis, University of Kansas, cited at p. 95, in Baldereschi, Czaja, Tosatti, Tosi (eds), *The Physics of Latent Image Formation in Silver Halides*.
73. M. T. Bennebroek, O. G. Poluektov, A. J. Zakrzewski, P. G. Baranov, J. Schmidt, *Phys. Rev. Lett.*, 74, 442 (1995).
74. M. T. Bennebroek, A. Amold, O. G. Poluektov, P. G. Baranov, J. Schmidt, *Phys. Rev.*, B53, 15607 (1996).
75. R. C. Baetzold, R. S. Eachus, *J. Phys.: Condens. Mater.*, 7, 3991 (1995).
76. E. Koch, C. Wagner, *Z. Physik. Chem.*, 38B, 295 (1937).
77. J. Teltow, *Ann. Physik.*, (6)5, 63 (1949).
78. J. Teltow, *Ann. Physik.*, (6)5, 71 (1949).
79. J. Teltow, *Z. Physik. Chem.*, 195, 197 (1950).
80. J. Teltow, *Z. Physik. Chem.*, 195, 213 (1950).
81. I. Ebert, J. Teltow, *Ann. Physik.*, (6)15, 268 (1955).
82. J. R. Hayes, W. Scockley, in *Report on a Conference on the Strength of Solids, University of Bristol, 1947* (London: The Physical Society, 1948), 151.
83. J. R. Hayes, W. Shockley, *Phys. Rev.*, 82, 935 (1951).
84. F. Hamilton, L. E. Brady, *J. Appl. Phys.*, 30, 1902 (1959).
85. J. F. Hamilton, L. E. Brady, *J. Phys. Chem.*, 66, 2384 (1962).
86. E. Brady, J. F. Hamilton, *J. Appl. Phys.*, 37, 6 (1966).
87. E. Brady, J. F. Hamilton, *J. Appl. Phys.*, 35, 1565 (1964).
88. J. W. Mitchell, *J. Photogr. Sci.*, 5, 49 (1957).
89. J. van Biesen, *J. Appl. Phys.*, 41(5), 1910 (1970).
90. S. Takada, *J. Appl. Phys.*, 12, 190 (1973).
91. S. Takada, *Photogr. Sci. Eng.*, 18, 500 (1974).
92. S. Takada, *Photogr. Sci. Eng.*, 19, 214 (1975).
93. S. Takada, *J. Soc. Photogr. Sci. Tech. Jpn.*, 44, 81 (1981).
94. H. A. Hoyen, Jr, *J. Appl. Phys.*, 47, 3784 (1976).
95. E. van Hulle, W. Maenhout-van der Vorst, *Phys. Stat. Sol.*(a), 39, 253 (1977).
96. M. E. van Hulle, W. Maenhout-van der Vorst, *Phys. Stat. Sol.*(a), 40, K57 (1977).
97. M. E. van Hulle, W. Maenhout-van der Vorst, *Phys. Stat. Sol.*(a), 52, 277 (1979).
98. F. Callens, W. Maenhout-van der Vorst, *Phys. Stat. Sol.*(a), 50, K175 (1978).
99. F. Callens, W. Maenhout-van der Vorst, *Phys. Stat. Sol.*(a), 71, K61 (1982).
100. F. Callens, W. Maenhout-van der Vorst, L. Ketellaper, *Phys. Stat. Sol.*(a), 70, 189 (1982).
101. M. Saito, *J. Photogr. Sci.*, 24, 205 (1976).
102. T. Tani, Y. Sano, S. Takada, M. Saito, *J. Photogr. Sci.*, 28, 28 (1980).
103. Sr. J. Feldott, A. P. Blose, B. D. Lounsberry, L. Slifkin, *J. Imaging Sci.*, 29, 39 (1985).
104. R. C. Baetzold, J. F. Hamilton, *Surface Sci.*, 33, 461 (1972).
105. R. C. Baetzold, J. F. Hamilton, *Surface Sci.*, 179, L85 (1987).
106. N. Starbov, A. Buroff, J. Malinowski, *J. Imaging Sci.*, 23, 44 (1975).
107. N. Starbov, A. Buroff, J. Malinowski, *Thin Solid Films*, 36, 175 (1976).
108. N. Starbov, A. Buroff, J. Malinowski, *Phys. Stat. sol.*(a), 37, 161 (1976).
109. N. Starbov, *J. Signalaufzeichnungsmateriallen*, 10, 17 (1982).
110. N. Starbov, *J. Inform. Record. Mater.*, 13, 305 (1985).
111. V. I. Saunders, R. W. Tyler, W. West, *Photogr. Sci. Eng.*, 12, 90 (1968).
112. F. Trautweiler, *Photogr. Sci. Eng.*, 12, 98 (1968).
113. N. Starbov, J. Malinowski, *J. Photogr. Sci.*, 32, 245 (1984).

114. G. Farlow, A. P. Blose, Sr. J. Feldott, B. O. Lounsberry, L. Slifkin, *Radiat. Effects*, 75, 1 (1983).
115. R. A. Hudson, G. C. Farlow, L. M. Slifkin, *Phys. Rev.*, B36, 4651 (1987).
116. R. A. Hudson, S. K. Wonnell, G. C. Farlow, L. M. Slifkin, *J. Imaging Sci.*, 34, 101 (1990).
117. S. K. Wonnel, L. M. Slifkin, *Phys. Rev.*, B48, 78 (1993).
118. S. Takada, T. Tani, *J. Appl. Phys.*, 45, 4767 (1974).
119. T. Tani, S. Takada, *Photogr. Sci. Eng.*, 18, 620 (1974).
120. T. Tani, T. Suzumoto, *J. Appl. Phys.*, 70, 3626 (1991).
121. T. Takada, *J. Soc. Photogr. Sci. Tech. Jpn.*, 42, 112 (1979).
122. T. Tani, *J. Imaging Sci. Technol.*, 51, 197 (2007).
123. K. Ohzeki, S. Urabe, T. Tani, *J. Imaging Sci.*, 34, 136 (1990).
124. J. F. Hamilton, L. E. Brady, *J. Appl. Phys.*, 35, 414 (1964).
125. J. F. Hamilton, *Photogr. Sci. Eng.*, 11, 57 (1967).
126. T. Funakubo, H. Mifune, T. Ohzeki, K. Ohzeki, K. Yamane, *J. Imaging Sci. Technol.*, 51, 212 (2007).
127. T. Tani, H. Mifune, S. Yamashita, S. Aiba, T. Ohzeki, K. Yamane, *J. Imaging Sci. Technol.*, 51, 202 (2007).
128. C. R. Berry, *Photogr. Sci. Eng.*, 20, 29 (1976).
129. G. Cochrace, N. H. Fletcher, *J. Phys. Chem.*, 32, 2557 (1971).
130. R. C. Baetzold, *Phys. Rev.*, B52, 11424 (1995).
131. J. Frenkel, in *Kinetic Theory of Liquids* (New York: Dover, 1955), 36.
132. J. Frenkel, *Z. Physik*, 35, 652 (1926).
133. K. Kliewer, K. Koehler, *J. Phys. Chem Solids*, 27, 705 (1966).
134. R. B. Poeppel, J. M. Blakely, *Surface Sci.*, 15, 507 (1969).
135. H. A. Hoyen, Jr, *Photogr. Sci. Eng.*, 17, 188 (1973).
136. Y. T. Tan, H. A. Hoyen, Jr, *Surface Sci.*, 36, 242 (1973).
137. Y. T. Tan, H. A. Hoyen, Jr, *J. Photogr. Sci.*, 26, 89 (1978).
138. H. A. Hoyen, Jr, Y. T. Tan, *J. Colloid Interface Sci.*, 79, 525 (1981).
139. V. I. Saunders, R. W. Tyler, W. West, *Photogr. Sci. Eng.*, 16, 87 (1972).
140. V. I. Arkhipov, H. Baessler, in W. Bruetting (ed.), *Physics of Organic Semiconductors* (Weinheim: Wiley-VCH, 2005), 183–233.
141. G. Le, Comber, W. E. Spear, *Phys. Rev. Lett.*, 25, 509 (1970).
142. F. C. Brown, K. Kobayashi, *J. Phys. Chem. Solids*, 8, 300 (1959).
143. W. von der Osten, in Baldereschi, Czaja, Tosatti, Tosi (eds), *The Physics of Latent Image Formation in Silver Halides*, 1.
144. R. Evrard, in Baldereschi, Czaja, Tosatti, Tosi (eds), *The Physics of Latent Image Formation in Silver Halides*, 57.
145. J. F. Hamilton, *Adv. Phys.*, 37, 359 (1988).
146. J. W. Hodby, G. P. Russell, F. M. Peeters, J. T. Devreese, D. M. Larsen, *Phys. Rev. Lett.*, 58, 1471 (1987).
147. D. C. Burnham, F. C. Brown, R. S. Knox, *Phys. Rev.*, 119, 1560 (1960).
148. S. van Heyningen, *Phys. Rev.*, 128, 2112 (1962).
149. K. Ahrenkiel, *Phys. Rev.*, 180, 859 (1969).
150. F. C. Brown, *Phys. Rev.*, 92, 502 (1953).
151. T. Masumi, R. K. Ahrenkiel, F. C. Brown, *Phys. Stat. Sol.*, 11, 163 (1965).
152. R. C. Hanson, F. C. Brown, *J. Appl. Phys.*, 31, 210 (1960).
153. J. F. Hamilton, L. E. Brady, *J. Appl. Phys.*, 30, 1893 (1959); an alternative explanation for the measured values of drift mobilities in this paper was given later by one of its authors (J. F. Hamilton, *Photogr. Sci. Eng.*, 18, 493 (1974)).
154. R. J. Deri, J. P. Spoonhower, *Appl. Phys. Lett.*, 43, 65 (1983).

155. R. J. Deri, J. P. Spoonhower, *J. Appl. Phys.*, 57, 2806 (1985).
156. A. Hirano, 'Highly Time Resolved Microwave Photoconductivity Technique Applied to Measurements of Drift Mobility and Capture Cross Sections in AgBr Microcrystals', in the preprint book of The SPSE Annual Conference, May, 1990, Rochester, NY.
157. M. Hohne, M. Stasiw, *Phys. Stat. Sol.*, 25, K55 (1968).
158. M. Hohne, M. Stasiw, *Phys. Stat. Sol.*, 28, 247 (1968).
159. Y. Toyozawa, *Prog. Theor. Phys., Kyoto*, 26, 29 (1961).
160. C. L. Marquart, R. T. Williams, M. N. Karler, *Solid State Commun.*, 9, 2285 (1971).
161. M. Fukui, Y. Hayashi, H. Kanzaki, S. Sakuragi, *J. Phys. Soc. Jpn.*, 30, 1510 (1971).
162. M. Fukui, Y. Hayashi, H. Yoshioka, *J. Phys. Soc. Jpn.*, 34, 1226 (1973).
163. E. Laredo, W. B. Paul, L. G. Rowan, L. Slifkin, *Phys. Rev.*, B27, 2470 (1983).
164. E. Laredo, W. B. Paul, S. E. Wang, L. G. Rowan, L. Slifkin, in Baldereschi, Czaja, Tosatti, Tosi (eds), *The Physics of Latent Image Formation in Silver Halides*, 35.
165. P. Mueller, S. Spenke, J. Teltow, *Phys. Stat. Sol.*, 41, 81 (1070).
166. H. Tamura, T. Masumi, *J. Phys. Soc. Jpn.*, 30, 897 (1971).
167. H. Tamura, *Solid State Commun.*, 10, 297 (1972).
168. J. W. Hodby, *Solid State Commun.*, 7, 811 (1969).
169. J. W. Hodby, *J. Phys. C*, 4, L8 (1971).
170. J. W. Hodby, J. G. Crowder, C. C. Bradley, *J. Phys. C*, 7, 3033 (1974).
171. R. K. Ahrenkiel, R. S. van Heyningen, *Phys. Rev.*, 144, 576 (1966).
172. M. Georgiew, J. Malinowski, *J. Phys. Chem. Solids*, 28, 931 (1967).
173. A. Buroff, J. Malinowski, *Phys. Stat. Sol.*, 26, 267 (1968).
174. R. C. Hanson, *J. Phys. Chem.*, 66, 2376 (1962).
175. W. West, B. H. Carroll, *J. Chem. Phys.*, 15, 529 (1947).
176. W. West, in J. W. Mitchell (ed.), *Fundamental Mechanisms of Photographic Sensitivity* (Proc. Bristol Symp., 1950) (London: Butterworths, 1951), 99.
177. Y. Toyozawa, A. Sumi, *Twelfth International Conference on Physics of Semiconductors* (Stuttgart: Teubner, 1974), 179.
178. H. Kanzaki, *Semicond. Insulators*, 3, 285 (1978).
179. J. W. Mitchell, *Sci. Ind. Photogr.* (2)28, 493 (1957).
180. J. F. Hamilton, *Photogr. Sci. Eng.*, 18, 493 (1974).
181. E. V. Baranov, I. A. Akimov, *Dokl. Akad. Nauk. SSSR*, 154, 7 (1964).
182. L. M. Kellogg, N. B. Liebert, T. H. James, *Photogr. Sci. Eng.*, 16, 115 (1972).
183. L. M. Kellogg, *Photogr. Sci. Eng.*, 18, 378 (1974).
184. J. Beutel, *Photogr. Sci. Emg.*, 19, 96 (1975).
185. T. Kaneda, *J. Imaging Sci.*, 33, 115 (1989).
186. T. Harada, T. Iijima, T. Koitabashi, *Photogr. Sci. Eng.*, 26, 137 (1982),
187. T. Nakayama, *J. Soc. Photogr. Sci. Tech. Jpn.*, 50, 459 (1987).
188. R. J. Deri, J. P. Spoonhower, J. F. Hamilton, *J. Appl. Phys.*, 57, 1968 (1985).
189. A. Hasegawa, O. Koseki, T. Sakaguchi, *J. Soc. Photogr. Sci. Tech. Jpn.*, 49, 509 (1986).
190. T. Oikawa, N. Saeki, T. Kaneda, A. Hirano, T. Tani, *J. Imaging Sci. Tech.*, 39, 233 (1995).
191. B. Levy, *Photogr. Sci. Eng.*, 15, 278 (1971).
192. K. C. Chang, F. P. Chen, B. Levy, *Photogr. Sci. Eng.*, 24, 55 (1980).
193. K. C. Chang, F. P. Chen, B. Levy, *Photogr. Sci. Eng.*, 25, 138 (1981).
194. G. Vlasov, P. V. Meiklyar, *Photogr. Sci. Eng.*, 17, 343 (1973).
195. J. E. Keevert, in the preprint book of *The 28th Annual Conference and Seminar on Quality Control, Denver, 1975* (Washington, DC: Society Photogr. Sci. Eng.), 186.
196. L. M. Kellogg, J. Hodes, J. Muenter, 'Measurement of Photocarrier Decay Times in Silver Halide Emulsion Grains Using Radiofrequency Techniques', in the preprint book of *The*

International East-West Symposium II on Factors Influencing the Efficiency of Photographic Imaging, Kona, Hawaii, October/November, 1988.
197. A. Hirano, *J. Imaging Sci. Technol.*, 42, 516 (1998).
198. K. Hosoi, A. Hirano, T. Tani, *J. Imaging Sci. Technol.*, 46, 76 (2002).
199. T. Tani, T. Tasaka, M. Murofushi, K. Hosoi, A. Hirano, *Imaging Sci. J.*,47, 1 (1999).
200. T. B. Brust, M. R. Mis, K. J. Lushington, 'Physical and Photographic Effects of Near-Surface Iodide Addition on {100}-Surface Tabular Silver Chloride Emulsions', in *The Proceedings of the Intern. Cong. Imaging Sci. September, 1998, Antwerp*, Vol 1, 11 (1998).
201. J. W. Mitchell, *J. Imaging Sci.*, 34, 113 (1990).
202. G. C. Farnell, R. B. Flint, J. B. Chanter, *J. Photogr. Sci.*, 13, 25 (1965).
203. R. Matejec, E, Moisar, *Photogr. Sci. Eng.*, 12, 133 (1968).
204. G. C. Farnell, *J. Photogr. Sci.*, 17, 116 (1969).
205. T. Tani, *J. Imaging Sci.*, 29, 165 (1985).
206. G. W. Luckey, *J. Phys. Chem.*, 57, 791 (1953).
207. G. W. Lukey, *J. Chem. Phys.*, 23, 882 (1955).
208. H. Kanzaki, T. Mori, *Smicond. Insul.*, 5, 401 (1983).
209. T. Tani, in the preprint of *The Autumn Meeting of Soc. Photogr. Sci. Tech. Japan, November, 1989, Kyoto.*
210. J. Malinowski, V. Platikanova, *Phys. Stat. Sol.*, 6, 885 (1964).
211. V. Platikanova, J. Malinowski, *Phys. Stat. Sol.*, 14, 205 (1966).
212. R. C. Brandt, F. C. Brown, *Phys. Rev.*, 181, 1241 (1969).
213. S. Sakuragi, H. Kanzaki, *Phys. Rev. Lett.*, 38, 1302 (1977).
214. H. Kanzaki, S. Sakuragi, *Photogr. Sci. Eng.*, 17, 69 (1973).
215. J. Z. Brescia, R. S. Eachus, R. Janes, M. T. Olm, *Cryst. Latt. Def. Amorph. Mater.*, 17, 165 (1987).
216. R. S. Eachus, R. Janes, M. T. Olm, in *Fifth Europhysical Topical Conference on Lattice Defect in Ionic Crystals, Madrid, 1986*, 259.
217. A. P. Marchetti, R. S. Eachus, in D. Volman, G. Hammond, D. Neckers (eds), *Advances in Photochemistry*, Vol. 17 (London: Wiley, 1992), 145.
218. K. K. Bajaj, T. D. Clark, *Phys. Stat. Sol.* (b), 52, 195 (1972).
219. M. H. Eigineer, N. Tozar, *Phys. Rev.*, B5, 3029 (1972).
220. M. Brauwers, R. Evrard, E. Kartheuser, *Solid State Commun.*, 19, 857 (1976).
221. M. Grinberg, S. Legowski, H. Meczynska, *Phys. Stat. Sol.* (b), 52, 195 (1972).
222. A. S. Malik, F. J. DiSalvo, R. Hoffman, J. T. Blair, *J. Imaging Sci. Technol.*, 42, 210 (1998).
223. A. S. Malik, J. T. Blair, W. A. Bernett, F. J. DiSalvo, R. Hoffmann, *J. Solid State Chem.*, 146, 516 (2000).
224. R. S. Eachus, Th. D. Pawlik, R. C. Baetzold, *J. Phys.: Condens. Mater.*, 7, 8893 (2000).
225. P. Suptitz, *Z. Elektrochem.*, 63, 400 (1959).
226. K. Hecht, *Z. Physik.*, 77, 235 (1932).
227. W. Lehfeldt, *Nachr. Ges. Wiss. Goettingen, Math.-Physik. Kl. Fachgruppe* 2, 1 (ns), 171 (1935).
228. K. Yamada, S. Oka, *Naturwissenschaften*, 43, 175 (1956).
229. R. Matejec, *Naturwissenschaften* 43, 533 (1956).
230. R. Matejec, *Z. Physik*, 148, 454 (1957).
231. H. Kanzaki, S. Sakuragi, *J. Phys. Soc. Jpn.*, 27, 109 (1969).
232. M. Lax, *Phys. Rev.*, 119, 1502 (1960).
233. R. M. Gibb, G. J. Rees, B. W. Thomas, B. L. H. Wilson, B. Hamilton, D. R. Wight, N. F. Mott, *Phil. Mag.*, 36, 1021 (1977).
234. T. Tani, M. Saito, *Nature*, 242, 47 (1973).
235. R. S. Eachus, M. T. Olm, *Cryst. Latt. Def. Amorph. Mater.*, 18, 297 (1989).

236. R. S. Eachus, in Baldereschi, Czaja, Tosatti, Tosi (eds), *The Physics of Latent Image Formation in Silver Halides*, 249.
237. R. S. Eachus, R. E. Graves, *J. Chem. Phys.*, 61, 2860 (1974).
238. H. Welzel, *Z. Wiss. Photogr.*, 58, 59 (1964).
239. J. Assa, E. Spasova, G. Danev, *J. Signalaufz.-Mater.*, 12, 41 (1984).
240. M. T. Olm, J. R. Niklas, J.-M. Spaeth, M. C. Symons, *Phys. Rev.*, B38, 4343 (1988).
241. H. Vercammen, T. Ceulemans, D. Schoemaker, F. Callens, D. Vandenbroucke, *J. Appl. Phys.*, 84, 4414 (1998).
242. H. V. F. Callens, P. Matthys, D. Vandenbroucke, *Solid State Commun.*, 110, 87 (1999).
243. K. Sabbe, H. Vrielinck, F. Callens, D. Vandenbroucke, *J. Phys.: Condens. Matter*, 12, 10611 (2000).
244. R. S. Eachus, R. E. Graves, M. T. Olm, *J. Chem. Phys.*, 69, 4580 (1978).
245. R. S. Eachus, R. E. Graves, *J. Chem. Phys.*, 65, 5445 (1976).
246. R. S. Eachus, R. E. Graves, M. T. Olm, *Phys. Stat. Sol.* (a) 57, 429 (1980).
247. J. R. Buntaine, R. S. Eachus, M. T. Olm, *Radiat Eff Defect Solid*, 119–121, 621 (1991).
248. W. G. McDugle, R. S. Eachus, M. T. Olm, in *44th Annual Conference of the SPSE, St Paul, 1991*.
249. R. S. Eachus, Th. D. Pawlik, R. C. Baetzold, D. A. Crosby, W. G. McDugle, *J. Phys.: Condens. Matter*, 12, 2535 (2000).
250. Th. D. Pawlik, R. S. Eachus, R. C. Baetzold, W. G. McDugle, *J. Phys.: Condens. Matter*, 12, 2555 (2000).
251. M. T. Bennebroek, J. Schmidt, R. S. Eachus, M. T. Olm, *J. Phys.: Condens. Matter*, 9, 3227 (1997).
252. R. C. Baetzold, *J. Phys. Chem. B*, 101, 1130 (1997).
253. M. T. Olm, R. S. Eachus, *Radiat Eff Deffect Solid*, 150, 71 (1999).
254. H. Vercammen, D. Schoemaker, D. Vandenbroucke, *J. Imaging Sci. Tech.*, 42, 416 (1998).
255. T. Tani, *Photographic Sensitivity: Theory and Mechanisms* (New York/Oxford: Oxford University Press, 1995), Chapter 6.
256. R. W. Gurney, N. F. Mott, *Proc. R. Soc., London A*, 164, 151 (1938).
257. N. F. Mott, R. W. Gurney, *Electronic Processes in Ionic Crystals* (Oxford: Clarendon Press, 1948), Chapter 7.
258. S. S. Collier, *Photogr. Sci. Eng.*, 20, 43 (1976).
259. T. Tani, M. Saito, *Photogr. Sci. Eng.*, 23, 323 (1979).
260. J. F. Hamilton, J. M. Harbison, D. L. Jeanmaire, *J. Imaging Sci.*, 32, 17 (1988).
261. L. M. Kellogg, J. Hodes, in the preprint book of *The 40th Annual Conference of SPSE, Rochester, New York, May, 1987*.
262. H. Kanzaki, Y. Tadakuma, *J. Phys. Chem. Solids*, 55, 631 (1994).
263. J. E. Keevert, V. V. Gokhale, *J. Imaging Sci.*, 31, 243 (1987).
264. Tadakuma, Y. Yoshida, H. Kanzaki, in the preprint book of *The Annual Conference of Soc. Photogr. Sci. Technol. Japan, May, 1990, Makuhari, Chiba*.
265. T. Tani, *J. Imaging Sci. Tech.*, 39, 386 (1995).
266. T. Tani, *J. Imaging Sci. Tech.*, 42, 135 (1998).
267. T. Tani, Y. Yoshida, *J. Imaging Sci. Tech.*, 44, 242 (2000).
268. H. E. Spencer, L. E. Brady, J. F. Hamilton, *J. Opt. Soc. Am.*, 57, 1020 (1967).
269. T. Tani, *Photogr. Sci. Eng.*, 15, 181 (1971).
270. T. Tani, *Photogr. Sci. Eng.*, 18, 569 (1974).
271. J. F. Hamilton, R. C. Baetzold, *Photogr. Sci. Eng.*, 25, 189 (1981).
272. T. Tani, M. Murofushi, *J. Imaging Sci. Tech.*, 38, 1 (1994).
273. T. Tani, *J. Imaging Sci. Tech.*, 41, 577 (1997).
274. T. Tani, T. Tasaka, M. Murofushi, K. Hosoi, A. Hirano, *Imaging Sci. J.*, 47, 1 (1999).

275. J. W. Mitchell, *Rep. Prog. Phys.*, 20, 433 (1957).
276. J. M. Harbison, H. E. Spencer, in James (ed.), *The Theory of the Photographic Process*, 152.
277. T. Tani, *J. Imaging Sci.*, 30, 41 (1986).
278. R. K. Hailstone, N. B. Leibert, M. Levy, R. T. McCleary, S. R. Gilolmo, D. L. Jeanmaire, C. R. Boda, *J. Imaging Sci.*, 32, 113 (1988).
279. C. R. Berry, *J. Photogr. Sci.*, 18, 169 (1970).
280. R. E. Marker, *J. Opt. Soc. Am.*, 44, 625 (1954).

4

Processes for Photographic Sensitivity

4.1 Framework of Current Mechanism of Latent Image Formation

As shown in Figure 1.3, the photographic processes for image formation are the capture of an incident photon by a silver halide (AgX) grain or a sensitizing dye molecule on the grain, the formation of a latent image center composed of a cluster of silver atoms on the grain, and the image formation by the catalytic action of a latent image center for the reduction of the grain by a developer to silver [1–3].

Figure 4.1 illustrates the framework of the current mechanism for the formation of a latent image center on an AgBr grain, being based on the proposals by many scientists including Sheppard et al. [4], Gurney and Mott [5], Burton and Berg [6–11], Mitchell et al. [12–15], and Hamilton et al. [1,16]. As described in Chapter 3, an AgBr grain contains many interstitial silver ions and shallow electron traps, which provide sites suitable for the formation of a latent image center. In particular, a latent image center is efficiently formed at one of electron traps provided by sulfur sensitization and sulfur-plus-gold sensitization (i.e., sulfur sensitization centers and sulfur-plus-gold sensitization centers) on an AgX grain. Details of those chemical sensitizations are given in Section 4.5.

As described in Chapter 3, a photoelectron and a positive hole are efficiently created as a result of the light absorption of an AgX grain. A photoelectron migrates and is then captured by one of the electron traps provided by chemical sensitization. This is known as the 'electronic process' for latent image formation. An interstitial silver ion migrates and reacts with the captured electron to form a silver atom. This is called the 'ionic process'. The repetition of the electronic and ionic processes one after another at the same site leads to the formation of a latent image center composed of a cluster of silver atoms.

Although a photolytically formed single silver atom, which is called a latent pre-image center, is unstable, it can trap a free electron and then capture a second interstitial silver ion to form Ag_2, which is stable and is called a latent sub-image center. The growth of a latent sub-image center by the repetition of the electronic and ionic processes leads to the formation of a latent image center. Namely, a latent image center forms through two distinct stages, which are termed nucleation and growth [6–11]. The formation of a stable latent sub-image center through the formation of an unstable latent pre-image center completes the nucleation stage and provides the foothold for latent image formation. The nucleation-and-growth model also proposes that the growth of a silver cluster may continue indefinitely by alternate repetition of the electron-trapping and silver ion events in the same sequence. It is estimated that the smallest latent image center is composed of three atoms on a sulfur-plus-gold-sensitized AgX grain and of four to five atoms on a sulfur-sensitized grain, as described in Section 4.4.1.

It is considered that the electrostatic charge of an initial electron-trapping center suitable for efficient latent image formation has a partial, preferably +1/2 charge in the case of a

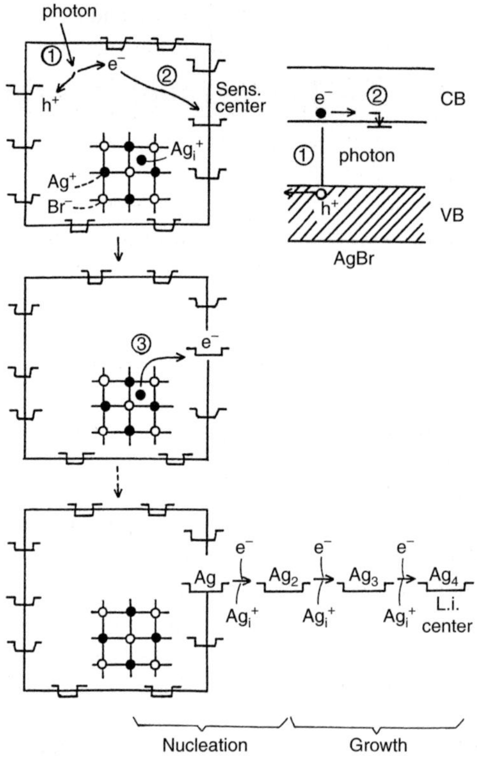

Figure 4.1 The framework of the current mechanism for the formation of latent image centers on an AgBrI grain, where Ag_i^+ is an interstitial silver ion, and CB and VB are the conduction and valence bands of AgBr [2,3].

surface kink site, and becomes a -1/2 charge when the center traps an electron. The electron-trapping by the center is followed by the ionic process, which causes the center to become +1/2 charged and to be 'reset' to trap another electron. The concept of the partially charged electron trap results in the charge sign alternating at each step in the latent image formation process, and minimizes the recombination probability by providing an electrostatic driving force for both the electronic and ionic processes of the latent image formation.

The reaction of a trapped electron with an interstitial silver ion is the ionic relaxation process leading to latent image formation and usually competes with the recombining of the trapped electron with a positive hole. As described in Sections 4.5.4, the most effective positive hole trap for preventing the recombination is a reduction sensitization center composed of Ag_2. The trapping of a positive hole by a reduction sensitization center is followed, not by the recombining of the trapped hole with a free electron, but by ionic relaxation with the release of a silver ion from the center into an interstitial position in an AgX grain, and furthermore by the release of a free electron from the relaxed center. It is noted that the ionic relaxation process, which follows the electronic processes, is very important for the efficient formation of a latent image center on AgX grain. As described earlier, a reduction sensitization center and a latent sub-image center, both of which are dimers of silver atoms, exhibit quite different functions, and have attracted the attention of many scientists, as described in Sections 4.3 and 4.4 as well as in Section 3.4.

A latent image center on an AgX grain has a deep electron-accepting level, and initiates photographic development by receiving an electron from a developer and acting as a catalyst for reducing the AgX grain to silver by a developer. The photographic development is a unique amplification process, whose degree of amplification is extremely large, and could be, for instance, more than 10^9 in the case of a cubic AgBr grain with edge length of 1 μm having a latent image center composed of four atoms, as illustrated in Figure 1.6. Details of photographic development are provided in Section 4.6.

4.2 Progresses in Mechanism of Latent Image Formation

Following Sheppard's recognition that Ag_2S specks could increase photographic sensitivity, Sheppard, Trivelli, and Loveland [4] first proposed in 1925 as the 'concentration-speck theory' the concept of the aggregation of silver atoms at an Ag_2S speck. Namely, a silver atom, which is formed as a result of the absorption of a photon by an AgX grain, migrates, and coagulates with other photolytically formed silver atoms to form a silver aggregate at an Ag_2S speck. It is noted that their theory was proposed too early to be influenced by the progress in solid-state physics. Although their detailed mechanism is not accepted now, the proposed concepts (i.e., a silver aggregate as a latent image center, and the concentration principle for its formation) are of great importance for the mechanism of photographic sensitivity.

Since then, remarkable progress has been made on solid-state physics including band theory with electrons in the conduction band and positive holes in the valence band in solids. In 1938, Gurney and Mott proposed the mechanism of photographic sensitivity [5] based on the progress of solid-state physics. The proposed idea, which was termed by Mitchell [17] the Gurney–Mott principle, was that the process of latent image formation consists of electronic and ionic stages. Namely, a photoelectron is liberated as a result of the absorption of a photon by an AgX grain. Photoelectrons migrate in the conduction band of an AgX grain, and are captured by an Ag_2S speck acting as a deep electron trap. This electronic process is followed by an ionic process. Namely, interstitial silver ions migrate and react with electrons captured by the speck to give an aggregate of silver atoms. They assumed that there were many interstitial silver ions, which were mobile and available for the latent image formation in an AgX grain. It was also assumed that positive holes were immobile, and not susceptible for recombination with electrons.

Other important features, which should be included in the mechanism of latent image formation, are the concept of an unstable and undevelopable latent pre-image center and that of a stable and undevelopable latent sub-image center. These concepts were evidenced and extended by Burton and Berg [6–11]. In this book, a latent pre-image center, a latent-sub-image center, and a latent image center are called image centers as a whole. They clearly showed by means of double-exposure techniques that a latent image center formed through two distinct stages, which they termed nucleation and growth. Namely, the inefficiency in the latent image formation for high-intensity exposure (i.e., high-intensity reciprocity law failure) could be eliminated by uniform post exposure with low intensity. This result indicates that the high-intensity exposure could form, not a latent image center, but several stable and undevelopable image centers, which can grow further to become latent image centers by the post-exposure. On the other hand, the inefficiency in the latent image formation for low-intensity exposure (i.e., low-intensity reciprocity law failure) could be eliminated by uniform pre-exposure with high intensity in order to form stable and undevelopable image centers. This result indicates the presence of an unstable image center prior to the formation of a stable and undevelopable image center. Taking into account the fact that the smallest latent image center is composed of three to five atoms, it is considered that the above-stated unstable image center (i.e., a latent pre-image center) is a single silver atom, and the stable and undevelopable image center (i.e., a latent sub-image center) is a dimer of silver atoms. The importance of the concept of the nucleation and growth processes was explored via computer simulation of the latent image formation by Hamilton, Hailstone, and co-workers [1,16,18–27]. Following the proposal of Seitz [28], the concept of a partially charged trap that will minimize the recombination probability by causing the charge sign to alternate at each step for latent image formation and smooth growth of image centers under illumination has been proposed [1].

Although the solid-state physics of AgX was not yet studied when Gurney and Mott proposed their theory in 1938, it has attracted keen interest from many physicists, and been subjected to vivid investigations, as described in Chapter 3. Based on the progress of large AgX crystals in solid-state physics, Mitchell et al. [12–15] pointed out that there were several problems in the Gurney–Mott mechanism that could not be supported by the physical properties of a large AgX crystal. Since it turned out that a positive hole in AgX is more mobile than an interstitial silver ion, Mitchell et al. considered that a positive hole should reach a negatively charged center with a trapped electron more rapidly than an interstitial silver ion. He also pointed out from the ionic conductivity of a large AgBr crystal at high temperature that there would be less than one interstitial silver ion per AgBr grain at room temperature [17]. Then, they proposed that a positive hole should be captured by a trap, such as a bromide ion at a surface kink site, before a photoelectron is captured. The hole-trapping should be followed by a prompt ionic relaxation, which makes it unlikely for the trapped hole to take part in the recombination. When a bromide ion at a kink site captures a positive hole, it bears a partial positive charge and repels its adjacent silver ion into an interstitial position to create an interstitial silver ion available for latent image formation and to leave Br_2^-, which cannot be a recombination center.

One of the most important achievements made by Mitchell et al. was the first-time observation of dislocations. Namely, they found that dislocations in AgBr are made visible, being decorated with photolytically formed silver atoms [29], leading to the observation of photolytic Ag specks formed along dislocations in AgX grains [30]. These observations indicated the importance of crystal defects as sites for the latent image formation. The role of crystal defects was not mentioned in the Gurney–Mott mechanism. It has turned out that the participation of crystal defects in the formation of latent image centers is important from the viewpoints of photographic science and technology [31,32], and is now one of the most important technologies for highly sensitive tabular AgX emulsion grains [32].

Mitchell also considered that deep electron traps should not be present before exposure to light; otherwise, they could initiate development to cause fog formation by taking an electron from a developer. According to Mitchell, a crystal defect should be a shallow electron trap, and could be a deep electron trap for capturing a photoelectron only in the presence of an interstitial silver ion in its vicinity. Then, a photoelectron and an interstitial silver ion are combined at the defect to form a silver atom.

It turned out that various kinds of shallow electron traps play important roles in photographic processes, and have been extensively studied and developed for highly sensitive AgX emulsion grains, as described in Section 3.4. They include sulfur and sulfur-plus-gold sensitization centers as sites suitable for latent image formation, dislocations for localizing photoelectrons, and metal-complex dopants for adjusting the time for latent image formation. In particular, sulfur sensitization centers and sulfur-plus-gold sensitization ones should be shallow enough to trap electrons temporarily in most trapping events, since as many as about $3000/\mu m^2$ are formed [33].

Although the Mitchell mechanism is based on the physical properties of a large AgX crystal, it has turned out, as described in Sections 3.2 and 3.3, that the ionic and electronic properties of AgX emulsion grains are quite different from those of a large AgX crystal. Recognition of this difference was initiated by Hamilton and Brady [34,35] in 1959, has been established since then, and has been included in the current mechanism of photographic sensitivity. One of the most important findings was that the ionic conductivity of AgBr grains was much larger than that of a large AgBr crystal. Namely, it was observed that a large number of interstitial silver ions were present and available for latent image formation in each AgBr grain in a photographic emulsion. It is quite probable that an interstitial silver ion can reach a trapped electron faster than a positive hole can, since it is judged that mobility multiplied by the concentration (i.e., conductivity) of interstitial silver ions is larger than that of positive holes [36]. The ionic property of silver halides has been studied in detail, and is described in Section 3.2.

Remarkable progress has also been made on information about the transport of electrons and positive holes in AgX grains. The measured lifetimes of electrons and positive holes in AgBr grains did not support Mitchell's hypothesis that a trap's capture of a positive hole should be faster than that of a photoelectron [35,37–40], as described in detail in Chapter 3. Although the mobility of electrons in AgX grains is about two orders of magnitude smaller than that in large AgX crystals, it is still large, and larger than the mobility of electrons in organic electronic devices, which are being developed as described in Chapters 9 and 10. On the other hand, it is judged that the mobility of positive holes is ordinarily smaller than that of photoelectrons in silver halides owing to the influence of self-trapped states for positive holes as described in Section 3.3.1. It is considered that this condition reduces the probability of recombination between electrons and positive holes, although positive holes are mobile, in contrast to the assumption made by Gurney and Mott.

As stated earlier, progress in the theory of photographic sensitivity has been initiated by progress in basic sciences and technologies, on which photographic processes are based. This includes the proposal of sulfur sensitization for the first mechanism of latent image formation by Sheppard et al., solid-state physics for the Gurney–Mott theory, physical properties of large single crystals of silver halides for the Mitchell theory, and physical properties of AgX grains leading to the current mechanism of latent image formation as described in detail in Chapter 3.

4.3 Efficiency of Latent Image Formation and Photographic Sensitivity

4.3.1 Efficiency of Latent Image Formation in Terms of Quantum Sensitivity

The sensitivity of a photographic emulsion layer primarily depends on the averaged sensitivity of the AgX grains in it. The photographic sensitivity of an AgX grain depends on the light absorbance of the grain, the efficiency of the latent image formation taking place in the grain, and the size of the smallest latent image center available for development of the grain [32,41]. Taking into account the influence of several different factors as stated above, it is rather difficult to define and measure accurately the efficiency of latent image formation. Instead, for many years the quantum sensitivity of AgX grains in an emulsion layer has usually been measured to judge the efficiency of latent image formation in the layer [42,25]. The quantum sensitivity is defined by the average number of absorbed photons per grain, which render half of the present grains developable. The measurement of quantum sensitivity is usually made for a thin emulsion layer, in which mono-dispersed AgX grains independently absorb incident photons, whereby accurate knowledge of the average number of photons absorbed per grain and the size distribution of the grains are needed in order to determine the fraction of developed grains in the layer as a function of the number of absorbed photon per grain.

The highest quantum sensitivity has been achieved for fine AgX emulsion layers, which were sulfur-plus-gold-sensitized and hydrogen-hypersensitized. Hydrogen hypersensitization was carried out by evacuating the layers and then keeping them in a hydrogen atmosphere to form hydrogen hypersensitization centers composed of dimers of silver atoms on the grains. Note that only less than three absorbed photons per grain were needed to render half the present grains developable [43,25]. Taking into account that the smallest latent image center is composed of three atoms on sulfur-plus-gold-sensitized grains [21–23], this result supported the idea that one absorbed photon could create two free electrons in the presence of hydrogen hypersensitization centers [25], as described in Sections 3.4.2 and 4.4.2.

The efficiency of latent image formation depends on the size of AgX emulsion grains as shown in Figure 4.2 [25]. From this viewpoint, AgX grains in emulsions could be classified into three groups: large (class I), small (class II), and very small (class III) grains, as shown in this figure [44]. The grains of class I are more than 0.5 μm in diameter, those of class II range between 0.5 and 0.1 μm, and those of class III are less than 0.1 μm. The highest quantum sensitivity (i.e., less than 3 absorbed photons per grain) has been achieved only for class II grains . It is likely that class I grains suffer from inefficiency in latent image formation because of the dispersion of image centers. Namely, grains are so large that some of photoelectrons cannot reach a growing image center on a grain, and take part in the formation of a new image center, resulting in the formation of more than one latent sub-image center and a latent image center not forming on some of the grains. It was observed that the intrinsic sensitivity of AgX grains was proportional to their volume for class II grains , while class I grains tended to deviate from the proportionality as grain size increased. The threshold size of the grains upon deviation was comparable with the values observed of the diffusion length of photoelectrons, as shown in Table 3.2, and supported the above-stated idea. Although the efficiency of latent image formation on class III grains is not clear due to the lack of accurate data, it is likely that the grains suffer from inefficiency due to the recombining of photoelectrons and positive holes, both of which are confined in such small volume.

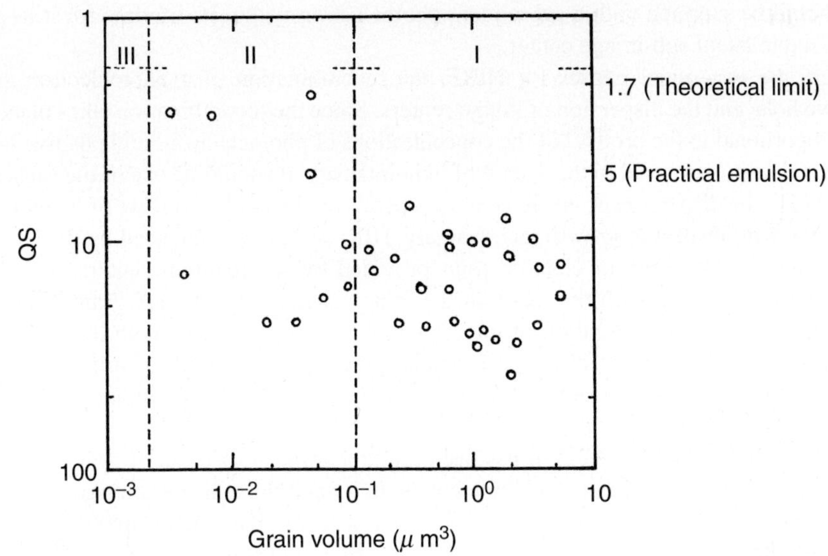

Figure 4.2 Grain volume dependence of the quantum sensitivity in terms of the number of absorbed photons per grain to render half of the AgX grains developable for various AgX grains [25] in addition to AgX grains for color films [76]. Here, AgX grains classified into three groups, I, II, III, with respect to grain size are also indicated [44]. This figure is modified with permission from reference [25], copyright 1988 IS&T, The Society for Imaging Science and Technology.

In order to understand the photographic process properly, it is necessary to analyze the factors influencing the photographic sensitivity of class II AgX grains. Octahedral and cubic AgBr grains of about 0.2 μm in diameter are considered standard, since they are single crystals and form only one latent image center per grain, regardless of exposure condition, due to the lack of competition for latent image formation among centers on the surface and interior of each grain. Chemically unsensitized grains usually exhibit very low sensitivity owing to vivid recombination between electrons and positive holes [for example, 45], and sometimes need about 5000 absorbed photons per grain to render half the present grains developable. On the other hand, only about 30, 30, and 10 absorbed photons per grains are needed to render half the grains developable on reduction-sensitized [44,45], sulfur-sensitized [44,46], and sulfur-plus-gold-sensitized AgBr grains. As stated earlier, the combination of sulfur-plus-gold sensitization with hydrogen hypersensitization only needed less than three absorbed photons per grain, although it was not practically useful owing to the instability of hydrogen hypersensitization centers [43,25].

Photographic sensitivity (*S*) is usually given by the reciprocal of exposure (*E*), which renders a certain fraction of AgX grains developable, while *E* is the product of light intensity (*I*) and exposure time (*t*). The reciprocity law stands when *S* is solely given by $1/E$ despite the variation of *I* [47]. Most AgX emulsions suffer from inefficiency in latent image formation under low-intensity light (i.e., LIRF; low-intensity reciprocity law failure) and/or from inefficiency under high-intensity light (i.e., HIRF; high-intensity reciprocity law failure). It is believed that LIRF arises from the instability of a latent pre-image center. Namely, a latent pre-image center lives only for several seconds [48,49], and should be supplied with a subsequent photoelectron within its lifetime for it to grow into a stable latent sub-image center.

There are two major causes for HIRF; the recombination of a photoelectron and a positive hole, and the dispersion of image centers. Since the recombination takes place at a rate proportional to the product of the concentrations of photoelectrons and positive holes, which is thus proportional to the square of light intensity, it should be one of the causes for HIRF [21]. The dispersion of image centers depends on the nucleation rate of latent image formation with respect to growth rate. Namely, HIRF should be enhanced by the increase in the number and depth of electron traps provided by sensitization centers, as will be described later. Along with the fact that the ionic conductivity of AgX grains, which is proportional to the concentration of interstitial silver ions, enhances both the nucleation and growth processes of latent image formation, the degree of HIRF of sulfur-sensitized AgX emulsions decreases, reaches the minimum, and then increases with increasing ionic conductivity of the grains [50].

As described in Section 3.2.2, it has been found that the adsorption of cyanine dyes to AgX emulsion grains increases the ionic conductivity and the concentration of interstitial silver ions in the grains [51,52]. Ihama and Tani observed the enhancement of HIRF in sulfur-sensitized cubic AgBr grains by their adsorption of cyanine dyes, and attributed the enhancement to the dispersion of image centers owing to the increase in concentration of interstitial silver ions in them [53,54].

One of the most important factors for the reciprocity law behavior of photographic emulsions is chemical sensitization. From this viewpoint, remarkable progress has been made on chalcogen sensitization. Namely, selenium and tellurium sensitizations have become practically useful in addition to sulfur sensitization [55]. These sensitization centers provide shallow electron traps acting as sites for latent image formation. It has been predicted that chalcogen sensitization centers depress LIRF and enhance HIRF more extensively as their trap depth increases. It has been found that the depth of traps is in the order of sulfur, selenium, and tellurium sensitizations, and that their influence on LIRF and HIRF was in accord with this prediction.

Recently, remarkable progress has also been made on shallow electron traps, which do not act as sites for latent image formation, as described in Section 3.4. They include dislocations [31,32] and doped metal complexes [56–59]. Since these traps repeatedly capture and release electrons, they increase the time for latent image formation, and improve HIRF.

4.3.2 Factors Influencing Photographic Sensitivity

Sensitivity is one of the most important capabilities of a photographic film. As stated earier, the sensitivity of a photographic film is primarily the averaged sensitivity of the AgX grains in it, and depends on the grain's light absorbance, the efficiency of the latent image formation taking place, and the size of the smallest latent image center for development of the grain [32,41]. The absorbance of a photographic emulsion grain is proportional to the grain's volume where blue light is absorbed by the grain itself [60,41,3], and to the grain's surface area where green and/or red light is absorbed by sensitizing dyes on the grain [61,41].

On the other hand, the sensitivity is proportional to the reciprocal of the resolving power for AgX emulsions with varied grain size [62]. As usual, the improvement in photographic film capability, for example, the increase in efficiency of latent image formation, can increase the sensitivity of a photographic film without deteriorating the image quality by keeping the size of AgX grains unchanged, or can improve the image quality of a photographic film without decreasing its sensitivity by decreasing the size of the grains. Therefore, it is essential to analyze the factors influencing photographic sensitivity for improvement in photographic performance.

There are various kinds of photographic films. They include color negative films, color reversal films, color papers, films for printing, and those for medical diagnosis. A systematic analysis of the factors influencing photographic sensitivity of a color negative film, which is regarded as representative of highly sensitive photographic films and described in Section 6.2.3, can be made, and these factors are briefly reviewed below.

Before the technology of tabular AgX grains [63,64] was developed, a highly sensitive photographic film could capture at most one-third of incident visible light [65]. Since a tabular AgX grain has a large specific surface, it can adsorb a large amount of sensitizing dye molecules on its surface, and make it possible for a color negative film to capture half of the incident visible light [66].

Taking into account that each tabular grain adsorbs a monomolecular layer of sensitizing dyes on its surface, which are not enough to capture all the incident light in the corresponding wavelength region, a technology that uses sensitizing dyes in amounts larger than that on the monomolecular layer on the grain surface is being developed. Several proposals have been made in the past, including a dye multi-layer by Gilman et al. [67], a dye/silver halide junction between dye and AgX crystals by House [68], and an antenna dye system with energy transfer among the dye molecules by Bird et al. [69]. Parton et al. described an antenna dye sensitization with bi-molecular layers of sensitizing dyes on the grain surface, which could be used for color films [70].

In order to evaluate the theoretical limit of the efficiency of latent image formation in terms of quantum sensitivity, it is necessary to take into account the following three facts. First, the smallest latent image center is composed of three atoms [21–23]. Second, one absorbed photon can create two free electrons [71,72,25,73,74], as described in Section 4.4.2. Third, the number of absorbed photons per grain is randomly distributed, obeying the Poisson distribution. As a result, it was determined that the theoretical limit of quantum sensitivity should be 1.7 absorbed photons per grain [25]. Although the highest quantum sensitivity is a little smaller than three absorbed photons per grain [75,25], the quantum sensitivity of highly sensitive AgX grains in color films was about five absorbed photons per grain [76].

As described in Section 4.4.2, it is known that one absorbed photon could create two electrons in the presence of reduction sensitization centers or hydrogen hypersensitization centers [71,72,25]. The technology to create two free electrons by one absorbed photon is sometimes known as two-electron sensitization. It was obvious that the application of hydrogen hypersensitization to color films was not possible owing to the instability of hydrogen hypersensitization centers under room atmosphere. It was also necessary to develop certain technologies to overcome the instability of reduction sensitization centers in order to apply reduction sensitization to color films. New types of two-electron sensitization technology were proposed. One depended on a photochemical reaction [74]. As illustrated in Figure 4.3, one absorbed photon creates an electron–hole pair. When a positive hole oxidize a certain compound, the compound undergoes a reaction to give an activated species, which can inject an electron into the conduction band of an AgX grain. It

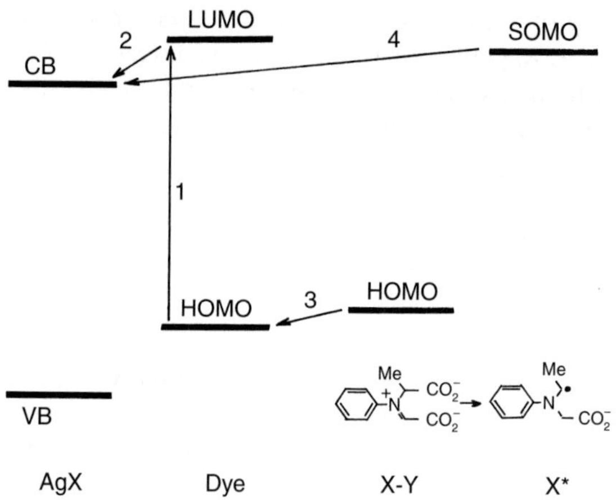

Figure 4.3 Scheme of a newly developed two-electron sensitization, which is initiated by the light absorption of an incident photon by a dye (1), and followed by the transfer of an electron from an excited dye to the conduction band of AgX (2), the transfer of a positive hole from the dye to compound X–Y, the formation of an activated species X* (3), and the transfer of an electron from X* to the conduction band of AgX (4) [74].

is still necessary to refine these technologies to really create two free electrons by one absorbed photon in color films without deteriorating their stability.

Hailstone used a computer simulation to analyze the processes for latent image formation, and predicted the quantum sensitivity of emulsion layers as a function of number and trap depth of sensitization centers, as shown in Figure 4.4. It was predicted that the decrease in trap depth should be associated with the increase in the number of sensitization

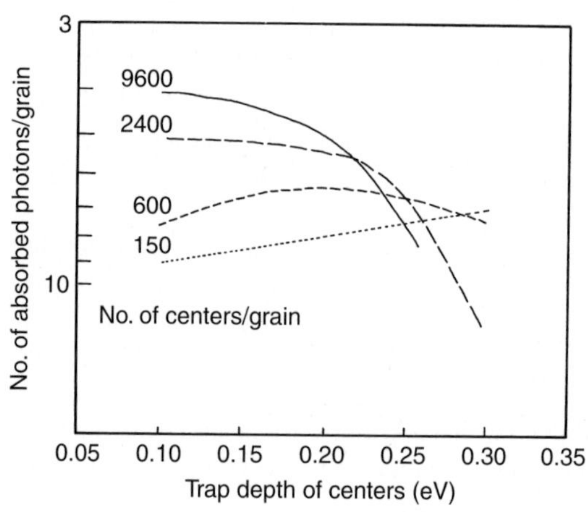

Figure 4.4 Quantum sensitivity as functions of the number and depth of the traps provided by chemical sensitization centers based on a simulation of latent image formation. Reproduced with permission from Reference [77], copyright 1991 IS&T, The Society for Imaging Science and Technology.

centers to maximize photographic sensitivity, and that the maximized sensitivity should increase with decreasing trap depth of the sensitization centers [77]. From this perspective, it is interesting to examine chalcogen sensitization because of the facts that selenium and tellurium sensitizations, in addition to sulfur sensitization, have become available for practical use, and that the depth of traps for electrons provided by sensitization centers is in the order of sulfur, selenium, and tellurium [55].

One of the most effective technologies for preventing recombination between photoelectrons and positive holes is the two-electron sensitization, by which a positive hole is used to create an additional free electron instead of recombining with an electron [71,72,25,73,74]. Another effective technology is introducing dense dislocations to the area near the edges of a dye-sensitized tabular AgX grain, where dislocations acting as shallow electron traps localize electrons in the area near the edges, and prevent them from recombining with positive holes, which are localized by sensitizing dyes on the main surfaces of the grain [32].

4.4 Latent Image Centers with Respect to Silver Clusters

4.4.1 Size of Latent Image Centers

Many attempts to determine the size of the smallest latent image centers have been made. Analyzing characteristic curves has given some indication [78–84]. By analyzing the ability of evaporated metal clusters to initiate physical development, Hamilton and Logel were able to estimate that the smallest latent image center was composed of four silver atoms or two gold atoms [85]. One of the most important attempts was to determine and analyze quantum sensitivity. Farnell and Chanter were the first to measure the quantum sensitivity of various emulsion grains with high sensitivity [42]. Marriage then analyzed their data from a statistical viewpoint [86]. It was found that the most sensitive grains were made developable by three absorbed photons per grain.

Hydrogen hypersensitization was invented by Babcock et al. [43] and enabled the most efficient latent image formation on fine AgBr grains in combination with sulfur-plus-gold sensitization. Babcock and James found that only two to three absorbed photons per grain were required to make more than a half of the grains developable [75]. By taking into account that an absorbed photon can create at best two free electrons in a grain in the presence of hydrogen hypersensitization centers or reduction sensitization centers, as described later in this section and in Section 3.4.2, Babcock and James were able to determine that the smallest latent image center was composed of three atoms of silver and gold. Computer simulation for latent image formation based on sensitometric data as stated above was helpful in estimating the size of the smallest latent image centers. Hailstone and Hamilton thus estimated that the smallest latent image was composed of three atoms on sulfur-plus gold-sensitized AgBr grains and of four to five atoms on sulfur-sensitized grains [21–23].

Fayet et al. measured the developability of silver cluster ions that were selected by size and deposited on AgBr grains in a mass spectrometer [87]. According to their measurements, the smallest developable silver cluster ion was found to be Ag_4^+, in accordance with the above-stated results. However, their subsequent result did not indicate that the size of the smallest developable gold cluster should be smaller than that of the smallest developable silver cluster.

Although the smallest latent image centers are too small to be observable by electron microscopy, it has been revealed that most latent image centers on AgBr grains, which provide the portion from around the middle to the shoulder of the characteristic curve, were large enough to be observable by electron microscopy [88]. It was confirmed that the efficiency of the growth of latent image centers could be unity when the activity of positive holes was depressed.

Since latent image centers are composed of silver clusters, the properties of silver clusters are of great importance. Konstantinov et al. prepared, by vapor deposition on carbon, silver clusters whose sizes were several nanometers [89–91]. They determined the size dependence of the redox potential of the clusters by an elaborated method. The redox potential of silver clusters was more negative than that of bulk silver, and the deviation (ΔE) was proportional to the reciprocal of the cluster diameter (d), obeying the Gibbs–Thomson equation. This result indicated that the electronic energy levels of the clusters studied were continuously distributed. Namely, the lowest unoccupied electronic energy level was situated just above the highest occupied one. It is noted that the smallest cluster studied still contained several hundred silver atoms. The extrapolation of ΔE to a cluster composed of four silver atoms in the Gibbs–Thomson equation did not give the value of ΔE for latent image centers actually formed on AgX grains [3]. This result indicates that the electronic energy levels of the smallest latent image centers are discrete and can be expressed by the molecular orbital method for organic molecules and the shell model for small metal clusters [92,93] instead of the Gibbs–Thomson equation.

Baetzold made extensive studies of the electronic structures of metal clusters by molecular orbital methods, and applied the results to the mechanism of latent image formation [94,95]. In order to get the general tendency of the electronic structure of small metal clusters, the electronic energy levels of linearly arrayed silver clusters were calculated by the Hueckel approximation molecular orbital method and is shown in Figure 4.5 [2,3]. It is noted that the electronic energy levels of the clusters are discrete, and in particular the highest occupied molecular orbital (HOMO) of a dimer is lower in electronic energy than those of larger clusters. It is known that, for molecules of photographic interests, the electronic energy level of HOMO and the lowest unoccupied molecular orbital (LUMO) (ϵ_{HO} and ϵ_{LU}, respectively) are related to oxidation and reduction potentials (E_{OX} and E_{R}, respectively) as follows [96].

$$\epsilon_{\mathrm{HO}} = -E_{\mathrm{OX}} + \text{constant} \tag{4.1}$$

$$\epsilon_{\mathrm{LU}} = -E_{\mathrm{R}} + \text{constant.} \tag{4.2}$$

Therefore, it can be predicted that E_{OX} of Ag_2 is more positive than those of larger silver clusters. This prediction has been confirmed, as described later in this section [45].

Other important feature of the clusters from the viewpoint of the molecular orbital methods is that the electronic structure of a cluster with an even number of silver atoms is quite different from that of a cluster with an odd number of silver atoms, and is experimentally observed as an odd–even alternation [94,95,97,98]. It is generally predicted that a cluster with an odd number of valence electrons has smaller ionization energy and larger electron affinity compared to that with an even number of valence electrons. However, it has been discovered that the formation of dimers of silver atoms adjacent to silver ions at surface kink sites on AgBr grains decrease the concentration of interstitial silver ions in the grains [99]. This observation indicates that a silver cluster is likely to combine with a silver ion at a kink site, and that the odd–even alternation is therefore overemphasized in Figure 4.5 for an explanation of the growth of image centers.

According to the shell model, each cluster is spherical in shape, and has discrete electronic energy levels such as 1*s*, 1*p*, 1*d* shells. According to this model, one of the most important and experimentally confirmed characteristics of metal nano-clusters is the magic number, which is the number of valence electrons at the closed shell configurations. Namely, the clusters with a closed shell such as Ag_2 with closed 1*s* shell, Ag_8 with closed 1*s* and 1*p* shells, and Ag_{18} with closed 1*s*, 1*p*, and 1*d* shells are more stable and less reactive than those with an open shell. The experimental confirmations were made on the stability of metal clusters with a magic number of valence electrons and on the anomaly of the ionization energy of the clusters with a magic number of silver atoms [97,100,101,93].

The Kubo effect is one of the most interesting subjects from the viewpoint of the relation between the size and electronic structure of metal clusters, and is described in Section 4.4.2.

The above-stated models that predict the characteristics of molecules or clusters of silver atoms as image centers are as follows:

(1) The electronic energy levels of image centers are discrete. This feature together with the Kubo effect contributes to the stability of small centers.
(2) While a single silver atom is unstable, Ag_2 is stable. This feature provides the threshold for the latent image formation, as illustrated in Figure 1.7. A silver dimer can also contribute to the photographic sensitivity from various aspects, as described later in this section.
(3) It is suggested that the clusters with an open shell or those with an odd number of electrons have deep electron-accepting levels suitable for initiating development.
(4) Since image centers are very small, it is quite probable that their property and behavior are influenced by the sites where they are formed.

Properties (1)–(3) are referred to as the 'quantum-size effect' and property (4) as the 'site effect' in this book.

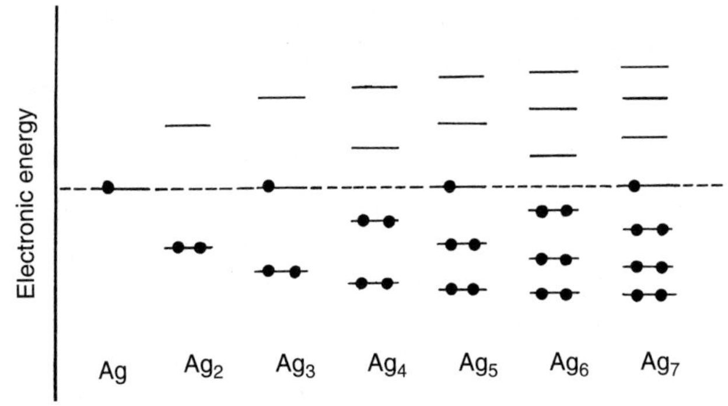

Figure 4.5 A schematic representation of electronic energy levels of silver clusters in a linear structure as calculated by the Huekel-approximation molecular orbital method [2, 3].

4.4.2 Properties and Behaviors of Silver Clusters

By using arrested development and latensification, which make various silver clusters observable by electron microscope, Spencer et al. [102, 3] observed two kinds of silver specks. During exposure of AgX grains to light, one of them appeared and grew, while the other disappeared. Namely, the former trapped a photoelectron, while the other trapped a positive hole during exposure to light. The presence of two kinds of silver clusters on AgX grains was supported by observing their effects on photographic performance and photoconductivity of the grains [103–110], as described in Section 3.4.2.

Among several proposals to explain the presence of the two kinds of silver clusters [107–112], this author has proposed that an electron-trapping silver cluster is formed at a positively charged site, while a hole-trapping center is formed at an electrically neutral site [104], and obtained the evidence to prove it [111]. Following this author's proposal and applying a molecular orbital method to models of Ag_2 on AgBr as shown in Figure 4.6, Hamilton and Baetzold have proposed that Ag_2 at a positively charged kink site is an electron trap, while Ag_2 at a neutral kink site is a positive hole trap [114]. Then, electron-trapping silver clusters, which are mostly formed during exposure to light, are called P centers, and hole-trapping silver clusters, which are usually formed by reduction sensitization, are called R centers.

This author has been able to observe the experimental results supporting the above-stated idea by studying the properties of silver clusters formed by reduction sensitization on AgBr grains [45,115–118]. As shown in Figure 3.32, the photographic sensitivity increases with two steps with increasing the amount of reduction sensitizer. The silver clusters that bring about the first step of sensitivity increase decrease the photoconductivity with positive holes as carriers; however, they have no influence on the photoconductivity of electrons. On the other hand, the silver clusters that bring about the second step of

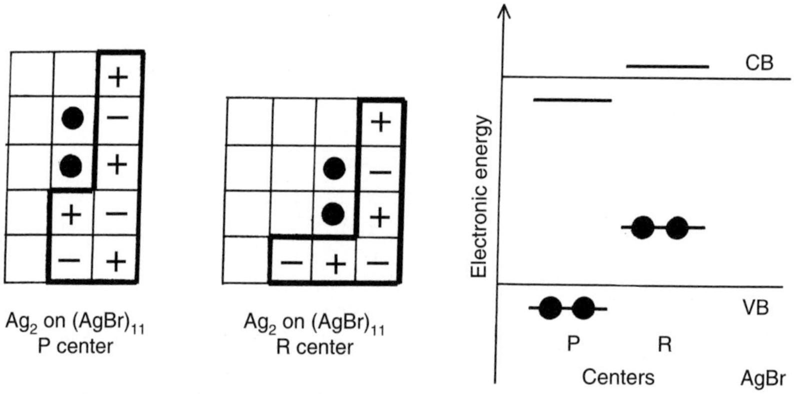

Figure 4.6 Models for the calculation of the electronic energy levels of Ag_2 at a positively charged kink site (i.e., P center) and Ag_2 at a neutral kink site (i.e., R center) on the surface of AgBr, and a schematic representation of the obtained electronic energy levels of the centers with respect to that of AgBr. Ions framed with thin lines form the first layer, and those framed with thick lines form the second partial layer. Reproduced with permission from Reference [114], copyright 1981 IS&T, The Society for Imaging Science and Technology.

sensitivity increase decrease the photoconductivity of electrons. This result proves the presence of two kinds of silver clusters on AgBr grains, indicating that the former is an R center, while the latter is a P center, and that the former is more likely to be formed than the latter when AgX grains are treated with reducing agents.

Those centers in emulsions were kept in equilibrium with redox buffer solutions. The oxidation potential (E_{OX}) of those centers is given by the redox potential of the solution, which eliminated half of the sensitizing effect of the centers and can be related to the HOMO energy level (ϵ_{HO}) of the centers on the basis of Eq. (4.1) [115]. As shown in Figure 4.7, the oxidation potential of P centers is more positive by about 0.4 V than that of R centers, indicating that the HOMO level of P centers is lower by about 0.4 eV than that of R centers. The oxidation potential of R centers in the interior of the grains is less positive by 0.05 V than that of P centers. Since positive holes must be injected from the

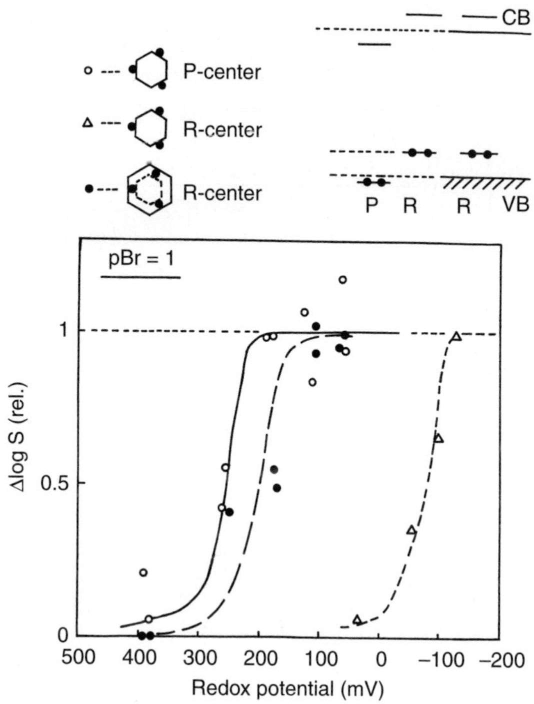

Figure 4.7 (Bottom) The residual proportions of P centers and R centers on the surface and R centers in the interior of octahedral AgBr grains with an average diameter of 0.2 μm after keeping them in redox buffer solutions with the redox potential indicated on the abscissa for 17 and 41 hours at 25°C. The residual proportion of the centers is expressed by the residual proportion of the logarithmic sensitivity increase caused by those centers. The oxidation potential of those centers is given by the midpoints of their curves. (Top) The heights of HOMO levels of the centers thus are obtained based on Eq. (4.1) with respect to the top of the valence band of AgBr (VB), and LUMO levels of the centers are shown with respect to the bottom of the conduction band of AgBr (CB) by taking into account their electron-trapping ability on the grains. Reproduced with permission from Reference [46], copyright 1994 IS&T, The Society for Imaging Science and Technology.

solutions to the valence band of AgBr grains to bleach R centers in the interior of the grains, the oxidation potential of R centers in the interior of the grains is thought to indicate the position of the top of the valence band of AgBr grains. The obtained electronic energy levels of those centers are in good accord with those described in Figure 4.6, supporting the 'site effect', due to which R centers are positive hole traps and P centers are electron traps.

Figure 4.8 shows the electronic energy levels of latent image centers as well as those of reduction sensitization centers on the basis of their oxidation potentials and Eq. (4.1). Since the oxidation potential of P centers was more positive than those of latent image centers, it is considered that the HOMO level of a P center (i.e., Ag_2) was lower than that

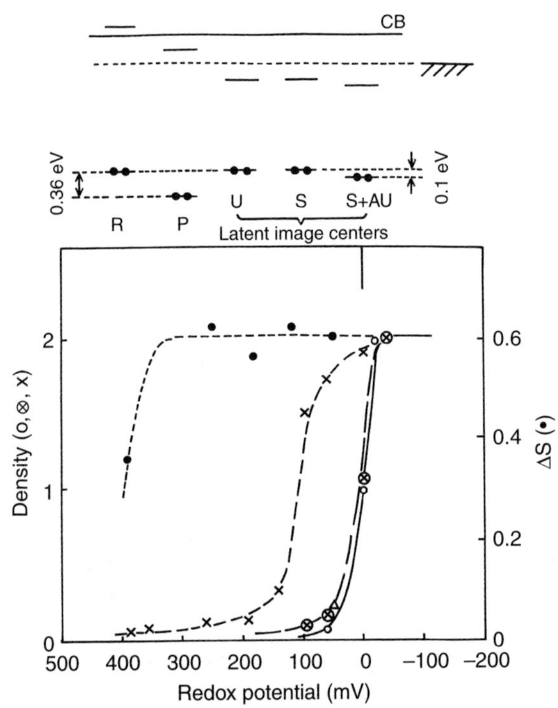

Figure 4.8 (Botttom) The residual proportions of P centers formed during reduction sensitization (●), and latent image centers formed on unsensitized (U;○), sulfur-sensitized (S;⊗), and sulfur-plus-gold-sensitized (S+Au;×) octahedral AgBr emulsion grains with an average diameter of 0.2 μm after keeping them in redox buffer solutions with the redox potential indicated on the abscissa for 17 hours at 25°C. The residual proportion of the centers is expressed by the residual proportion of the logarithmic sensitivity increase caused by the centers (ΔS) and the residual density of the exposed and developed emulsion layer. The oxidation potentials of those centers is given by the midpoints of their curves. (Top) The height of HOMO levels of those centers thus is estimated on the basis of Eq. (4.1) and the height of LUMO levels of those centers with respect to the bottom of the conduction band (CB) of AgBr and to the redox potential of a developer by taking into account their electron-trapping ability on the grains and their developability. Reproduced with permission from Reference [45], copyright 1994 IS&T, The Society for Imaging Science and Technology.

of latent image centers (i.e., Ag_n; $n > 2$) on the basis of Eq. (4.1). It is considered that LUMO level of a P center is higher than those of latent image centers, since the former cannot initiate development. The electronic energy levels of P centers and latent image centers thus estimated could support the 'quantum size effect' as shown in Figure 4.5.

It has been found from the above-stated examinations that two kinds of silver dimers (R and P centers) contribute through different ways to the processes of photographic sensitivity. As to the behavior of R centers, the proposals by Mitchell [12] and Lowe [71] are important. According to Mitchell's mechanism for photographic sensitivity, the hole-trapping by an R center is followed by the ionic relaxation (i.e., the release of a silver ion into an interstitial position), preventing the recombination of the trapped hole at an R center with a photoelectron.

Including the above-stated process, Lowe proposed an hypothesis for the behavior of positive holes in the presence of R centers [71], which is composed of the processes of light absorption of an AgX emulsion grain,

$$h\nu \rightarrow e^- + h^+, \tag{4.3}$$

trapping of a positive hole by an R center,

$$Ag_2(R) + h^+ \rightarrow Ag_2{}^+, \tag{4.4}$$

ionic relaxation of the positive hole trapping,

$$Ag_2{}^+ \rightarrow Ag + Ag^+, \tag{4.5}$$

and dissociation of a single silver atom and liberation of an extra free electron,

$$Ag \rightarrow Ag^+ + e^-, \tag{4.6}$$

In total,

$$Ag_2(R) + h\nu \rightarrow 2Ag^+ + 2e^-, \tag{4.7}$$

where $Ag_2(R)$ is a R center. Namely, in the presence of R centers, one absorbed photon can produce at most two free electrons available for the latent image formation.

This author confirmed Lowe's hypothesis for the first time by observing the growth of latent sub-image centers after exposing reduction-sensitized grains to light in the presence of hole-trapping dyes [72]. Latent sub-image and image centers are formed rapidly. When positive holes migrate on the grain surface in the presence of many sensitizing dye molecules acting as deep positive hole traps, it takes a long time (e.g., several minutes, depending upon the dyes used) for positive holes to bleach the latent image centers as well as the R centers [72,119–121], in order to bring about the liberation of extra free electrons available for the growth of the latent sub-image centers.

$$h^+ + \text{Dye} \rightarrow \text{Dye}^+, \tag{4.8}$$

$$\text{Dye}^+ \rightarrow \text{Dye} + h^+. \tag{4.9}$$

Then, it was observed that in the presence of a sensitizing dye, the number of latent image centers decreased after exposure to light in the absence of R centers, while the number of latent image centers increased after exposure to light in the presence of R centers, as shown in Figure 4.9. The time constants of these two reactions were similar to each other

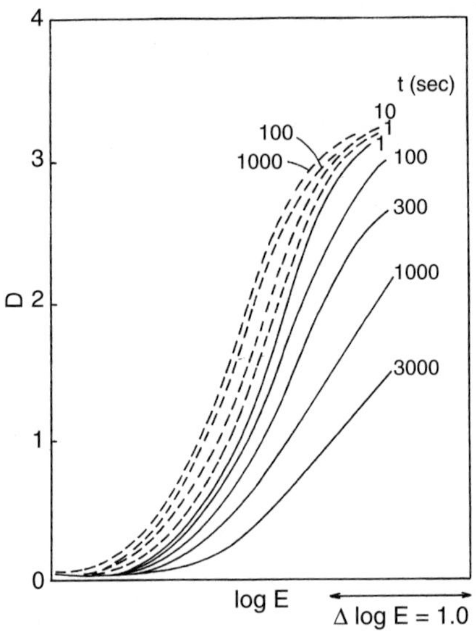

Figure 4.9 Time-dependent characteristic curves of unsensitized (solid line) and hudrogen-hypersensitized (broken line) octahedral AgBr emulsion grains (0.2 μm) with 3.3'-disulfopropyl-5,5'-dichloro-9-ethyl-thiacarbocyanine adsorbed. The emulsions were exposed for 0.001 s, kept for *t* s, immersed in water to deactivate positive holes trapped by the dye, and developed by a surface developer (MAA-1). Reproduced with permission from Reference [72], copyright 1986 IS&T, The Society for Imaging Science and Technology.

and to the time constant of the decay of Dye^+ as measured by means of an electron paramagnetic spectrometer [72,119–121]. Namely, positive holes reacted with latent image centers with a time constant of several minutes to bring about their fading in the absence of R centers, while positive holes mostly reacted with R centers with a similar time constant to create extra electrons, which were used for the growth of sub-image centers into latent image centers.

Other evidence for Lowe's hypothesis has been provided by Hailstone and co-workers [25], who observed that a large fraction of latent image centers can be formed by only two absorbed photons per grain on hydrogen-hypersensitized and sulfur-plus-gold-sensitized fine AgBr grains (Figure 4.10), despite the fact that the smallest latent image center was estimated to be composed of three atoms. These results cannot be explained without taking into account the extra electrons created according to Lowe's hypothesis.

As stated above [88], it was possible to observe by electron microscope such small silver clusters as latent image centers, which gave the portion from around the middle to the shoulder of characteristic curves of photographic emulsions, and fog centers formed by excessive reduction sensitization. They ranged from less than 2 to more than 8 nm in diameter, and should be subjected to the Kubo effect, according to which they should be electrically neutral on keeping since it needed large energy to add an electron to or remove

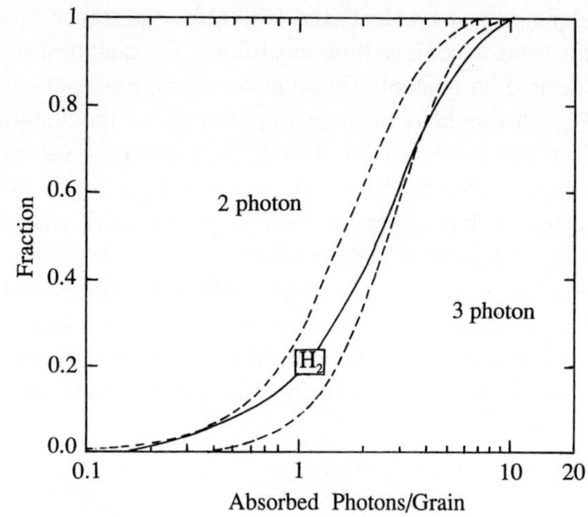

Figure 4.10 Observed *D*–log *E* curve of a hydrogen-hypersensitized and sulfur-plus-gold-sensitized octahedral AgBr emulsion (solid line) and theoretical curves of the corresponding emulsions, in which a latent image center is formed by two and three absorbed photons per grain (broken lines). Reproduced with permission from [25], copyright 1988 IS&T, The Society for Imaging Science and Technology.

it from a cluster [122,123]. Namely, the Kubo effect explains why there is high stability of latent image centers and fog centers at the temperature of photographic interest.

It has been clearly demonstrated that only one latent image center forms on an AgBr grain of 0.2 μm in diameter, while as many as 200 fog centers form on the grain [123]. This result indicates that the concentration principle actually takes place for latent image formation as described in Sections 4.1 and 4.2, while it does not for the formation of fog centers by excessive reduction sensitization. It has also been revealed that the efficiency of the growth of latent image centers could be unity when positive holes are prevented from attacking latent image centers [88]. This result is in contrast to the fact that the efficiency of latent image center formation is considerably smaller than unity on the grains, as described in Section 4.3.1.

Metal clusters, to which the Kubo effect is applicable, can be classified into two groups; even and odd. In photographic films, even clusters are composed of an even number of silver atoms and are therefore diamagnetic. On the other hand, odd clusters are composed of an odd number of silver atoms and therefore paramagnetic. Latent image centers give electron spin resonance (ESR) signals, being in accord with the idea that nearly half of the latent image centers formed are odd clusters. On the other hand, most fog centers formed by excessive reduction sensitization are diamagnetic, indicating that almost all the fog centers are even clusters. This result indicates that reduction sensitization directly forms only dimers of silver atoms, which are stabilized at surface kink sites to act as reduction sensitization centers, and coagulate to form fog centers when the sites are saturated with dimers of silver atoms [123].

In summary, the 'quantum size effect' and 'site effect' of silver clusters are important for AgX photographic films to achieve high sensitivity. The quantum size effect provides a stable silver dimer despite an unstable single silver atom, giving the threshold for latent image formation. This threshold is of great importance for the stability of AgX photographic films, as illustrated in Figure 1.7. The discrete energy levels could be the reason for high stability of a small silver cluster. A cluster with an open shell or with an odd number of valence electrons has a deep electron-accepting level, and might be the reason for high developability of a small silver cluster.

The site effect provides two kinds of silver dimers; R centers and P centers. Being assisted by ionic relaxation processes, a P center (i.e., a latent sub-image center) captures a photoelectron to grow during exposure to light, while a R center captures a positive hole to prevent the recombination of a photoelectron and a positive hole and the creation of an additional electron.

4.5 Chemical Sensitization and Stabilization

4.5.1 Introduction

In order to achieve latent image formation in photographic emulsions with high efficiency, many attempts have been made to modify AgX grains in emulsions. While these attempts include chemical sensitization, hypersensitization, and latensification, chemical sensitizations are most effective, and widely used in practical photographic films and papers [124,125]. Chemical sensitization is a method of putting small centers on the surface or in the interior of AgX grains via chemical reactions taking place during the ripening of emulsions (i.e., agitation of emulsions at elevated temperature for a certain time period). Hypersensitization is a method of treating coated emulsion layers before exposure to light in order to improve the efficiency of the latent image formation. Latensification is a method of treating coated emulsion layers after exposure to light but before development in order to increase the developability of latent sub-image centers and small latent image centers.

There are three major chemical sensitizations, i.e., chalcogen sensitization including sulfur, selenium, and tellurium sensitizations; gold sensitization; and reduction sensitization. Chemical species that are provided by chemical sensitization and effective for improving the efficiency of latent image formation are called chemical sensitization centers in this book. Figure 4.11 illustrates the mechanism of latent image formation. Chalcogen and chalcogen-plus-gold sensitization centers provide shallow electron traps where photoelectrons are preferentially trapped, and silver clusters (latent image centers) are thereby formed. An important role of gold sensitization centers is to increase the developability of small silver clusters by incorporating them with gold atoms. The role of reduction sensitization centers is to prevent the recombination of photoelectrons and positive holes by trapping and deactivating positive holes. Furthermore, the reaction of a reduction sensitization center with a positive hole in an AgX grain provides a free electron available for latent image formation [124].

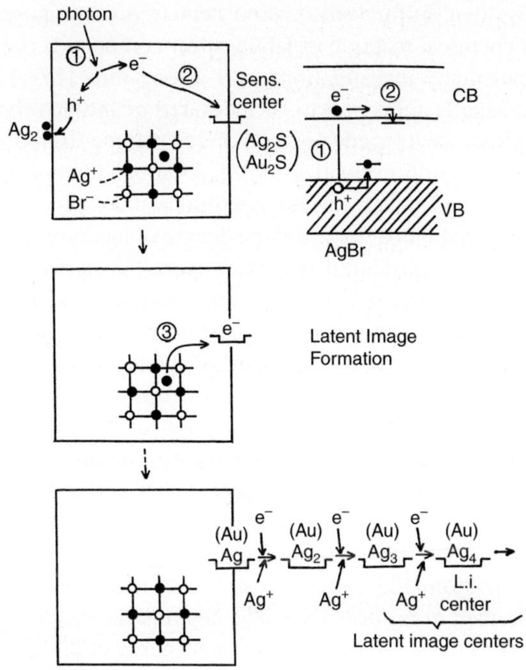

Figure 4.11 The formation mechanism of a latent image center on a AgBr grain with emphasis on the roles of chemical sensitization centers, where e^- and h^+ are a photoelectron and a positive hole, respectively, and CB and VB are the conduction and valence bands of AgBr, respectively. An electron-trapping sensitization center is composed of Ag_2S and/or Au_2S, while Ag_2 is a hole-trapping reduction sensitization center (i.e., R center), and the smallest latent image center is composed of three atoms containing a gold atom.

Stabilizers and antifoggants are also important from a practical viewpoint, since chemical sensitizations tend to make emulsions unstable and to bring about fog formation during storage and development.

4.5.2 Chalcogen Sensitization

Since gelatin inherently contains some sulfur compounds available for sulfur sensitization, it is believed that sulfur sensitization was introduced into AgX photographic materials when Maddox used gelatin for those materials in 1871. Paying attention to the fact that photographic sensitivity remarkably depended upon the kind of gelatin, Sheppard analyzed the chemical contents of gelatin, and discovered in 1925 that sulfur compounds could increase photographic sensitivity [126,127]. Among many sulfur sensitizers, thiosulfates and allylthiourea are well known and widely used for sulfur sensitization. Sheppard proposed that Ag_2S specks were sulfur sensitization centers, and that Ag_2Se and Ag_2Te specks should function similarly to selenium and tellurium sensitization centers, respectively. Although it took a long time to practically develop selenium and tellurium sensitization, sulfur sensitization soon became a central technology for enhancing photographic sensitivity.

It is now doubtless that sulfur sensitization centers are composed of Ag_2S clusters formed as a result of chemical reaction of labile sulfur compounds (i.e., sulfur sensitizers) owing to the results of many investigations over a long time [124,125]. The amount of Ag_2S thus formed is usually too small to be measured quantitatively by most analytical methods. Frieser and Ranz developed a method of quantifying the amount of Ag_2S using a radioactive thiosulfate as a sulfur sensitizer [128]. Although many investigators measured the amount of Ag_2S formed under various conditions, they could not find any definite correlation between the amount of Ag_2S and the degree of increase in sensitivity caused by sulfur sensitization. It was considered that there were two types of Ag_2S, which were useful and useless for increasing sensitivity, respectively [e.g., 129–131].

The function of sulfur sensitization centers have been studied for a long time by many groups of investigators, as described in Section 3.4.2 and in this author's previous work [124]. Gurney and Mott proposed that sulfur sensitization centers were deep electron traps [5]. Mitchell proposed that sulfur sensitization centers were positive hole traps [12, 132], emphasizing that deep electron traps on AgX grains before being exposed to light should act as fog centers. It is now accepted that sulfur sensitization centers are shallow electron traps, where latent image centers are preferably formed [124]. Evidences for this theory include the observation of the decrease in lifetime of photoelectrons in AgX grains by sulfur sensitization [133,134], indicating that sulfur sensitization centers act as electron traps, as described in Section 3.4.2. The depth of the electron traps was estimated by Hamilton et al. as 0.33 eV based on the activation energy of the spectral sensitization of the photographic process by sulfur sensitization centers [135]; by Kellogg and Hodes as 0.39 eV with a thermally stimulated current technique [136]; and by Kanzaki and Tadakuma as 0.3 eV with luminescence-modulation spectroscopy [137].

Keevert analyzed the rearrangement process of Ag_2S based on Smoluchowski's equation [138], and concluded that sulfur sensitization centers were dimers of Ag_2S and numbered as many as $1000/\mu m^2$. Kanzaki and Tadakuma also identified sulfur sensitization centers as dimers of Ag_2S, which were formed at a concentration of 2800 dimers/μm^2 on the surface of cubic AgBr grains [139]. As shown in Figure 4.12, Mifune et al. succeeded in observing sulfur sensitization centers formed on AgBr grains by electron microscope. They decorated them with silver atoms by use of a specially prescribed physical developer, and found sulfur sensitization centers of about $3000/\mu m^2$ on the grains [140]. They compared the concentration of sulfur sensitization centers with the amount of Ag_2S formed during sulfur sensitization, and supported the idea that a sulfur sensitization center was composed of a dimer of sulfide ions.

Based on results accumulated over many years, a model of the structure and electronic energy levels of sulfur sensitization centers and fog centers was proposed [124,141,142]. In Figure 3.31, two versions of the conditions of Ag_2S on AgBr grains are proposed, i.e., a solid solution type composed of substitutional sulfide ions and a cluster type. In the former case, a sulfide ion replaces a bromide ion at a lattice site and is associated with an interstitial silver ion for the charge compensation. In the latter, Ag_2S clusters are formed on the surface of AgBr grains, being separated from the lattice of AgBr. An interstitial silver ion, which a substitutional sulfide ion carries, provides a very shallow trap with a hydrogen-like orbital for a free electron in AgBr. Two hydrogen-like orbitals, which the dimers of substitutional sulfide ions carry, interact with each other to give bonding and nonbonding orbitals. It was proposed that the bonding orbital should act as a sulfur

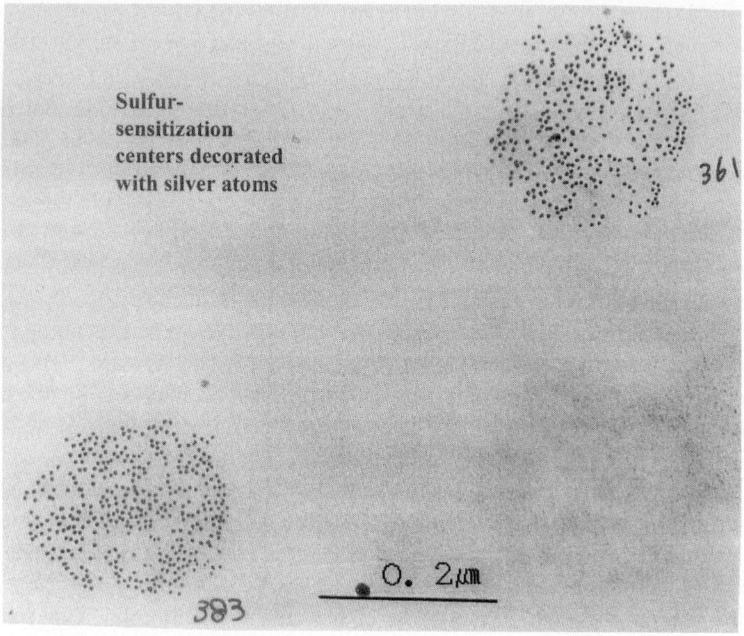

Figure 4.12 Electron micrograph of sulfur sensitization centers, which were formed on AgBr grains (0.2 μm) and decorated by silver atoms. When AgBr grains are dissolved, sulfur sensitization centers decorated with Ag atoms are observed in the gelatinate shells, which are slightly darker than the background, as seen in the lower right part. The centers in the upper right and lower left parts are marked with ink in order to make them clear. Reproduced with permission from Reference [33], copyright 2002 IS&T, The Society for Imaging Science and Technology.

sensitization center by providing a deeper trap level for a free electron compared to that of a single interstitial silver ion. Effective sulfur sensitization centers and fog centers are assigned to the dimers of substitutional sulfide ions and Ag_2S clusters, respectively, as shown in Figure 3.31.

Baetzold made a theoretical calculation of various centers containing one or two substitutional sulfide ions with silver ions at interstitial and kink sites for the compensation of the excess negative charges on the sulfide ions in order to discover a model of a sulfur sensitization center with electron-trapping ability [143]. It was indicated that not only an interstitial silver ion, but also a silver ion at a kink site, should be available for the compensation of an excess negative charge of a substitutional sulfide ion, and that some sort of lattice relaxation, in which silver ions at surface sites were involved, should take place for a sulfur sensitization center to stabilize and render a trapped electron available for the formation of a latent image center. The importance of the lattice relaxation process in photographic sensitivity was already proposed by Hamilton [16,144] and by Hoffmann et al. [145,146].

The above-stated results indicate that tremendous numbers of sulfur sensitization centers are formed during sulfur sensitization, and act as shallow electron traps [124,141,142]. As indicated in Section 4.3, optimizing the number and depth of shallow electron traps acting as the sites for latent image formation is of great importance for enhancing the efficiency of latent image formation, and had a good chance when chalcogen sensitization was practically expanded from sulfur to selenium and tellurium sensitizations [147,148,55]. The sites for chalcogen sensitization centers are considered to be the same among sulfur, selenium, and tellurium sensitizations, and are about $3000/\mu m^2$. By use of the method developed by Hamilton et al. [135], Morimura et al. determined the depths of electron traps provided by sulfur, selenium, and tellurium sensitization centers as 0.28, 0.17, and 0.12 eV, respectively [55].

This result could be explained by expanding the model for a sulfur sensitization center with a dimer of substitutional sulfide ions to selenium and tellurium sensitizations in such a way that sulfide ions are replaced by selenide and telluride ions. It is noted that both a selenide ion and a telluride one are larger than a sulfide ion. The increase in size of the ions in a sulfur sensitization center should decrease the interaction between the two hydrogen-like orbitals and therefore decrease the depth of the electron trap provided by it [149].

It is considered [149] that the above-stated result is not consistent with the idea that the temperature coefficient of the long-wavelength sensitivity is attributed to the energy for the hole injection from a photo-excited chalcogen sensitization center to the valence band of an AgX grain [e.g., 150], since the height of the filled level of a chalcogenide ion is on the order of its atomic number.

Also note that a sulfur sensitization center is not silver sulfide itself, but a dimer of substitutional sulfide ions in AgX. In contrast, it has been proposed that clusters of Ag_2S itself act as fog centers [142]. Electron-microscopic observation revealed as many as $3000/\mu m^2$, and that the developability of each cluster was so small that only one among several hundreds clusters on a fine AgBr grain could initiate development when all the latent image centers could do it. In contrast to substitutional sulfide ions, which constitute sulfur sensitization centers, most clusters of Ag_2S contribute neither to the increase in sensitivity nor to the increase in fog density. This result could explain the existence of two kinds of Ag_2S, which was proposed based on the relation between sensitivity and the amount of Ag_2S in sulfur-sensitized emulsions [e.g., 129–131].

In summary, the model illustrated in Figure 3.31 can explain the mechanisms of chalcogen sensitization and related phenomena, as itemized below [149]:

(1) The model can explain why an electron trap available for latent image formation is provided by chalcogen sensitization without assuming that the electron affinity of Ag_2S is larger than that of AgX.
(2) The model can explain the significant difference in the amount of a sulfur sensitizer for the increase in sensitivity from that for the appearance of fog density.
(3) The model can explain the existence of two kinds of Ag_2S, which are effective and ineffective, respectively, for the increase in photographic sensitivity.
(4) As described in the next section, the model can explain why a sulfur-plus-gold sensitization center provides a shallower electron trap and is therefore more suitable for latent image formation than a sulfur sensitization center.
(5) The model can explain why the increase in size of chalcogenide ions causes the reduction in depth of electron traps provided by chalcogen sensitization centers and in the degree of HIRF in chalcogen-sensitized emulsions.

4.5.3 Gold Sensitization and Gold Latensification

In collaboration with Mueller [151–153], Koslowsky discovered gold sensitization in 1936 by use of $AuCl_4^-$ and SCN^-. Considerable efforts to clarify the chemistry of gold sensitization have been made by many investigators. It is now accepted that Au^+ rather than Au and Au^{3+} is a gold sensitization center [124,125]. The chemistry of the formation of gold sensitization centers has been studied in detail by many groups of investigators. The studies have been focused on the reduction of Au^{3-} to Au^-, stabilization of Au^- by appropriate ligands, and incorporation of Au^- into AgX grains in competition with their binding by gelatin. The binding of ligands to Au^+ should be strong enough to prevent Au^+ from combining with gelatin and from undergoing their disproportionation to Au and Au^{3+}, while it should be weak enough to have an appropriate probability for complexes between Au^+ and ligands to dissociate to allow the incorporation of Au^+ into AgX grains. Examples of appropriate ligands are SCN^- and $S_2O_3^{2-}$. On the other hand, CN^- is not appropriate since its binding with Au^- is too strong for Au^+ to be allowed to enter into an AgX grain.

In practice, since gold sensitization is usually used with sulfur sensitization, it is appropriate to study sulfur-plus-gold sensitization instead of gold sensitization alone. The following three important proposals were made to explain the mechanism of the increase in sensitivity brought about by sulfur-plus-gold sensitization centers [124]:

(1) A sulfur sensitization center and a gold sensitization center are combined to form a sulfur-plus-gold sensitization center, which is composed of a dimer of substitutional sulfide ions associated with an interstitial silver ion and an interstitial gold ion.
(2) A gold atom should be incorporated into a latent image center to deepen its electronic energy levels, and thus reduce the size of the smallest latent image center [154].
(3) An electron trap provided by a sulfur-plus-gold sensitization center should be more suitable for an efficient latent image center than that provided by a sulfur sensitization center.

Proposal (1) is supported by the amount of sulfur sensitizers and gold sensitizers used to maximize photographic sensitivity, as shown in Figure 4.13 [155]. Namely, the amount of sulfur sensitizer for maximum sensitivity increased with increasing the amount of gold sensitizer, the former being nearly proportional to the latter.

Proposal (2) is supported by the evaluation that the smallest latent image center is composed of three atoms on sulfur-plus-gold-sensitized AgBr grains and four atoms on sulfur-sensitized ones [21–23]. The incorporation of a gold atom into a latent image center can be repeatedly proved by the fact that the oxidation potential of latent image centers formed on sulfur-plus-gold-sensitized grains is more positive than that of latent image centers on sulfur-sensitized and unsensitized grains [156–159]. It is therefore evident that Proposal (1) contributes to the increase in sensitivity caused by sulfur-plus-gold sensitization. Proposal (2) is consistent with Proposal (1), according to which a sulfur-plus-gold sensitization center contains Au^+ available for the incorporation of a gold atom into a latent image center formed at the center.

There have been discussions on the adequacy of Proposal (3). As described in the previous section, the trap depth of sulfur sensitization centers was evaluated to be 0.33 eV

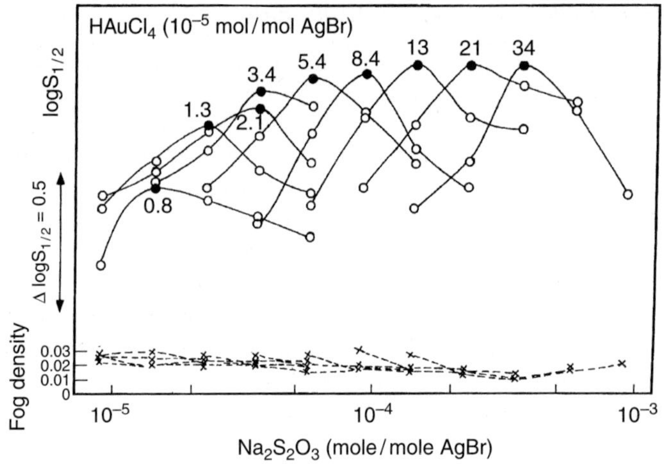

Figure 4.13 Sensitivity and fog density of sulfur-plus-gold-sensitized octahedral AgBr emulsion grains with an equivalent diameter of 0.2 μm as functions of the amounts of a sulfur sensitizer ($Na_2S_2O_3$) and a gold sensitizer ($HAuCl_4$). Emulsions were digested for 60 min at 60°C. Reproduced with permission from [155], copyright 2000 IS&T, The Society for Imaging Science and Technology.

by Hamilton and Harbison [135]. They also evaluated the trap depth of sulfur-plus-gold sensitization centers as 0.19 eV. Kellogg and Hodes obtained similar result [136]. Namely, an electron trap provided by a sulfur-plus-gold sensitization center was shallower than that by a sulfur sensitization center. Since the latter is deep enough to induce the dispersion of latent image centers, it is considered that the former is more suitable for efficient latent image formation.

According to the simulation of latent image formation as shown in Figure 4.4, Hailstone suggested the tendency that the number of electron trapping sensitization centers should increase with decreasing trap depth and that the sensitivity achieved by the centers should increase with decreasing trap depth [77]. The result shown in Figure 4.13 is consistent with these predictions from the viewpoints that the number of the sensitization centers should increase with increasing amount of sulfur sensitizer, and that the trap depth of the centers should decrease with increasing amount of gold sensitizer. Since a decrease in the depth of electron traps should result in an increase in the mobility and diffusion length of photoelectrons, it should enhance the formation of a latent image center on the grain surface in competition with the formation of a latent image center in the interior [160,161]. These results and considerations support the idea that gold sensitization is achieved, not only by Proposal (2), but also by Proposal (3).

The reason why the electron trap provided by a sulfur-plus-gold sensitization center is shallower than that provided by a sulfur sensitization center is explained on the basis of the model for sulfur sensitization center, as illustrated in Figure 3.31. Several experiments supported the idea that one of two interstitial silver ions is replaced by a monovalent gold ion to form a sulfur-plus-gold sensitization center [124]. Since a gold ion is larger than a silver ion, the interaction between the two hydrogen-like orbitals associated with two interstitial ions should be weaker in sulfur-plus-gold sensitization center than in a sulfur sensitization center, giving a shallower electron trap in the former than in the latter.

4.5.4 Reduction Sensitization

Carroll and Hubbard observed chemical sensitization using sodium sulfite as early as 1933 [162]. Paying attention to the difference from sulfur sensitization, Lowe, Roberts, and Jones introduced the concept of reduction sensitization in 1951 [163]. Many kinds of reducing compounds and methods have been proposed as reduction sensitizers [124] including stannous chloride, dimethylamineborane, aminoiminomethanesulfinic acid, formamidine sulfinic acid (thiourea dioxide), and hydrazine derivatives. It is known that the digestion of emulsions at low pAg (silver digestion) [164,165] and in the presence of hydrogen gas [166] could also produce reduction sensitization centers.

The proposed mechanism of reduction sensitization is described in detail in Section 4.4.2, and summarized below:

(1) Reduction sensitization centers are composed of dimers of silver atoms on the surface and in the interior of AgX grains.
(2) There are two types of reduction sensitization centers. The first (R center) is a positive hole trap, and is formed at an electrical neutral kink site. The other (P center) is an electron trap, and is formed at a positively charged kink site, as described in Section 4.4.2.
(3) Proposal (2) is supported by the observation that radio frequency photoconductivity with positive holes as carriers in AgBr grains can be reduced by R centers, while microwave photoconductivity with photoelectrons as carriers can be reduced by P centers.
(4) Proposal (2) is supported by the observation that the oxidation potential of R centers is less positive than that of P centers.
(5) While one absorbed photon creates a free electron and a positive hole, the capture of a positive hole by an R center creates an additional free electron to achieve so-called two-electron sensitization.
(6) Excessive reduction sensitization forms fog centers, which are aggregates of reduction sensitization centers and diamagnetic in contrast to latent image centers, which are mostly paramagnetic. Namely, all the fog centers are even nano-clusters, while many latent image centers are odd nano-clusters.

Additional results have been reported as shown below.

Further characterization of reduction sensitization centers have been made. By applying theoretical calculation to reduction sensitization centers, Baetzold supported the formation of P centers by reduction sensitization, and proposed the possibility that reduction sensitization P centers were different from photolytically formed ones [167]. Kuge et al. also supported the formation of reduction sensitization P centers by observing their enhancement of the dispersion of latent image centers [168]. By use of diffuse reflection spectra of reduction-sensitized emulsions, Tani and Murofushi observed not only plasmon-resonance absorption spectra of fog centers composed of silver nano-clusters, but also absorption spectrum of reduction sensitization centers with HOMO–LUMO transition for the first time (Figure 8.10) [45], providing photographic scientists with a new method of studying silver clusters in photographic films [169–171,115,117,172–175]. Marchetti et al. were able to observe the absorption bands of two kinds of reduction sensitization centers and the low-temperature fluorescence bands of reduction sensitization centers [173].

As stated above, reduction sensitization is effective for increasing sensitivity by bringing about functions that other chemical sensitizations do not. However, there are some problems in applying reduction sensitization to practical photographic films. The formation of P centers should be depressed since P centers disturb the function of chalcogen-plus-gold sensitization by competing against them for trapping photoelectrons and by forming fog centers in gold-sensitized emulsions. It has been noted that hydrogen hypersensitization does not disturb the function of sulfur-plus-gold sensitization. It has been pointed out that P centers formed on the grain surface can be R centers in the interior when they are buried by AgX [176].

Although reduction sensitization R centers and hydrogen-hypersensitization centers are useful for increasing sensitivity, they are unstable since their HOMO is high and apt to be oxidized by oxygen and water. In particular, hydrogen hypersensitization centers are less stable than reduction sensitization R centers because introducing water into photographic films changes the condition for the former [43], and are not practically useful. One way to overcome the instability of reduction sensitization centers is to incorporate reduction sensitization centers into the interior of AgX grains [177]. Although it is difficult to incorporate the centers into the interior of the grains with perfect crystal structure, crystal defects such as dislocations should enhance the incorporation of the centers into the interior of the grains.

The rate of the two-electron sensitization caused by R centers is slow because its main process, which is the dissociation of a single silver atom on an AgX grain to create an additional free electron, takes several seconds at room temperature [48]. As described in Section 4.4.2, proposals on the technologies for new two-electron sensitization have been made [73,74,178] and should be subjected to improvements for their practical use.

Although reduction sensitization centers are formed by use of reduction sensitizers, they are also formed by the electron transfer from gelatin to AgX grains. As shown in Figure 4.14, the redox potential of an aqueous gelatin solution decreases with increasing pH and becomes more negative than the silver potential, which increases with increasing concentration of silver ions in emulsions. Thus, the reduction sensitization of AgX emulsions could be realized by silver digestion, according to which the emulsions are digested in the presence of high concentraions of silver ions at elevated temperature [164,165]. As described in Section 2.2.2, it is difficult to avoid the portion in the reaction solution in a vessel, where the concentration of silver ions is high enough to form reduction sensitization centers under neutral and alkaline condition during the precipitation of AgX grains [179]. In order to depress the formation of reduction sensitization centers during the precipitation of AgX grains, it is necessary to keep the reaction solution under acidic conditions, as indicated in Figure 4.14. In a dried photographic film, the electron transfer between silver halide grains and gelatin is determined by the relative heights of their Fermi levels, which could be measured by ultraviolet photoelectron spectroscopy (UPS) [180]. It has been revealed that the Fermi level of gelatin is higher than that of silver bromide. In accord with this result, it has been observed that reduction sensitization centers form on AgBr grains in dried photographic films during storage, bringing about an increase in the

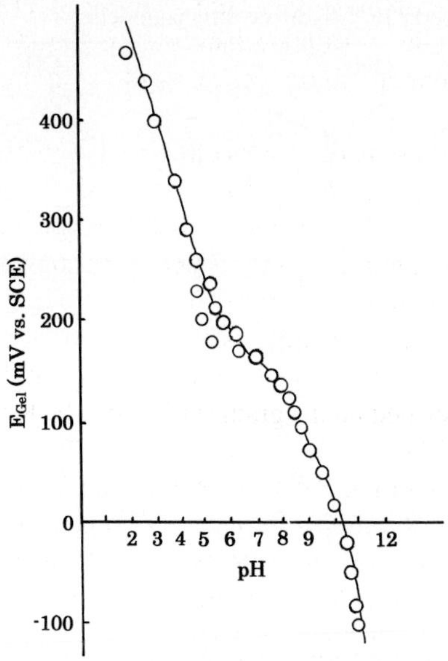

Figure 4.14 Redox potential of an aqueous gelatin solution with concentration of 3 wt.% as a function of pH of the solution. Reproduced with permission from Reference. [180], copyright 2003 IS&T, The Society for Imaging Science and Technology.

sensitivity of unsensitized emulsions and an increase in fog density in sulfur-plus-gold-sensitized emulsions. Extensive efforts to replace gelatin in photographic films with synthetic polymers have been made [e.g., 181]. Although some synthetic polymers behave similarly to gelatin for the precipitation of AgX grains and are not able to form reduction sensitization centers on AgX grains [182], they have not been put to practical use because they were co-polymers made of several different monomers and could not be synthesized reproducibly. On the other hand, it is expected that the phenomena caused by reducing ability of gelatin is depressed by stabilizers and antifoggands, which are described in next section [180].

4.5.5 Stabilizers and Antifoggants

The increases in sensitivity caused by chemical sensitizations are more or less associated with the decrease in stability of photographic films during their manufacturing, storage, and development. Namely, chemically sensitized emulsion layers tend to suffer from fog formation and/or change in sensitivity during storage and development processes. For this reason, most photographic films contain stabilizers and/or antifoggants for their stabilization. In this section, stabilizers and antifoggands for improving the stability of photographic films in this respect are focused on among various kinds of photographic additives.

Stabilizers are agents that decrease the changes in developable fog and/or in other sensitometric characteristics of photographic films that occur during storage. Antifoggants are agents that decrease the rate of fog density growth during development to a larger

Table 4.1 Structure and property of 4-hydroxy-tetraazaindenes[a]

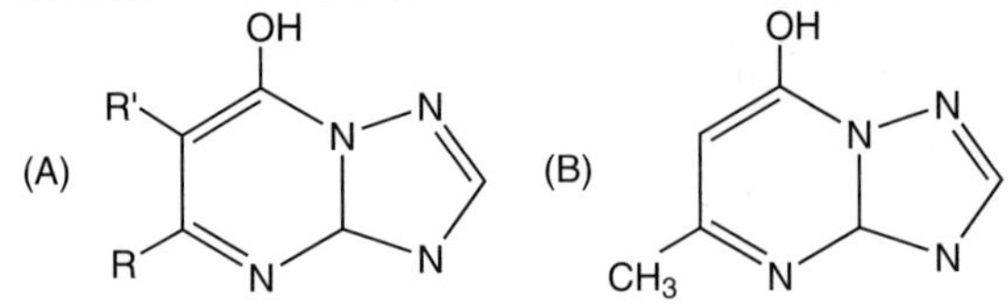

Compound	Structure	R	R'	pK_a	pK_{sp}
1	A	CH_3	H	6.27	10.5
2	A	H	H	5.86	10.0
3	A	Ph	H	5.32	9.3
4	A	CF_3	H	3.38	—
5	A	$t\text{-}B_u$	H	6.41	—
6	A	—$(CH_2)_3$—		6.10	10.2
7	B			6.53	10.8

[a]Although the structure and names of the agents are expressed in the traditional way here, it is known that a dissociative proton is attached to one of the nitrogen atoms in each agent.

degree than they decrease the rate of development of latent images. Tremendous numbers of stabilizers and antifoggants have been studied [183–187]. Among these, some detailed descriptions have been made on the fundamental properties of the following three families: 4-hydroxy-tetraazaindenes (Table 4.1), benzotriazoles (Table 4.2), and mercaptoazoles (Table 4.3) [187]. Among stabilizers, most extensive studies have been made on 4-hydroxy-6-methyl-1,3,3a,7-tetraazaindene (TAI), which was invented by Birr [188–193]. Among antifoggants, the most extensive studies have been made on 1-phenyl-5-mercaptotetrazole (PMT).

The first step for the stabilization and antifogging action by these compounds is their adsorption to AgX grains. The adsorptivity of these compounds to AgX grains is obvious

Table 4.2 Structure and property of benzotriazoles

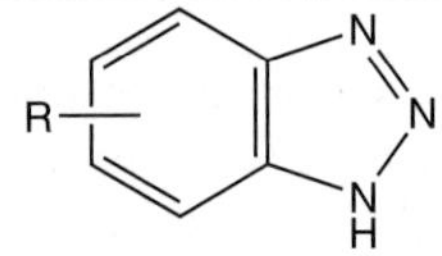

Compound	R	pK_a	pK_{sp}
8	Br	7.49	12.8
9	Cl	7.54	12.9
10	H	8.30	13.6
11	CH_3	8.50	13.7
12	NO_2	6.40	11.9

Table 4.3 Structure and property of mercaptoazoles

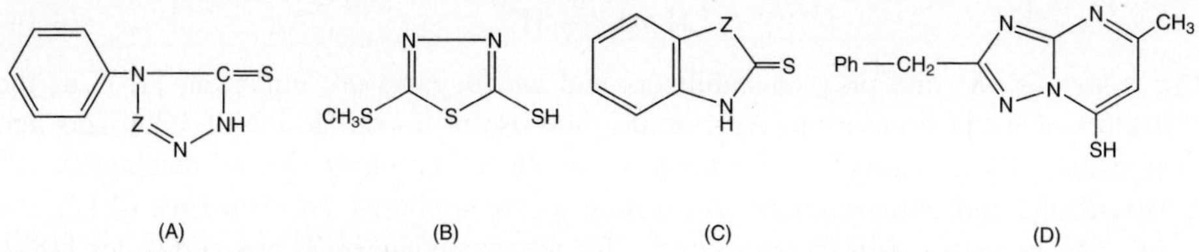

Compound	Structure	Z	pK_a	pK_{sp}
13	A	N	2.95	16.1
14	A	CCh_3	7.68	16.1
15	A	CH	7.28	15.9
16	B		4.81	16.5
17	C	S	6.94	16.5
18	C	NH	10.07	16.5
19	D		4.31	16.8

in the absence of gelatin [187], and was not necessarily obvious for tetraazaindene derivatives in the presence of gelatin [188–193]. However, as described in Section 3.2.2, the adsorptivity of all these compounds including tetraazaindene derivatives to AgX grains in the presence of gelatin is obvious when it is observed that they can definitely decrease the grains' ionic conductivity, which is proportional to the concentration of interstitial silver ions in the grains, and which are in equilibrium with silver ions at surface kink sites [194–201]. The dielectric loss method has often been used to learn about the adsorption of these compounds to AgX grains in photographic films and the change in the grains' ionic conductivity caused by the adsorption of these compounds.

As stated above, the formation of silver salts is essential to the adsorption of these compounds to AgX grains, and is expressed for tetraazaindene and benzotriazole derivatives as

$$-\mathrm{NH}- + \mathrm{Ag}^+ \leftrightarrows -\mathrm{NAg}- + \mathrm{H}^+, \tag{4.10}$$

and for mercaptoazoles as

$$-\mathrm{NH{-}CS}- + \mathrm{Ag}^+ \leftrightarrows -\mathrm{N} = \mathrm{CSAg}- + \mathrm{H}^+, \tag{4.11}$$

and their important property is the solubility product of silver salt (K_{sp}; $K_{sp} = -\log K_{sp}$) as expressed

$$K_{Sp} = [\mathrm{Ag}^+][\mathrm{L}^-], \tag{4.12}$$

where $[Ag^+]$ and $[L^-]$ are the concentrations of silver ions and anions of a stabilizer or an antifoggant (LH), which is characterized by the dissociation constant (K_a; $pK_a = -\log K_a$) as

$$K_a = [\mathrm{L}^-][\mathrm{H}^+]/[\mathrm{LH}] \tag{4.13}$$

for their acid–base equilibrium:

$$LH \leftrightharpoons L^- + H^+. \tag{4.14}$$

The values of pK_a and pK_{sp} of stabilizers and antifoggants are important [187] as the indicators of its adsorptivity to AgX grains, and useful for the design of stabilizers and antifoggants. The adsorptivity is primarily in the order of groups of mercaptoazole, benzotriazole, and tetraazaindene. According to the similarity between Eqs. (4.12) and (4.13), pK_a increases with increasing pK_{sp} for tetraazaindenes and benzotriazoles [187]. However, the adsorption of each agent to the grains is realized only after its dissociation according to the process of silver formation of silver salts in Eqs. (4.10) and (4.11). For example, an agent with large pK_a is strongly adsorbed to the grains only in a solution with high pH, and is hardly adsorbed to the grains in a solution with low and medium pH [187].

In view of the fact that unstable conditions are usually brought about by chemical sensitizations, it is now reasonable to consider the mechanisms of photographic stabilization in relation to the mechanisms of chemical sensitizations as reviewed in Sections 4.5.2–4.5.4. In relation to the mechanism of chalcogen sensitization, one of the most probable mechanisms for stabilization is the prevention of the formation of sensitization centers and fog centers during storage. As proposed in Section 4.5.2, chalcogen sensitization centers are dimers of substitutional calcogenide ions, which occupy lattice sites of AgX [124,141,142], and fog centers are clusters of silver chalcogenide [142]. It has also been demonstrated that TAI could effectively retard the aggregation of Ag_2S for sulfur sensitization [202]. It is thus expected that stabilizers retard the formation of fog centers by preventing the aggregate formation of silver chalcogenide during storage.

In relation to reduction sensitization, it was demonstrated that reduction sensitization centers are also formed on AgX grains in the presence of gelatin during storage for a long time. They are likely to become to be fog centers, especially when they interact with gold sensitizers and/or gold sensitization centers in emulsions [116]. As described by Eqs. (4.4)–(4.6), the oxidation of reduction sensitization centers on AgX grains during storage creates free electrons responsible for the formation of fog centers especially in the presence of gold sensitization centers. It was therefore expected and actually observed that the stabilizers decreased the concentration of silver ions available for the formation of reduction sensitization centers, and stabilized the photographic properties of emulsions by preventing the formation of reduction sensitization centers during storage [116].

4.6 Development-Amplification of Latent Image Centers

4.6.1 Introduction

Photographic development is a process that amplifies a latent image center formed according to the mechanisms described in the previous sections, and has been reviewed in detail [203–206]. Photographic development is important in relation to photographic sensitivity and image quality from the following viewpoints.

(1) As described in Chapter 1, photographic development is a unique and effective amplification process that makes it possible for AgX photographic materials to achieve such a high sensitivity and image quality. In general, the increase in degree

of amplification brings about the increase in sensitivity and the deterioration of image quality.

(2) The initiation of development of a latent image is the electron transfer from developing agents to latent image centers and should thus determine photographic sensitivity, since the development condition might determine the size of the smallest detectable latent image centers, which is one of the most important factors influencing sensitivity.

(3) The development process is based on the redox reaction, which involves AgX grains and developing agents with the aid of the catalyzing effect of latent image centers, and thus provides opportunities for variety of image processing techniques for AgX photographic materials.

The centers that initiate development are called development centers, and include latent image centers and fog centers. A development center grows to a mass of silver during photographic development, being repeatedly supplied with electrons and silver ions one after the other. The compound that provides a latent image center with an electron during development is called the developing agent. From the viewpoint of process, photographic development is classified into two types: conventional and thermal. Conventional development is carried out by putting exposed photographic films into a developer, which is an aqueous solution containing developing agents. Thermal development is carried out by heating exposed photographic films that contain developing agents and silver salts of organic compounds and are called photothermographic films [207–209]. On heating, the silver salts release silver ions available for development. From the viewpoint of mechanism, photographic development is classified into two types: chemical and physical. Silver ions for development are supplied from the interior of an AgX grain in the case of chemical sensitization and from a developer in the case of physical development. Thermal development belongs in principle to physical development. Although photothermographic films are popular for their diagnostic use [208–211], chemical development is mainly discussed in this chapter, since most practical photographic films are chemically developed.

4.6.2 Electron Transfer in Development

Figure 4.15 schematically shows the processes of chemical development of an AgBr grain bearing a latent image center on its surface. In comparison with Figures 4.11 and 4.1, it is understood that the development process is regarded as the extension of the formation and growth of silver clusters on AgX grains during exposure to light. In both cases, silver clusters grow through the capture of electrons and interstitial silver ions one after the other. In the case of the growth of silver clusters during exposure, electrons are provided by AgX grains as a result of the light absorption of the grains or sensitizing dyes adsorbed by the grains. In the case of development, electrons are provided by a developer, since a latent image center is such a deep electron trap as to readily accept an electron from a developing agent.

In the first step of development processes, a development center on an AgX grain receives an electron from a developing agent in a developer (i.e., electronic process). Next, the center attracts and combines with an interstitial silver ion (i.e., ionic process). The repetition of the above-stated electronic and ionic processes finally reduces all the silver

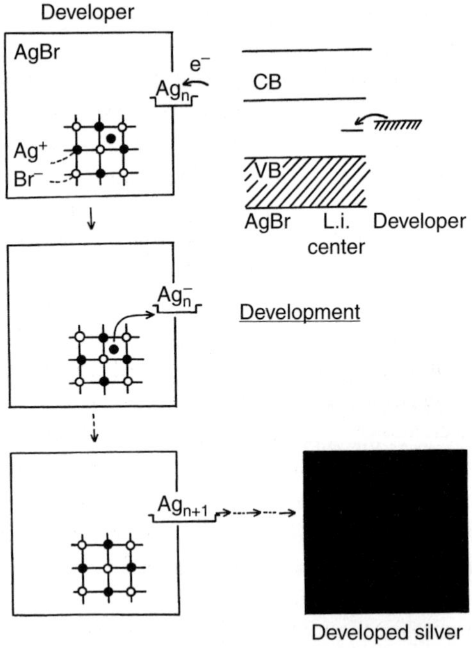

Figure 4.15 Schematic illustration of the development process on an AgBr grain, where Ag_n is a latent image center, and CB and VB are the conduction and valence bands of AgBr, respectively.

ions in a grain to silver atoms, releasing halogen ions into a developer solution. Interstitial silver ions initially present in a grain is limited in number, and should be rapidly supplied during development via the process described in Section 3.2.2.

The development process is usually expressed by an electrochemical way [203–205]. The general reaction of the development can be expressed as

$$\text{Red} + n\text{Ag}^+ \leftrightarrows \text{nAg} + \text{Ox} + \text{mH}^+, \tag{4.15}$$

where Red is a developing agent, Ox is an oxidized form of the developing agent, n is the number of electrons given up by a molecule of the developing agent, and m is the number of protons produced. The general cell with Eq. (4.15) can be separated into two half-cells, i.e., the silver half-cell with electrochemical potential of E_{Ag},

$$\text{Ag}^+ + e^- \leftrightarrows \text{Ag}, \tag{4.16}$$

and the developer half-cell with electrochemical potential of E_{Dev},

$$\text{Ox} + m\text{H}^+ + ne^- \leftrightarrows \text{Red}. \tag{4.17}$$

The cell potential E_{cell} for (4.15) is the difference between the above-stated half-cell potentials,

$$E_{\text{cell}} = E_{\text{Ag}} - E_{\text{Dev}}. \tag{4.18}$$

When E_{cell} is positive, Reaction (4.15) proceeds to the right, and when E_{cell} is negative, Reaction (4.15) proceeds to the left.

Since photographic development is initiated by the electron transfer from developing agents to latent image centers, ϵ_{HO} of a developing agent should be high enough for the electron transfer. General rules have been proposed for the molecular structure of developing agents by many investigators [203,204,212], including Kendall [213] and Pelz [214]. It is thus helpful for the understanding of the mechanism of development to examine the relation between molecular and electronic structures of developing agents. Among the above-stated rules, Kendall's and Pelz's rules are simple and cover most developing agents. According to Kendall's rule, compounds are developing agents if they have such a structure as a-(C=C)$_m$–a′, where a and a′ are –OH, $-NH_2$, –NHR, $-NR_1R_2$ and $m = 0$ or a whole number. According to Pelz's rule, compounds that have such a structure as a-(C=N)$_m$–a′ are developing agents. The examination in terms of the alternate nature of π-electron conjugated systems in molecular orbital theory indicated that the HOMOs of the compounds, which obey Kendall's and Pelz's rules, are non-bonding, while those of the compounds that do not obey them are bonding, and that the former are higher in nature than the latter [215,203]. This result supports the idea that photographic development is initiated by the electron transfer from a developing agent in a developer to a latent image center on an AgX grain.

The rate of development depends upon several factors including the overpotential of the development Reaction (4.15), the concentration of developing agents, the surface area of development centers, and diffusion of the chemical species involved. Kinetic studies of photographic development have been carried out by many investigators from various viewpoints, as exemplified by some references [203,206].

The most important process for the rate of development is the electron transfer from a developing agent to a development center on an AgX grain, whose rate depends upon the overpotential of the development Reaction (4.15), i.e., Eq. (4.18). The value of E_{Dev} is related to the polarographic half-wave oxidation potential ($E_{1/2}$) of a developing agent. Although it was demonstrated by Tong et al. that the potential for one-electron oxidation reaction (E_1) was more closely related to the rate of development than $E_{1/2}$ [216], $E_{1/2}$ is still useful for studying the rate of development among developing agents, whose molecular structures are closely related to each other.

The most useful developing agents for color photography are found among 4-amino-N-dialkylanilines, and were therefore studied extensively from various viewpoints. Figure 4.16 shows the result of the extensive analysis of 4-amino-N-dialkylanilines [217] as developing agents by Bent et al., indicating the relation between their development rates ($1/t'$) and $E_{1/2}$ where $t' = t - t_o$, t_o is the induction period, and t is the time required to give optical density of 0.2 above fog. The linear relationship between the logarithm of the development rate and $E_{1/2}$ indicates that $E_{1/2}$ is approximately proportional to the activation energy of the rate of development by those compounds. Compounds with potentials more negative than about + 170 mV gave development rates that were nearly independent of $E_{1/2}$. The authors considered that the rate of development by these compounds was so high that the diffusion rather than the chemical reaction became the rate-determining step.

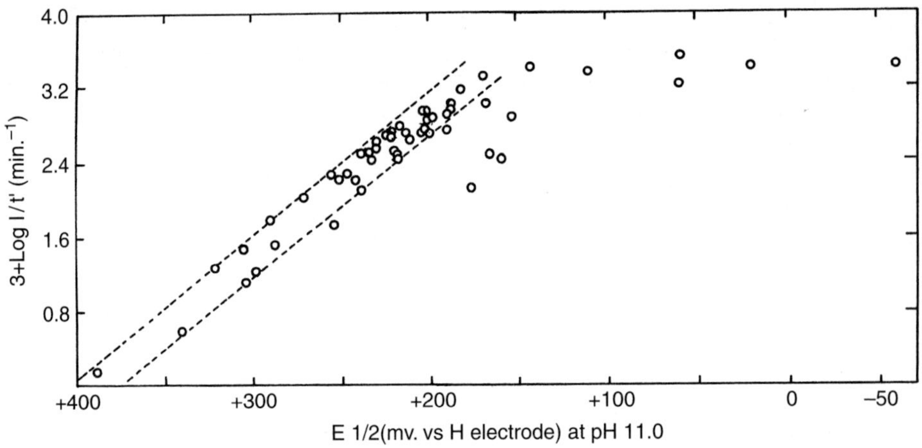

Figure 4.16 Relation between development rates and half-wave oxidation potentials of 4-amino-N-dialkylanilines, where *t'* is development time less induction period. Reproduced with permission from Reference [217], copyright 1951 American Chemical Society.

4.6.3 Photographic Sensitivity and Image Quality in Relation to Development

Photographic sensitivity should depend upon the development conditions. However, it has been confirmed by several groups that photographic sensitivity does not depend upon the kind of developers when each developer is used under optimum conditions [206,218–221]. It is considered from the above-stated result that each developer under optimum conditions can detect the same smallest latent image centers on emulsion grains, whereas it can hardly detect latent sub-image centers. On the other hand, photographic sensitivity should be increased by the technologies for decreasing the size of the smallest development centers. As stated in Section 4.5.3, gold sensitization decreases the size of the smallest latent image centers by incorporating gold atoms into image centers, and brings about the fact that the smallest latent image center formed in sulfur-plus-gold-sensitized emulsions is composed of three atoms, while the smallest latent image center formed in sulfur-sensitized emulsions is composed of four or five atoms. It is known that latent sub-image centers are stable and become developable after some latensification treatment [222,223] or enforced development conditions are applied [222–225]. These facts imply some possibility for further increase in photographic sensitivity by finding the development conditions that could detect an image center composed of two atoms (i.e., a latent sub-image center under usual development conditions), discriminating from fog formation. Namely, the discrimination of the development of image centers from fog formation might be one of the key technologies for future increases in sensitivity.

The development process is based on the redox reaction, which involves AgX grains and developing agents with the aid of the catalyzing effect of development centers. In addition, there are various chemical interactions among AgX grains during development. Development processes thus provide the opportunities for various image-processing techniques for AgX photographic films, as briefly described below for example. One of them is chemical adjacency effects, which are based on the finite diffusion and mass transfer rates of development byproducts and developer [226]. They take place among emulsion layers in multilayer films, as well as within a single emulsion layer, and are controlled to improve image quality. In color development, colored couplers are widely used to provide image-forming dyes with pure colors [227,228]. Other examples are development-inhibitor-releasing (DIR) couplers [229] and development-accelerator-releasing (DAR) couplers [230,231]. It is expected that DIR couplers provide fine images at the expense of sensitivity and gamma, while DAR couplers provide high sensitivity at the expense of graininess.

References

1. J. F. Hamilton, in T. H. James (ed.), *The Theory of the Photographic Process* (New York: Macmillan, 1977), 105.
2. T. Tani, *Physics Today* (September 1989), 36.
3. T. Tani, *Photographic Sensitivity: Theory and Mechanisms* (New York/Oxford: Oxford University Press, 1995), Chapter 4.
4. S. E. Sheppard, A. P. H. Trivelli, R. P. Loveland, *J. Franklin Inst.*, 200, 51 (1925).
5. R. W. Gurney, N. F. Mott, *Proc. R. Soc. London, Ser. A*, 164, 151 (1938).
6. P. C. Burton, W. F. Berg, *Photogr. J.*, 86B, 2 (1946).
7. P. C. Burton, *Photogr. J.*, 86B, 62 (1946).
8. W. F. Berg, P. C. Burton, *Photogr. J.*, 88B, 84 (1948).
9. P. C. Burton, *Photogr. J.*, 88B, 13 (1948).
10. P. C. Burton, *Photogr. J.*, 88B, 123 (1948).
11. W. F. Berg, *Rep. Prog. Phys.*, 11, 248 (1948).
12. J. W. Mitchell, *Recent Prog. Phys.*, 20, 433 (1957).
13. J. W. Mitchell, *J. Phys Chem.*, 66, 2359 (1962).
14. J. W. Mitchell, *Photogr. Sci. Eng.*, 22, 249 (1078).
15. J. W. Mitchell, *Photogr. Sci. Eng.*, 25, 170 (1981).
16. J. F. Hamilton, *Adv. Phys.*, 37, 359 (1988).
17. J. W. Mitchell, *J. Photogr. Sci.*, 5, 49 (1957).
18. J. F. Hamilton, *Photogr. Sci. Eng.*, 14, 102 (1970).
19. J. F. Hamilton, *Photogr. Sci. Eng.*, 18, 371 (1974).
20. R. K. Hailstone, N. B. Liebert, M. Levy, *J. Imaging Sci.*, 34, 169 (1980).
21. R. K. Hailstone, J. F. Hamilton, *J. Imaging Sci.*, 29, 125 (1985).
22. R. K. Hailstone, N. B. Liebert, M. Levy, J. F. Hamilton, *J. Imaging Sci.*, 31, 185 (1987).
23. R. K. Hailstone, N. B. Liebert, M. Levy, J. F. Hamilton, *J. Imaging Sci.*, 31, 255 (1987).
24. R. K. Hailstone, N. B. Liebert, N. B. Levy. J. F. Hamilton, *J. Photogr. Sci.*, 36, 2 (1988).
25. R. K. Hailstone, N. B. Leibert, M. Levy, R. T. McCleary, S. R. Gilolmo, D. L. Jeanmaire, C. R. Boda, *J. Imaging Sci.*, 32, 113 (1988).
26. R. K. Hailstone, N. B. Liebert, M. Levy, J. F. Hamilton, *J. Imaging Sci.*, 32, 150 (1988).
27. R. K. Hailstone, N. B. Liebert, M. Levy, *J. Imaging Sci.*, 34, 169 (1990).
28. F. Seitz, *Rev. Mod. Phys.*, 23, 328 (1951).
29. J. M. Hedges and J. W. Mitchell, *Phil. Mag.*, 44, 223 (1953).
30. G. C. Farnell, *J. Photogr. Sci.*, 27, 160 (1979).
31. J. W. Mitchell, *J. Soc. Photogr. Sci. Technol. Jpn.*, 54, 248 (1991).
32. T. Tani, *J. Imaging Sci. Tech.*, 51, 110 (2007).

33. H. Mifune, M. Mizuno, Y. Toyama, T. Shiozawa, J. Okuda, *J. Imaging Sci. Technol.*, 46, 262 (2002).
34. J. F. Hamilton, L. E. Brady, *J. Appl. Phys.*, 30, 1893 (1959).
35. J. F. Hamilton, L. E. Brady, *J. Appl. Phys.*, 30, 1902 (1959).
36. J. F. Hamilton, L. E. Brady, *Photogr. Sci. Eng.*, 8, 189 (1964).
37. L. M. Kellogg, J. Hodes, J. Muenter, in the preprint of *The International East-West Symposium II: Factors Influencing The Efficiency of Photographic Imaging, Kona, Hawaii, October/November 1988.*
38. A. Hirano, *J. Imaging Sci. Tech.*, 42, 516 (1998).
39. K. Hosoi, A. Hirano, T. Tani, *J. Imaging Sci. Tech.*, 46, 76 (2002).
40. T. Tani, T. Tasaka, M. Nurofushi, K. Hosoi, A. Hirano, *Imaging Sci. J.*, 47, 1 (1991).
41. T. Tani, *J. Imaging Sci.*, 29, 93 (1985).
42. G. C. Farnell, J. B. Chanter, *J. Photogr. Sci.*, 9, 73 (1961).
43. T. A. Babcock, P. M. Ferguson, W. C. Lewis, T. H. James, *Photogr. Sci. Eng.*, 19, 49 (1975).
44. T. Tani, *Radiation Measurements*, 44, 733 (2009).
45. T. Tani, M. Murofushi, *J. Imaging Sci. Technol.*, 38, 1 (1994).
46. K. Ohzeki, *J. Imaging Sci. Tech.*, 40, 591 (1966).
47. J. F. Hamilton, in James (ed.), *The Theory of the Photographic Process*, 133.
48. J. H. Webb, *J. Opt. Soc. Am.*, 40, 3 (1950).
49. C. R. Berry, *J. Photogr. Sci.*, 18, 169 (1970).
50. S. Takada, *J. Soc. Photogr. Sci. Technol. Jpn.*, 42, 112 (1979).
51. S. Takada, T. Tani, *J. Appl. Phys.*, 45, 4764 (1974).
52. T. Tani, S. Takada, *Photogr. Sci. Eng.*, 18, 620 (1974).
53. M. Ihama, T. Tani, *J. Imaging Sci.*, 31, 157 (1987).
54. T. Tani, M. Ihama, in F. Granzer, E. Moisar (eds), *Progress in Basic Principles of Imaging Systems, Proceedings of the International Congress of Photographic Science, Koeln, 1986* (Braunschweig/Wiesbaden: Friedr. Vieweg & Sons 1987), 335.
55. K. Morimura, H. Sasaki, T. Kojima, S. Yamashita, T. Tani, H. Mifune, in the preprint of *The Beijing International Conference, 'Technology and Application for 21st Century', Beijing, China, May 2005.*
56. W. G. McDugle, R. S. Eachus, M. T. Olm, in Paper Summaries in 44th Annual Conference of the SPSE, St Paul, 1991.
57. R. S. Eachus, Th. D. Pawlik, R. C.Baetzold, D. A. Crosby, W. G. McDugle, *J. Phys.: Condens. Matter*, 12, 2535 (2000).
58. Th. D. Pawlik, R. S. Eachus, R. C. Baetzold, W. G. McDugle, *J. Phys.: Condens. Matter*, 12, 2555 (2000).
59. M. T. Bennebroeck, J. Schmidt, R. S. Eachus, M. T. Olm, *J. Phys.: Condens. Matter*, 9, 3227 (1997).
60. G. C. Farnell, *J. Photogr. Sci.*, 17, 117 (1969).
61. H. Zwicky, *Z. Wiss. Photogr.*, 53, 68 (1959).
62. Tani, *Photographic Sensitivity: Theory and Mechanisms*, Chapter 1.
63. Anonymous, *Res. Discl.*, 225, 20 (1983).
64. J. T. Kofron, R. E. Booms, *J. Soc. Photogr. Sci. Tech. Jpn.*, 49, 15 (1986).
65. G. R. Bird, *Photogr. Sci. Eng.*, 18, 562 (1974).
66. J. C. Marchant, in F. Granzer and E. Moisar (eds), *Progress in Basic Principles of Imaging Systems* (Braunschweig, Germany: Vieweg, 1987), 9.
67. T. I. Penner, P. B. Gilman, *Photogr. Sci. Eng.*, 20, 97 (1976).
68. G. I. House, in Granzer and Moisar (eds), *Progress in Basic Principles of Imaging Systems*, 348.
69. G. R. Bird, R. C, Jones, A. E. Ames, *Appl. Opt.*, 8, 2389 (1969).

70. R. L. Parton, T. L. Penner, W. J. Harrison, J. C. Deaton, A. Muenter, in *AgX 2004: International Symposium on Silver Halide Technology* (Apringfield, VA: IS&T, 2004), 2.
71. M. Harbison, H. E. Spencer, in James (ed.), *The Theory of the Photographic Process*, 152.
72. T. Tani, *J. Imaging Sci.*, 30, 41 (1986).
73. J. Belloni, M. Treguer, H. Remita, R. de Keyzer, *Nature (London)*, 402, 865 (1999).
74. I. R. Gould, J. R. Lenhard, A. A. Muenter, S. A. Godleski, S. Y. Farid, *J. Am. Chem. Soc.*, 122, 11934 (2000).
75. T. A. Babcock, T. H. James, *J. Photogr. Sci.*, 24, 19 (1976).
76. J. D. Bologa, in *PICS Conference of IS&T, Portland, 1998.*
77. R. Hailstone, in B. Levy (ed.), *IS&T Proceedings. Symposium on Electronics and Ionic Properties of Silver Halides-Common Trends with Photocatalysis. Part of the 44th Annual Conference* (Soc. Imaging Sci. Technol., 1991), 102.
78. L. Silverstein, *Phil. Mag.*, 44, 257, 956 (1922).
79. L. Silverstein, *Phil. Mag.*, 45, 1062 (1923).
80. A. E. Ames, *Photogr. Sci. Eng.*, 17, 154 (1973).
81. R. Shaw, *J. Imaging Sci.*, 21, 25 (1973).
82. M. Kawasaki, S. Fujiwara, H. Hada, *Photogr. Sci. Eng.*, 22, 290 (1978).
83. H. Hada, M. Kawasaki, H. Fujimoto, *Photogr. Sci. Eng.*, 24, 232 (1980).
84. H. Hada, M. Kawasaki, *J. Imaging Sci.*, 29, 51 (1985).
85. J. F. Hamilton, P. C. Logel, *Photogr. Sci. Eng.*, 18, 507 (1974).
86. A. Marriage, *J. Photogr. Sci.*, 9, 93 (1961).
87. P. Fayet, F. Granzer, G. Hegenbart, E. Moisar, B. Pischel, L. Woeste, *Phys. Rev. Lett.*, 55, 3002 (1985)
88. T. Tasaka, M. Murofushi, T. Tani, *J. Imaging Sci. Tech.*, 47, 463 (2003).
89. I. Konstantinov, A. Panov, J. Malinowski, *J. Photogr. Sci.*, 21, 250 (1973).
90. I. Konstantinov, J. Malinowski, *J. Photogr. Sci.*, 23, 1 (1975).
91. I. Konstantinov, J. Malinowski, *J. Photogr. Sci.*, 23, 45 (1975)
92. M. G. Mayer, J. H. Jensen, *Elementary Theory of Nuclear Shell Structure* (New York: Wiley, 1955).
93. S. Sugano, *Microcluster Physics* (Berlin: Springer-Verlag, 1991).
94. R. C. Baetzold, *J. Solid State Chem.*, 6, 352 (1973).
95. R. C. Baetzold, *Photogr. Sci. Eng.*, 17, 78 (1973).
96. T. Tani, *J. Imaging Sci.*, 34, 143 (1990).
97. W. D. Knight, K. Clemenger, W. A. de Heer, W. A. Saunders, M. Y. Chou, M. L. Cohen, *Phys. Rev. Lett.*, 52, 2141 (1984).
98. M. Kawasaki, Y. Tsujimura, H. Hada, *Phys. Rev. Lett.*, 57, 2796 (1986).
99. T. Tani, *J. Imaging Sci. Tech.*, 41, 577 (1997).
100. I. Katakuse, T. Ichikawa, Y. Fujita, T. Matsuo, T. Sakurai, H. Matsuda, *Intern. J. Mass. Spectrom. Ion Proc.*, 67, 229 (1985).
101. I. Katakuse, in S. Sugano, Y. Nishina, S. Ohnishi (eds), *Microclusters,* Springer Ser. Mat. Sci., Vol. 4 (Berlin/Heidelberg: Springer, 1987), 10.
102. H. E. Spencer, L. E. Brady, J. F. Hamilton, *J. Opt. Soc. Am.*, 57, 1020 (1967).
103. T. Tani, *Photogr. Sci. Eng.*, 15, 28 (1971).
104. T. Tani, *Photogr. Sci. Eng.*, 15, 181 (1971).
105. T. Tani, *Photogr. Sci. Eng.*, 16, 35 (1972).
106. T. Tani, S. Takada, *Photogr. Sci. Eng.*, 26, 111 (1982).
107. E. Moisar, F. Granzer, D. Dantrich, E. Palm, *J. Photogr. Sci.*, 25, 12 (1977).
108. E. Moisar, E. Palm, F. Granzer, D. Dantrich, *J. Photogr. Sci.*, 25, 19 (1977)
109. D. Dantrich, F. Granzer, E. Moisar, E. Palm., *J. Photogr. Sci.*, 25, 169 (1977).
110. E. Moisar, F. Granzer, D. Dantrich, E. Palm, *J. Photogr. Sci.*, 28, 71 (1980).

111. T. Tani, *Photogr. Sci. Eng.*, 18, 569 (1974).
112. H. E. Spencer, *Photogr. Sci. Eng.*, 11, 352 (1967).
113. H. E. Spencer, *J. Photogr. Sci.*, 20, 143 (1972).
114. J. F. Hamilton, R. C. Baetzold, *Photogr. Sci. Eng.*, 25, 189 (1981).
115. T. Tani, *J. Imaging Sci. Technol.*, 41, 577 (1997).
116. T. Tani, N. Muro, A. Matsunaga, *J. Imaging Sci. Technol.*, 42, 349 (1998).
117. T. Tani, *J. Imaging Sci. Technol.*, 42, 402 (1998).
118. T. Tani, T. Tasaka, M. Murofushi, K. Hosoi, A. Hirano, *Imaging Sci. J.*, 47, 1 (1999).
119. T. Tani, *Photogr. Sci. Eng.*, 19, 356 (1975).
120. T. Tani, *J. Appl. Phys.*, 62, 2456 (1987).
121. T. Tani, Y. Sano, *J. Appl. Phys.*, 69, 4391 (1991).
122. R. Kubo, *J. Phys. Soc. Jpn.*, 30, 41 (1962).
123. T. Tani, *J. Appl. Phys.*, 91, 4595 (2002).
124. Tani, *Photographic Sensitivity: Theory and Mechanisms*, Chapter 6.
125. J. M. Harbison, H. E. Spencer, in James (ed.), *The Theory of the Photographic Process*, 149.
126. S. E. Sheppard, *Photogr. J.*, 65, 380 (1925).
127. S. E. Sheppard, *Photogr. J.*, 66, 399 (1926).
128. H. Frieser, E. Ranz, *Ber. Bunsenges. Phys. Chem.*, 68, 389 (1964).
129. D. J. Cash, *J. Photogr. Sci.*, 20, 19 (1972).
130. D. J. Cash, *J. Photogr. Sci.*, 20, 77 (1972).
131. D. J. Cash, *J. Photogr. Sci.*, 20, 223 (1972).
132. J. W. Mitchell, *J. Photogr. Sci.*, 6, 57 (1958).
133. L. M. Kellogg, N. B. Liebert, T. H. James, *Photogr. Sci. Eng.*, 16, 151 (1970).
134. L. M. Kellogg, *Photogr. Sci. Eng.*, 18, 378 (1974).
135. J. F. Hamilton, J. M. Harbison, D. L. Jeanmaire, *J. Imaging Sci.*, 32, 17 (1988).
136. L. M. Kellogg, J. Hodes, in the preprint of *The 40th Annual Conference of SPSE, Rochester, New York, May 1987.*
137. H. Kanzaki, Y. Tadakuma, *J. Phys. Chem. Solids*, 55, 631 (1994).
138. J. E. Keevert, V. V. Gokhale, *J. Imaging Sci.*, 31, 243 (1987).
139. Y. Tadakuma, Y. Yoshida, H. Kanzaki, *The Annual Conference of Soc. Photogr. Sci. Tech. Japan, May 1990, Makuhari, Chiba.*
140. H. Mifune, M. Mizuno, Y. Toyama, T. Shiozawa, J. Okuda, *J. Imaging Sci. Technol.*, 46, 262 (2002).
141. T. Tani, *J. Imaging Sci. Technol.*, 39. 386 (1995).
142. T. Tani, *J. Imaging Sci. Technol.*, 42, 135 (1998).
143. R. C. Baetzold, *J. Imaging Sci. Tech.*, 45, 247 (2001).
144. J. F. Hamilton, in A. Balderaschi, W. Czaja, M. Tosatti (eds), *The Physics of Latent Image Formation in Silver Halides* (Singapore: World Scientific, 1984), 203.
145. A. S. Malik, F. J. DiSalvo, R. Hoffmann, J. T. Blair, *J. Imaging Sci. Tech.*, 42, 210 (1999).
146. A. S. Malik, J. T. Blair, W. A. Bernett, F. J. DiSalvo, R. Hoffmann, *J. Solid State Chem.*, 146, 516 (2000).
147. E. Spencer, R. E. Atwell, M. Levy, *J. Photogr. Sci.*, 31, 158 (1983).
148. R. K. Hailstone, J. French, R. de Keyzer, *Imaging Sci. J.*, 51, 33 (2003).
149. T. Tani, *Imaging Sci. J.*, 55, 65 (2007).
150. R. K. Hailstone, T. Zhao, A. G. DiFrancesco, M. Tyne, *J. Imaging Sci. Tech.*, 45, 76 (2001).
151. R. Koslowsky, *Z. Wiss. Photogr.*, 46, 65 (1951).
152. F. W. H. Mueller, *J. Opt. Soc. Am.*, 39, 494 (1049).
153. F. W. H. Mueller, *Photogr. Sci. Eng.*, 10, 338 (1966).
154. E. T. Larson, *PSA J (Photogr. Sci. Tech.)*, 17B, 19 (1951).
155. T. Tani, Y. Yoshida, *J. Imaging, Sci. Tech.*, 44, 242 (2000).

156. P. J. Hillson, *J. Photogr. Sci.*, 6, 97 (1958).
157. J. Bourdon, A. Bonnerot, *Sci. Ind. Photogr.*, (2)30, 205 (1959).
158. J. Bourdon, A. Bonnerot, *Sci. Ind. Photogr.*, (2)32, 369 (1961).
159. T. Tani, *Photogr. Sci. Eng.*, 27, 75 (1983).
160. G. C. Farnell, L. R. Solman, *J. Photogr. Sci.*, 24, 303 (1976).
161. G. C. Farnell, L. R. Solman, *J. Photogr. Sci.*, 28, 185 (1980).
162. B. H. Carroll, H. Hubbard, *J. Res. Nat. Bur. Stand.*, 11, 743 (1933).
163. G. W. Lowe, J. E. Jones, H. E. Roberts, in J. W. Mitchell (ed.), *Fundamentals of Photographic Sensitivity (Proc. Bristol Symposium, 1950)* (London: Butterworths, 1951), 112.
164. H. W. Wood, *J. Photogr. Sci.*, 1, 163 (1953).
165. H. W. Wood, *J. Photogr. Sci.*, 6, 33 (1958).
166. E. Moisar, B. Schmitt, F. Granzer, *J. Imaging Sci.*, 29, 233 (1985).
167. R. C. Baetzold, *J. Phys. Chem.*, 105, 3577 (2001).
168. K. Kuge, H. Shimabukuro, T. Tsutsumi, M. Kato, K. Sakashita, H. Kumagai, N. Aoi, A. Hasegawa, M. Mii, *Imaging Sci. J.*, 48, 107 (2000).
169. S. Guo, R. K. Hailstone, *J. Imaging Sci. Tech.*, 40, 210 (1996).
170. A. G. DiFrancesco, M. Tyne, C. Pryor, R. K. Hailstone, *J. Imaging Sci. Tech.*, 40, 576 (1996).
171. R. K. Hailstone, J. Dai, A. G. DiFrancesco, in the preprint of *ICPS'98, Intern. Cong. Imaging Sci., Antwerp, September 1998*, 253.
172. Y. Oku, M. Kawasaki, *J. Imaging Sci. Tech.*, 42, 346 (1998).
173. A. P. Marchetti, A. A. Muenter, R. C. Baetzold, R. T. McCleary, *J. Phys. Chem. B*, 102, 5287 (1998).
174. M. Treguer, H. Remita, J. Belloni, R. de Keyzer, in the preprint of *ICPS'98, Intern. Cong. Imaging Sci., Antwerp, September 1998*, 73.
175. M. Treguer, R. de Keyzer, H. Remita, J. Belloni, in the preprint of *ICPS'98, Intern. Cong. Imaging Sci., Antwerp, September 1998*, 243.
176. D. Dautrich, F. Granzer, E. Moisar, E. Palm, *J. Photogr. Sci.*, 25, 169 (1977).
177. X. Zhi, Y. Shi, J. Cao, P. Xia, *Imaging Sci. J.*, 54, 53 (2006).
178. W. Ma, H. Guo, P. Xia, in the preprint of *2004 Intern. Symposium on Silver Halide Technology, September 13–15, 2004, Ventura, California, USA*.
179. H. Nakatsugawa, T. Tani, *J. Imaging Sci. Tech.*, 47, 78 (2003).
180. T. Tani, N. Muro, A. Matsunaga, *J. Imaging Sci. Technol.*, 42, 349 (1998).
181. For example, T. Yano, *Photogr. Korr.*, 104, 121 (1968).
182. S. Urabe, T. Sano, *J. Soc. Photogr. Sci. Tech. Jpn.*, 62, 295 (1999).
183. E. J. Birr, *Stabilization of Photographic Silver Halide Emulsions* (London/New York: Focal Press, 1974).
184. T. H. James, in James (ed.), *The Theory of the Photographic Processes*, 396.
185. F. W. H. Mueller, in S. Kikuchi (ed.), The Photographic Image (Invited papers, International Congress of Photographic Science, Tokyo, 1967) (London: Focal Press, 1970), 91.
186. G. Fischer, *J. Inf. Rec. Mater.*, 16, 91 (1988).
187. Tani, *Photographic Sensitivity: Theory and Mechanisms*, Chapter 6.
188. E. J. Birr, *Z. Wiss. Photogr.*, 47, 72 (1952).
189. E. J. Birr, *Z. Wiss. Photogr.*, 48, 103 (1953).
190. E. J. Birr, *Z. Wiss. Photogr.*, 49, 1 (1954).
191. E. J. Birr, *Z. Wiss. Photogr.*, 49, 261 (1954).
192. E. J. Birr, *Z. Wiss. Photogr.*, 50, 107 (1955).
193. E. J. Birr, *Z. Wiss. Photogr.*, 50, 124 (1955).
194. J. F. Hamilton, L. E. Brady, *J. Phys. Chem.*, 66, 2384 (1962).
195. J. van Biesen, *J. Appl. Phys.*, 41, 1910 (1970).
196. S. Takada, *Jpn. J. Appl. Phys.*, 12, 190 (1973).

197. H. A. Hoyen, Jr, *J. Appl. Phys.*, 47, 3784 (1976).
198. M. E. van Hulle, W. Maenhout-van der Vorst, *Phys. Stat. Sol.*(a), 39, 253 (1977).
199. F. Callens, W. Maenhout-van der Vorst, *Phys. Stat. Sol.*(a), 50, K175 (1978).
200. M. Saito, *J. Photogr. Sci.*, 24, 205 (1976).
201. T. Tani, Y. Sano, S. Takada, M. Saito, *J. Photogr. Sci.*, 28, 28 (1980).
202. D. Corbin, A. Gingello, G. MacIntyre, B. H. Caroll, *Photogr. Sci. Eng.*, 24, 45 (1980).
203. Tani, *Photographic Sensitivity: Theory and Mechanisms*, Chapter 7.
204. W. E. Lee, E. R. Brown, in James (ed.), *The Theory of the Photographic Process*, Chapter 11.
205. T. H. James, in James (ed.), *The Theory of the Photographic Process*, Chapter 13.
206. T. H. James, W. E. Lee, in James (ed.), *The Theory of the Photographic Process*, Chapter 14.
207. D. H. Klosterboer, in J. M. Sturge, V. Walworth, A. Shepp (eds), *Imaging Processes and Materials, Neblette's Eighth Edition* (New York: Van Norstrand Reinhold, 1989), Chapter 9.
208. D. R. Whitcomb, *Photogr. Sci. Photochem.*, 21, 1 (2003).
209. M. R. V. Sahyun, *J. Imaging Sci. Technol.*, 49, 337 (2005).
210. H. Mifune, T. Ohzeki, K. Yamane, F. Nariyuki, K. Watanabe, M. Yoshikane, M. Nakanishi, T. Maekawa, T. Funakubo, *J. Imaging Sci. Tech.*, 51, 207 (2007).
211. T. Funakubo, H. Mifune, T. Ohzeki, K. Ohzeki, K. Yamane, *J. Imaging Sci. Tech.*, 51, 212 (2007).
212. J. Eggers, in H. Frieser, G. Haase, E. Klein (eds), *Die Grundlagen der Photographischen Prozesse mit Silberhalogeniden* (Frankfurt: Akademische Verlag, 1968), 813.
213. J. D. Kendall, in L. P. Clere (ed.), *IX Congress Intern. de Photographie Scientifique et Appliquee, Paris, 1935* (Paris: Edition Rev. d'Optique, 1936), 277.
214. W. Pelz, *Angew. Chem.*, 66, 231 (1954).
215. T. Tani, *Photogr. Sci. Eng.*, 15, 374 (1971).
216. L. K. J. Tong, M. C. Glesman, *Photogr. Sci. Eng.*, 8, 326 (1964).
217. R. L. Bent, J. C. Dessloch, F. C. Duennebier, D. W. Fassett, D. B. Glass, T. H. James, D. B. Julian, W. R. Ruby, J. M. Snell, J. H. Sterner, J. R. Thirtle, P. W. Vittum, A. Weissberger, *J. Am. Chem. Soc.*, 73, 3100 (1951).
218. V. I. Sheberstov, Yu. I. Bukin, *Kino-Foto Prom.*, 1, 101 (1932).
219. T. H. James, *J. Franklin Inst.*, 239, 41 (1945).
220. V. I. Sheberstov, *Usp. Nauch. Fotogr.*, 4, 210 (1955).
221. P. J. Hillson, *J. Photogr. Sci.*, 22, 31 (1974).
222. H. E. Spencer, *J. Imaging Sci.*, 32, 40 (1988).
223. M. Kawasaki, H. Hada, *J. Imaging Sci.*, 35, 129 (1991).
224. R. W. Swenson, *Photogr. Sci. Eng.*, 1, 119 (1958).
225. P. J. Hillson, *Photogr. Sci. Eng.*, 23, 38 (1979).
226. M. A. Kriss, in James (ed.), *The Theory of the Photographic Process*, Chapter 21.
227. W. T. Hanson, Jr., P. W. Vittum, *J. Photogr. Soc. Am.*, 13, 94 (1947).
228. W. T. Hanson, Jr., *J. Opt. Soc. Am.*, 40, 171 (1950).
229. C. R. Barr, J. R. Thirtle, P. W. Vittum, *Photogr. Sci. Eng.*, 13, 74 (1969).
230. H. Kobayashi, K. Mihayashi, in *SPSE Annual Conference, Atlantic City, May, 1985*.
231. J. R. Jarvis, P. J. Twist, in *The Intern. East-West Symposium II: Factors Influencing The Efficiency of Photographic Imaging, October/November, 1988, Kona, Hawaii.*

5

Dye Sensitization

5.1 Introduction

As described in Chapter 1, the capability to form color images is one of the most important properties of photosensitive films. At least three different photosensitive layers with sensitivity to the three primary colors (i.e., blue, green, and red), are required in order to form color images. As described in Section 3.1.1, silver halide (AgX) itself absorbs ultraviolet and blue light, and is thus sensitive only to blue light in the visible region. In 1873, Vogel observed for the first time the phenomenon that rendered silver halides sensitive to light of wavelength longer than that of blue light and to light of wavelength that sensitizing dyes could absorb [1]. This phenomenon has been called spectral sensitization in photographic science and technology, and dye sensitization in some cases [2–4]. Now, this phenomenon is widely used for other systems, especially solar cells, and is called dye sensitization, as described in Chapter 9. In this book, it is referred to as dye sensitization when dyes are used as sensitizers, and as spectral sensitization when materials other than dyes are used as sensitizers.

For the mechanism of dye sensitization, the electron transfer mechanism was proposed by Gurney and Mott in 1938 [5], and the energy transfer mechanism by Mott in 1948 [6]. Figure 5.1 illustrates these mechanisms. According to the electron transfer mechanism, an electron is photo-excited from the highest occupied molecular orbital (HOMO) to the lowest unoccupied one (LUMO) in a sensitizing dye molecule, and transferred to the conduction band of AgX. According to the energy transfer mechanism, the energy in the photo-excited dye is transferred by Foerster's resonance energy transfer model to AgX to excite an electron at intermediate level in the forbidden band to the conduction band. It is now accepted [4,7,8] that dye sensitization takes place according to the electron transfer mechanism, owing to the result of discussions taking place over many years about the two mechanisms, as described in Section 5.8. In this book, the knowledge related to dye sensitization of photographic films is described on the basis of the electron transfer mechanism.

Silver halide grains, which are used in highly sensitive layers in color negative films, are tabular in shape with large aspect ratios and therefore with large surface-to-volume ratios. Namely, they are poor in absorbing incident blue light and have large surface areas for the adsorption of sensitizing dyes. Sensitizing dyes are therefore used as sensors even in blue-sensitive layers in color films [7]. In this chapter, the sensitivities on the excitation of silver halides and sensitizing dyes are called intrinsic and spectral, respectively. Photographic performance of sensitizing dyes on AgX grains is characterized by the absorption of light by dyes, the efficiency of the electron transfer, and the desensitization, which refers to the decrease in intrinsic sensitivity of AgX grains by dyes. The shortage in efficiency of the electron transfer can be improved by supersensitization. The mechanisms of supersensitization and desensitization are also described in Section 5.9.

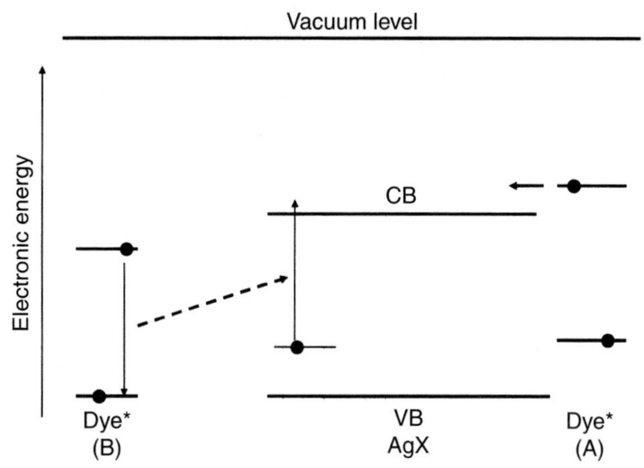

Figure 5.1 Electronic energy level diagram of AgX and an optically excited dye (Dye*) indicating the electron transfer (A) and energy transfer (B) mechanisms.

5.2 Sensitizing Dyes

Since dye sensitization takes place according to the electron transfer mechanism, sensitizing dyes are required to have the capabilities to strongly absorb light at a desired wavelength region and to transfer electrons to the conduction band of silver halides from their excited states. Comprehensive reviews of sensitizing dyes have been published by several authors [9–12].

The essential molecular structure in sensitizing dyes is the polymethine chain with an odd number of methine groups and with hetero atoms at its ends. As shown in Figure 5.2, there are three major systems, i.e., amidinium, carboxyl-ion, and amidic, which are exemplified by groups (1), (2), and (3), respectively in Table 5.1 [9,13–16]. Group (1) is an example of cyanine dyes, among which the dyes with *n* of 0, 1, 2, and 3 are called cyanine, carbocyanine, dicarbocyanine, and tricarbocyanine dyes, respectively. Group (2) is an example of oxonol dyes, and group (3) is an example of merocyanine dyes. Most sensitizing dyes used for photographic films are cyanine dyes, while merocyanine dyes are also used in some cases.

Molecular structures of sensitizing dyes have ben studied by X-ray crystallography [9, 17–30]. Most typical sensitizing dyes are nearly planar in molecular structure with

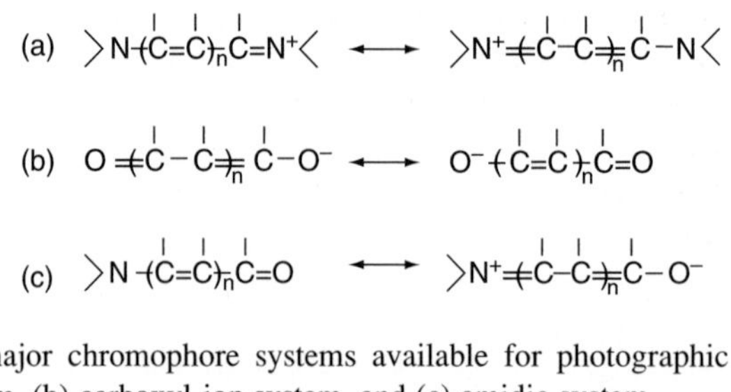

Figure 5.2 Three major chromophore systems available for photographic sensitizing dyes: (a) amidinium-ion system, (b) carboxyl-ion system, and (c) amidic system.

Table 5.1 Wavelength of absorption peak (λ_{max}) and molecular extinction coefficient (ϵ) of typical cyanine dyes (a), oxonol dyes (b), and merocyanine dyes (c), which represent amidinium-ion, carboxyl-ion, and amidic systems, respectively

Dyes	λ_{max}, nm($\epsilon \times 10^{-4}$, $lm^{-1}cm^{-1}$)			
	$n = 0$	$n = 1$	$n = 2$	$n = 4$
(1) 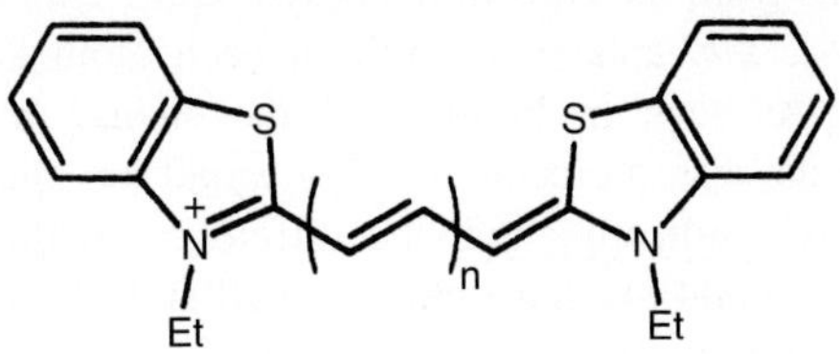	423 (8.5)	557 (15)	650 (23)	758 (25)
(2) 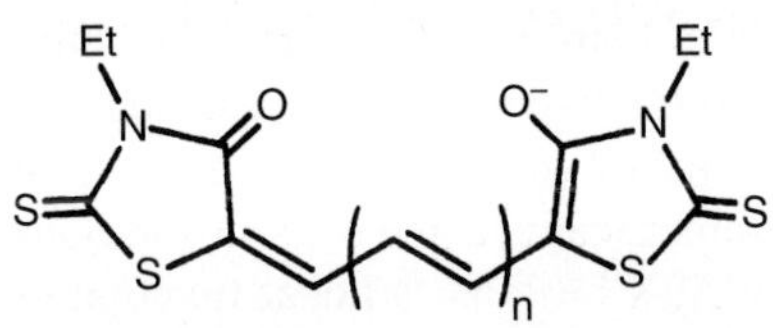	542 (8.2)	613 (11)	714 (14)	
(3) 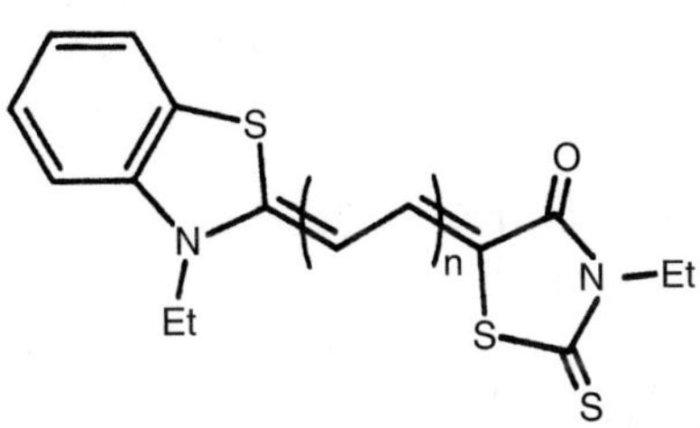	432 (6.0)	528 (9.3)	605 (8.5)	635 (6.9)

bond lengths in agreement with those observed for simple molecules with theoretical bond orders. However, steric effects have been observed for some cyanine dyes [26–29].

Several theoretical approaches for predicting the absorption wavelengths of dyes have been proposed. They include the free electron model [31–35] and molecular orbital (MO) methods [36–42]. The free electron model treats π electrons in the polymethine chain in a cyanine dye molecule as a free electrons in a one-dimensional well and has successfully predicted the absorption wavelengths of cyanine dyes. The systematic study of the electronic transitions of various kinds of cyanine dyes by the MO method was made by Tani and Kikuchi with the Hueckel molecular orbital (HMO) method [36]. The wavelength of the absorption maximum (λ_{max}) of a dye is related to the transition energy (E_{TR}) from HOMO to LUMO,

$$\varepsilon_{LU} - \varepsilon_{HO} = E_{TR}(\text{calc}) \tag{5.1}$$

$$E_{TR}(\text{obs}) = hc/\lambda_{max}, \tag{5.2}$$

where ϵ_{HO} and ϵ_{LU} are the electronic energy levels of HOMO and LUMO, respectively. The calculated and observed transition energies are in fairly good agreement except for monomethine dyes with steric effects [40]. Some MO methods with higher order approximation were also used to calculate the electronic structures of sensitizing dyes [9].

As shown in Figure 5.3, MO calculation indicates π-electron density distribution in the frontier MO (i.e., HOMO and LUMO) of cyanine dyes, where π-electrons are concentrated and delocalized within the polymethine chain with nitrogen atoms at the ends, and a π-electron in HOMO is localized on odd atoms from the ends of the polymethine chain and a π-electron in LUMO on even atoms [4]. The area of a circle at each atom is proportional to the π-electron density there. As seen in Figure 5.3, the change in distribution of π-electron density on a cyanine dye molecule is very small on the electronic transition from its HOMO to LUMO, indicating that the difference in the equilibrium nuclear configuration between ground and excited states is small and likely to give a sharp and intense absorption band (Figure 9.3), which is suitable for dye sensitization of color films.

The value of E_{TR} of a molecule is usually a function of both the electrostatic dielectric constant (D) and the optical constant (n^2, where n is the refractive index) of its surrounding according to the theory of solvent effects of absorption spectra such as MacRae's equation [43]. However, it has been determined that E_{TR} of symmetrical cyanine dyes in solutions [44–46], and even on the surface and in the interior of AgX [46], has a linear function of only $(n^2 - 1)/(2n^2 + 1)$ of solvents and AgX. This result indicates that the shift of E_{TR} resulted solely from the electronic polarization of the solvents and the AgX for the excited dye molecules and that the interaction of the dye molecules with solvents and AgX was

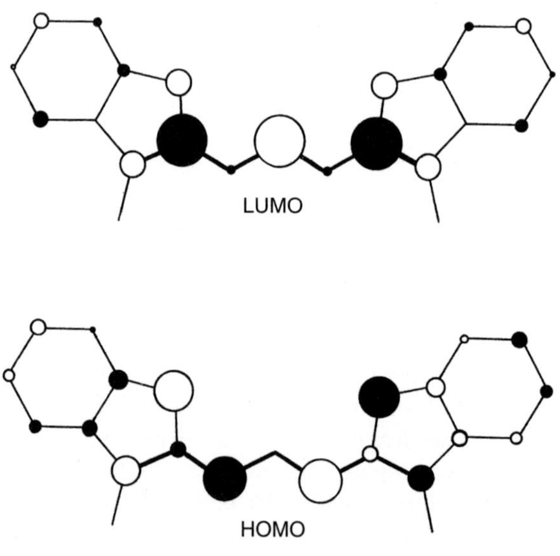

Figure 5.3 LUMO and HOMO electron clouds of a thiacarbocyanine dye molecule as calculated by the Hueckel-approximation molecular orbital method [36]. The area of the circle represents the electron density of the molecular orbital at each atom. The open and closed circles represent the difference in sign of each atomic orbital contributing to the LUMO and HOMO of the dye.

based on the van der Waals force. Although this general shift was also expected to take place for E_{TR} of merocyanine dyes, it was usually superimposed by other, more powerful effects resulting from the interaction between dye–solvent permanent dipoles and the formation of hydrogen bonding in solutions and coordinate bonding between merocyanine dye molecules with >C=S groups and silver ions on the surface of silver halides, as described in detail in Section 5.4.

5.3 Aggregate Formation of Sensitizing Dyes

Many cyanine dyes form aggregates in aqueous solutions and especially in AgX emulsions, bringing about large spectral shifts and distinct changes in absorption band shape. Three kinds of aggregates were observed for cyanine dyes adsorbed on AgX grains in photographic films. They are J-aggregate, H-aggregate, and aggregate with herring-bone structure (herring-bone aggregate), which give the absorption bands named J-band, H-band, and double band, respectively [4,9]. Figure 5.4 shows the absorption spectrum of 3,3'-disulfopropyl-5,5'-dichloro-9-ethyl-thiacarbocyanine (dye 1) in methanol and those of the dye in photographic emulsions composed of octahedral AgBr grains [4]. The dye showed an M-band corresponding to its monomeric state in methanol and a sharp and bathochromatically shifted J-band corresponding to its J-aggregate. J-aggregates of dyes were observed for the first time by Jelley and Scheibe in 1936 [47–49], and became

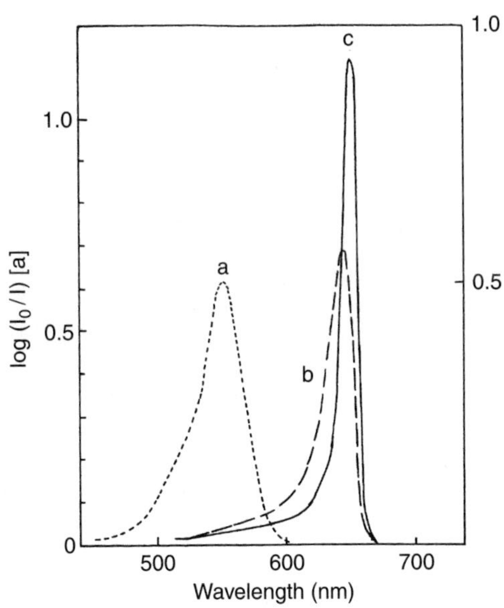

Figure 5.4 Absorption spectra of dye 1 in methanol (a) and in octahedral AgBr emulsions, which were agitated for 20 min at 40°C (b) and at 70°C (c) [4]. It was believed that the elevated temperature for the adsorption enhanced the growth of J-aggregates of the dye. The diffuse reflectance spectra of emulsions were treated by the Kubelka–Munk equation to give the absorption spectra of J-aggregates of dye 1.

indispensable for color films. The J-band became sharper and appeared at longer wavelength, as the growth of J-aggregates was enhanced by the increase in temperature and time for the dye adsorption to AgX grains. As indicated by this result, the sharpening and shift of absorption band are enhanced by the increase in size of J-aggregates [50–52]. Figure 5.5 shows the absorption spectrum of a 9-methyl-thiacarbocyanine in methanol and that of the dye in a photographic emulsion composed of cubic AgBr grains. While the dye in methanol exhibited an M-band, the dye on the cubic grains exhibited a double band with two absorption peaks. It is noted that the wavelength of one of the bands was longer than that of the M-band, while that of the other was shorter.

Brooker et al. classified the molecular structures of cyanine dyes into three groups, i.e., 'loose', 'compact', and 'crowded', among which the π-conjugated system of 'compact' dyes is the most rigid and planar, and found that compact cyanine dyes were the most likely to form J-aggregates [53]. Several groups of scientists observed the general tendency that J-aggregates are formed by planar cyanine dyes with such protruding substituents as to cause the slipping of the pile of dye molecules [54], and had an empirical idea that slipped piles of planar dye molecules with protruding substituents are J-aggregates, while parallel piles of planar dye molecules without protruding substituents are H-aggregates, as illustrated in Figure 5.6.

Several structural models were proposed for dye aggregates [55–60,19,9,4]. As shown in Figure 5.7, Kasha classified the structure of molecular alignments into three groups, i.e.,

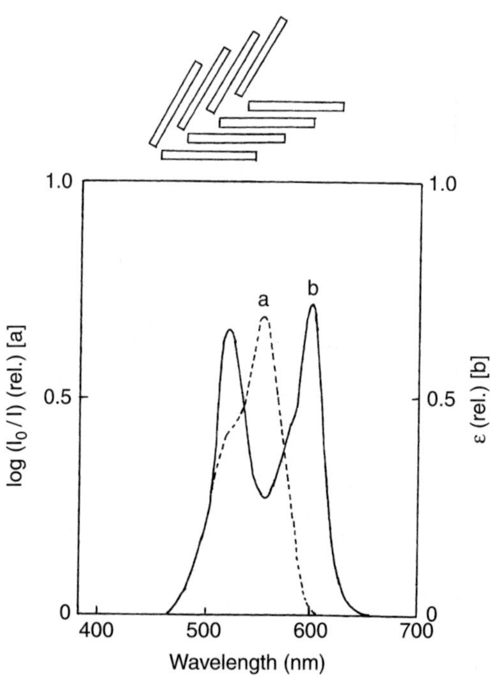

Figure 5.5 Absorption spectra of dye 2 in the monomeric state in a gelatin layer (a) and in an emulsion layer containing cubic AgBr grains with edge length of 0.65 μm and dye 2 of 0.23 mmole/mole AgBr [4].

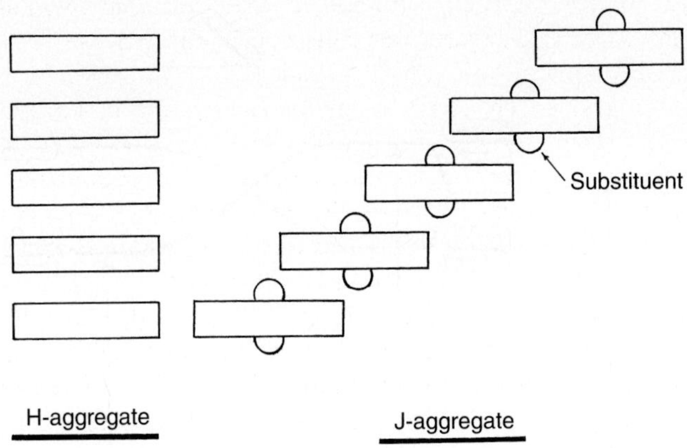

Figure 5.6 The structures of an H-aggregate with flat molecule and of a J-aggregate with molecules with protruding substituents.

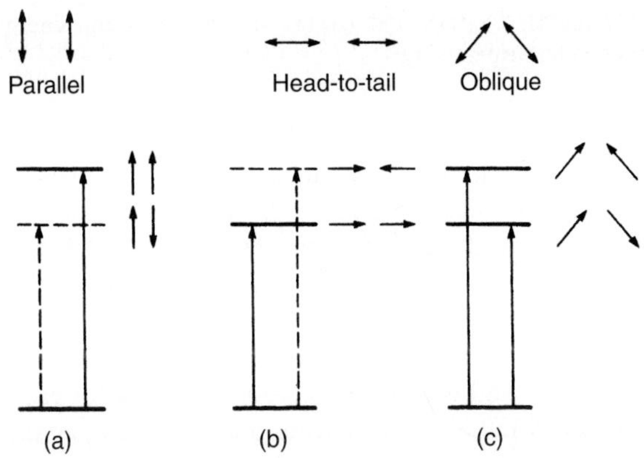

Figure 5.7 The structure and energy levels with transition moments for three kinds of aggregates of organic molecules named parallel (a), head-to-tail (b), and oblique (c) after Kasha et al. [55–57]. The transitions to the levels represented by solid line and broken one are allowed and forbidden, respectively.

parallel, head-to-tail, and oblique [55–57], which should correspond to H-aggregate, J-aggregate, and herring-bone aggregate, respectively. Parallel and head-to-tail alignments are considered to correspond to the extremes of the aggregates with one molecule in a unit cell, while oblique alignment corresponds to the aggregate with two molecules in a unit cell. As to the aggregate with one molecule in a unit cell, the slip angle between the long axis of a dye molecule and that of an aggregate (α) is general and most important, indicating that J-aggregates and H-aggregates have small and large values of α, respectively [55–57,59,60], as illustrated in Figure 5.8 with the electronic transition energy (E_{TE}) as a function of α for a monomer and its dimers on the basis of the molecular exciton theory [55–58,61,62,4].

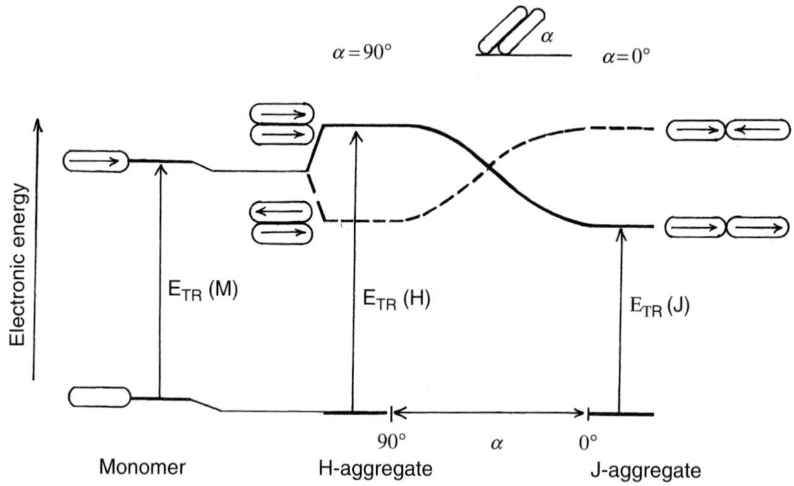

Figure 5.8 The electronic transitions in a monomer, H-aggregate, and J-aggregate on the basis of the interaction between molecular excitons in aggregates, where α is the slip angle between molecules in an aggregate, and $E_{TR}(M)$, $E_{TR}(H)$, and $E_{TR}(J)$ are the transition energies of a monomer, H-aggregate, and J-aggregate, respectively.

The relation between E_{TE} and α was evaluated for dimers on the basis of the point dipole model [55–57] and the transition dipole model [59,60]. As seen in Figure 5.7, the transitions to the excited states of aggregates with transition dipoles in-phase and out-of-phase arrangements are allowed and forbidden, respectively [4,57]. The energy of the allowed transition is larger than that of the transition of the monomer for the aggregate with large α (i.e., H-aggregate), and smaller than that for the aggregate with small α (i.e., J-aggregate). The angle for the absence of exciton splitting is 54.7° according to the point dipole model [55–57], and was quantitatively evaluated for each dye according to the transition density model [59,60]. The herring-bone structure, as shown in Figure 5.5, has been proposed for the molecular arrangement of the aggregate that gives the double band [63–65]. It is known that 9-methyl-thiacarbocyanine dye derivatives adsorbed on the {100} face of AgX exhibit the double band.

It is noted that adjacent plane dye molecules in J-aggregates, H-aggregates, and herring-bone aggregates are in face-to-face arrangements. Although a herring-bone aggregate is similar to an oblique alignment in that there are two molecules in a unit cell, it does not seem that adjacent molecules in an oblique alignment are always in face-to-face arrangement. Many organic semiconductor molecules give an absorption band with two peaks owing to Davidov splitting [61]. However, molecules of aromatic hydrocarbons, which are typical organic semiconductors, are in edge-to-face arrangements due to repulsive forces between electrons in adjacent molecules [66], as also described in Section 10.3.

The above-stated models were described for dimers, and should be extended for large aggregates including those with two and three dimensions. For two-dimensional aggregates, Kuhn et al. examined brick stonework arrangement, ladder arrangement, and staircase arrangement, and proposed the first one for J-aggregates formed in Langmuir–Blodgett (LB) membranes [67,68]. However, analysis of the molecular packing in the crystals of J-aggregating cyanine dyes by Smith supported the staircase arrangement [19]. Consistent with the staircase arrangement, analysis by Asanuma and Tani of the desorption process of sensitizing dye molecules from J-aggregates on AgBr grains revealed that the J-aggregates were in linear arrangement, although they formed two-dimensional layers on AgX grains, as described in the next section [69]. This result indicates that the interaction among dye molecules within a linear J-aggregate was more important than that among dye molecules in adjacent linear J-aggregates, as described in the next section, and is in accord with the fact that most J-aggregates of cyanine dyes on AgX grains exhibit absorption bands with a single peak.

Since J-aggregated cyanine dye molecules form mono-molecular layers on AgX grains, as described in the next section, attention should be paid to the effect of the AgX grain surface structure on the J-aggregates' molecular alignment. Although one of the molecular alignments proposed is based on the coordinate-bonding of sulfur atoms in thiacyanine dye molecules with silver ions on the surface of AgX grains [59,60,64], subsequent examinations from various viewpoints have not confirmed this [19,22,70–72]. The difference in absorption spectra of many cyanine dyes on AgBr grains between {100} and {111} faces on the grain surface indicates the influence of the AgX grain surface structure, and can be explained by the difference between them in such physical interactions as those involving electrostatic and van der Waals's forces working between the cyanine dyes and AgX grains, as described in the next section.

Direct evidence of the relation between the molecular arrangements and absorption spectra of dyes have been obtained for single crystals of cyanine dyes by X-ray diffraction analysis and microscopic reflection spectra with Kramers-Kronig transformation [19, 22–25]. A single crystal of a simple cyanine dye was confirmed to have a molecular arrangement with a small slip angle [73]; a popular J-aggregate-forming dye (5,5'-dichloro-9-ethyl-thiacarbocyanine derivative) [30] gave a sharp J-band, the direction of the transition moment of the J-band coinciding with that of the transition moment of each molecule, in agreement with the molecular exciton theory [55,68,61,62]. However, the molecular arrangement of J-aggregates of a dye on AgX grains is not necessarily the same as that of J-aggregates of the same dye in its single crystal, and should be subjected to subsequent studies.

Marchetti et al. compared the absorption spectra of solution aggregates and of a crystal of 2, 2'-cyanine, and found that the interactions between the molecules in the aggregate and in the crystal of the dye were the same [22]. On large tabular AgX grains, Maskasky formed mono-molecular layers of J-aggregates of cyanine dyes, which were as large as several micrometers, and directly observed them via polarized fluorescence microscopy to determine their molecular arrangement [74,75]. It is important to note that these J-aggregates are judged to be single crystals because all the dye molecules in each aggregate are aligned in the same direction. He also confirmed the herring-bone structure for the aggregate of a 9-methyl-thiacarbocyanine dye on the {111} face of AgBr.

The structure of J-aggregates of sensitizing dyes on AgX is important for the design of color films. Attempts to observe J-aggregates and their molecular arrangement on the surface of AgX by such new techniques as scanning tunnel microscope (STM) and atomic force microscope (AFM) have been made. While AFM can be used to observe the shape of J-aggregates of cyanine dyes on AgX emulsion grains [76], its resolution is not enough to observe the J-aggregates' molecular arrangement on the AgX surface. While STM can have the lowest resolution for observing the molecular arrangement in J-aggregates, it cannot be used for J-aggregates on the surface of the AgX layer since the electric conductivity is not large enough to prevent the charge-up of the layer during the STM measurement. Then, STM has been used to confirm the above-stated models by observing the molecular arrangement in aggregates of cyanine dyes adsorbed to pseudo-AgX monolayers on Ag substrates [77,78] and to graphite [79],

5.4 Adsorption of Sensitizing Dyes to Silver Halide Grains

Since dye sensitization takes place as a result of the electron transfer from excited dyes to the conduction band of AgX, knowledge about the adsorption of sensitizing dyes to AgX grains and the nature of the interaction are of great importance for understanding the mechanism of dye sensitization. Adsorption isotherms and heats of adsorption were measured for various sensitizing dyes and AgX grains by many working groups [80–82]. The amount of dye adsorbed to AgX grains and its adsorption isotherm are measured by a conventional centrifugal method and an optical method with the Kubelka–Munk equation [83]. The latter technique could be applied, not only to liquid emulsions, but also to dried emulsion layers [84].

Although it is known that the adsorptivity of sensitizing dyes to AgX grains varies to a large extent [80–85], many sensitizing dyes are adsorbed on the grains, forming monomolecular layers with Langmuir-type adsorption isotherms, as shown in Figure 5.9. In particular, the adsorption of cyanine dyes to AgX grains was strengthened to enhance the realization of the Langmuir-type adsorption when the dyes formed aggregates, indicating that a dye molecule was stabilized by its interaction with other dye molecules as well as with the grain surface [86,87]. Although dye adsorption to AgX grains with multimolecular layers has been reported in the past [82,88], careful experiments could not

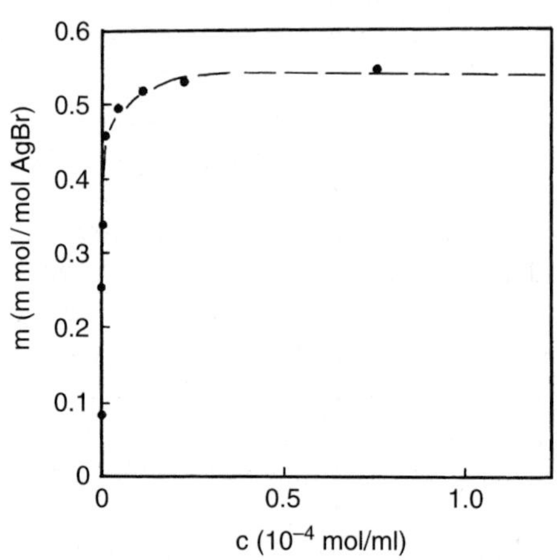

Figure 5.9 Adsorption isotherm of dye 2 to cubic AgBr grains in an emulsion, with the amount of adsorbed dye (m) as a function of the concentration of free dye (c) as given by a phase-separation method (●) [84].

prove it [89], revealing the necessity for special molecular designs for enhancing attractive interaction between dye layers in order for multi-molecular layers to form on the silver halide surface [90,91].

The area occupied by a dye molecule on the grain surface as given by the amount of saturation is usually in agreement with the product of the length and thickness of the dye molecule, indicating the edge-on orientation for the adsorption of J-aggregated dye molecules to the grain surface instead of flat-on and end-on orientations [82–85]. It has been noted that the area occupied by a dye molecule with edge-on orientation (0.68 nm^2 for thiacarbocyanine dye molecule) is much smaller than that by a dye molecule with flat-on orientation (1.55 nm^2 for the same dye molecule) [92,17,18], and that the formation of J-aggregates makes it possible to increase the amount of dye molecules per unit area of the grain surface owing to attractive interaction among dye molecules.

Although gelatin tends to reduce the rate and heat of dye adsorption, its influence diminishes with increasing dye levels, and the saturated amount of adsorbed dyes has shown to be independent of the presence of gelatin for various cyanine dyes [80–82, 93–95]. Namely, at high concentrations of cyanine dyes with strong adsorptivity to AgX grains in liquid emulsions, almost all gelatin can ultimately be displaced from the surface of the grains except for a small part of gelatin molecule in contact with the grain surface. However, Tani and Suzumoto observed that the competition of sensitizing dyes with gelatin for the adsorption to AgX grains was more severe in dried emulsion layers than in liquid emulsions, and that the saturated amount of a cyanine dye adsorbed to AgBr grains was diminished by gelatin in dried emulsion layers [84].

The heats of adsorption of sensitizing dyes to AgX grains were measured by several methods and were as large as around 10 kcal/mole for J-aggregated cyanine dyes [4,82,96,69]. This result indicates the presence of strong van der Waals's interaction between a dye molecule and AgX along with that among dye molecules, and does not indicate the presence of any chemical interaction between a dye molecule and AgX. This idea was supported by the following facts.

The adsorption behavior of 3,3'-disulfopropyl-9-methyl-thiacarbocyanine (dye 2) to cubic and octahedral AgBrI grains in emulsions was studied comparatively [96]. The heats of its adsorption to octahedral and cubic AgBrI grains as measured by a microcalorimeter were 9 and 11 kcal/mole, respectively. The reversibility of the adsorption was confirmed by observing the movement of all the dye molecules on the octahedral grains to the cubic ones, when an emulsion of undyed cubic grains was mixed with an emulsion of dyed octahedral grains. The heat of the desorption of the dye from the octahedral grains was 26 kcal/mole. The potential energy profile for the dye in an aqueous gelatin solution, on {111}, and {100} surfaces of AgBr grains, is illustrated in Figure 5.10. Thus, the dye molecules were being rearranged on the grain surfaces through their repeated desorption and adsorption, and the rate-determining step was their desorption from the grain surface. The desorption of the dye molecules in J-aggregates on the octahedral grains was a zero-order reaction, indicating that the J-aggregates were linear in structure and that only the molecules at the ends of the aggregates could be desorbed. Similarly, it was observed that a dye molecule in a large J-aggregate was more stable than that in a small J-aggregate, and that a dye molecule tended to move from a small J-aggregate to a large one with large activation energy,. This result indicates that the growth of J-aggregates could be enhanced by increasing the temperature and time for the dye adsorption [69].

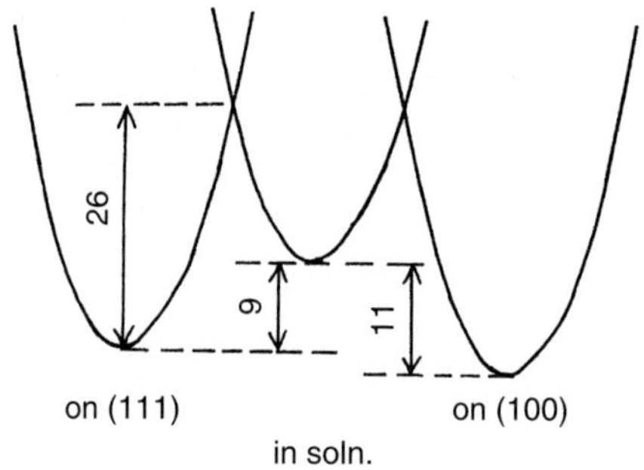

Figure 5.10 Potential energy profile of dye 2 on {111} and {100} faces of AgBrI grains and in an aqueous gelatin solution with enthalpy differences in kcal/mole. Reproduced with permission from Reference [96], copyright 1985 IS&T, The Society for Imaging Science and Technology.

Infrared absorption spectra of sensitizing dyes were studied by several working groups [3,97]. The spectra of cyanine dyes were not subjected to any essential change when they were adsorbed to AgX grains [97,98]. This result indicates that the molecular structures of cyanine dyes were not changed by their adsorption to the grains. The wavelengths of the peaks of the visible absorption spectra of cyanine dyes in the monomeric state on AgX grains were longer by 20–40 nm than those of the corresponding dyes in solution. As described in Section 5.2, this bathochromatic shift of the absorption bands of cyanine dyes due to their adsorption to AgX grains was within the range of the solvent effect on the absorption spectra of cyanine dyes based on the van der Waals interaction between the dyes and the grains [43–46,98].

As stated in Section 3.2.2, the adsorption of cyanine dye molecules with positively charged chromophores increases the concentration of interstitial silver ions in cubic AgBr grains by repelling silver ions at the surface kink sites into interstitial positions [99,100], regardless of the change in electric charges on the substituents attached to nitrogen atoms in a cyanine dye [4]. In accord with this result, the adsorption to AgX grains of positively charged cyanine dyes increases the concentration of free silver ions in an liquid emulsion by repelling silver ions at surface sites into solutions [101–103]. In agreement with those results, the adsorption of cyanine dyes to AgX grains is enhanced by elevated pAg values [80,81,89,94,95,104–108]. These phenomena indicate that Coulombic interaction, i.e., the minimization of the electric energy of a positive charge on the chromophore of a cyanine dye molecule by association with negative ions on the grain surface, is a necessary condition for the adsorption of cyanine dyes to emulsion grains, and that the van der Waals interaction between the conjugated π-electrons in a dye molecule and AgX, along with that among dye molecules, was the driving force for the adsorption, since Columbic interaction also took place when the dye molecule was free in a liquid emulsion.

In the case of the adsorption of merocyanine dyes to AgX, there has been some experimental evidence to suggest the presence of a chemical interaction between dye molecules and the AgX surface. For merocyanine dyes with longer polymethine chains, the infrared absorption wave number corresponding to the stretching vibration of >C=O bond decreased on their adsorption to AgBr grains [109]. West, Carroll, and Whitcomb found that merocyanine dyes with >C=S groups were adsorbed to AgX grains more strongly than those without C=S [80,81]. Seki, Tani, and co-workers observed the $1s$-π^* transition in X-ray absorption near edge structure (XANES) spectra at the K edge of sulfur in >C=S groups in merocyanine dyes disappeared on their adsorption to AgCl grains. This result was attributed to a coordinate bonding forming between a sulfur atom in a >C=S group in a merocyanine dye and a silver ion on the AgCl surface [110,111].

5.5 Electronic Energy Levels of Sensitizing Dyes Adsorbed on Silver Halide

The light-induced electron transfer process in dye sensitization is illustrated in Figure 5.11 [4]. An electron is excited from HOMO to LUMO in a dye molecule, and then transferred to the conduction band of AgX. It is therefore considered that the electronic energy levels of LUMO and HOMO of a dye (ϵ_{LU} and ϵ_{HO}, respectively) with respect to the bottom of the conduction band and the top of the valence band of AgX are important in understanding the photo-induced electron transfer in dye sensitization.

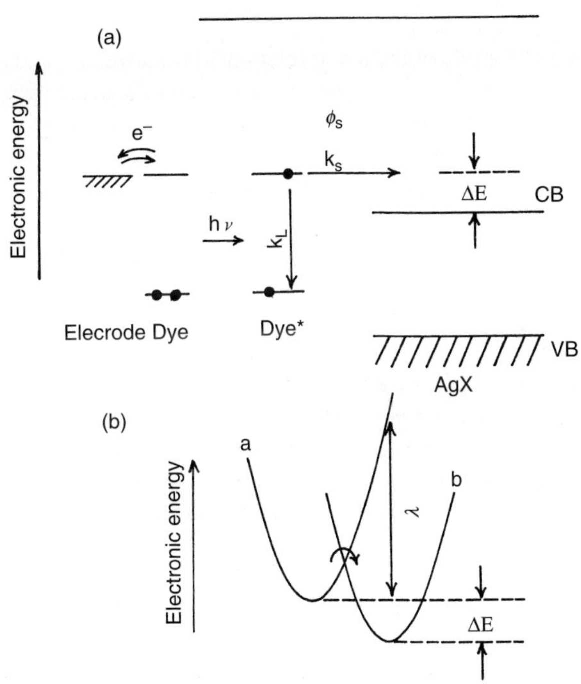

Figure 5.11 (A) The electron transfer process in dye sensitization. The excitation of an electron from HOMO to LUMO to give a dye molecule in the lowest excited singlet state (Dye*) is followed by the transfer of the excited electron to the conduction band of AgX (CB) with efficiency of ϕ_s, rate constant of k_s, and energy gap of ΔE. The reduction potential (E_R) of a dye could be used as the measure for the height of LUMO and for ΔE on the basis of Eq. (5.5). (B) The potential energy profiles of the initial state (a) and the final state (b) of the electron transfer, where λ is the rearrangement energy of solvents as expressed in Eq. (5.8).

In general, ϵ_{LU} and ϵ_{HO} are related to electron affinity (EA) and ionization energy (I), respectively,

$$EA = -\varepsilon_{LU} \tag{5.3}$$

$$I = -\varepsilon_{HO}, \tag{5.4}$$

and were studied by several methods including photoelectron spectroscopy, electrochemical methods including polarography, and molecular orbital calculations [4]. Sheppard first recognized the photographic significance of the polarographic potentials of dyes by showing that the degree of decrease in intrinsic sensitivity of AgX grains by dyes (i.e., desensitization) increased with decreasing negative values of their reduction potentials (E_R) [112]. The measurement of E_R and E_{OX} of sensitizing dyes was made by Stanienda [113–116] and then by several working groups including Daehne [117], Tamura and Hada [118], and Honda and co-workers [119]. The systematic examination of the relationship of ϵ_{LU} and ϵ_{HO} with E_R and E_{OX} of various kinds of sensitizing dyes was studied by Tani and Kikuchi [46,119–122] according to the relationships

$$\varepsilon_{LU} = -E_R + \text{constant} \tag{5.5}$$

$$\varepsilon_{HO} = -E_{OX} + \text{constant}. \tag{5.6}$$

Figure 5.12 shows an example of the linear relationships between ϵ_{LU} and E_R and between ϵ_{HO} and E_{OX} of cyanine dyes with varying polymethine chain length and types of heterocyclic rings [40]. The linear relationships between polarographic potentials and electronic energy levels as given by the HMO method were so good that deviations from the linear relationships characterizes the steric effects in their molecules [40], and can be

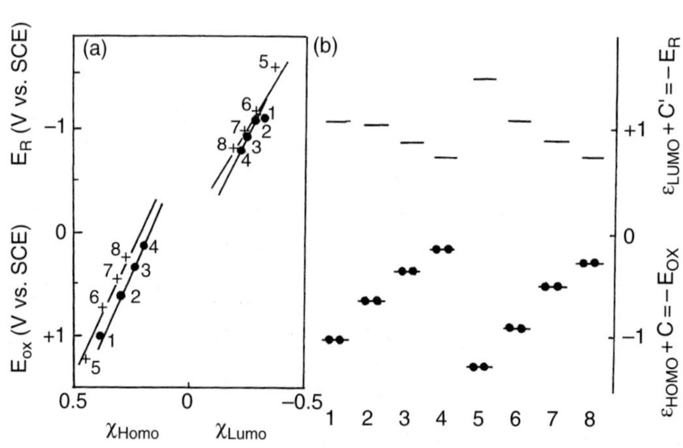

Figure 5.12 (a) Relations of E_R and E_{OX} with calculated electronic energy levels of 2,2'-quinocyanine dyes (1–4) and thiacyanine dyes (5–8), where χ_{LU} and χ_{HO} are $(\epsilon_{LU} - \alpha)/\beta$ and $(\epsilon_{HO} - \alpha)/\beta$, respectively, and α and β are Coulomb and resonance integrals of $2p$ atomic orbitals of carbons in π-conjugated hydrocarbons, respectively. The numbers of methane groups are 1 (1 and 5), 3 (2 and 6), 5 (3 and 7), and 7 (4 and 8). (b) Electronic energy levels of the dyes on the basis of Eqs. (5.5) and (5.6) [40].

used to predict the effects of substituents on the wavelength of absorption peak and electronic energy levels [4,9,39].

On the basis of Eqs. (5.5) and (5.6), the electronic energy levels of various dyes with respect to the conduction band and valence band of silver halides were estimated and used to study desensitization and then dye sensitization by several working groups, including Daehne [117], Tani and Kikuchi [122,36], Berriman and Gilman [123], and James [124].

The accuracy of the values of E_{OX} and E_R as measured by conventional polarography is deteriorated by secondary reactions of the oxidized and reduced dyes, respectively [125]. Lenhard introduced into this field phase-selective second harmonic (PSSH) voltammetry, which measures the second-harmonic current for the AC voltage superimposed on the DC voltage as a function of the AC voltage's frequency [126]. Since the current measured by this method is responsible for the ratio of the concentration of dye molecules to that of their ionized species caused by the AC voltage, the measured values of E_{OX} and E_R by means of PSSH voltammetry can escape from secondary reactions when the AC voltage's frequency is too large for detecting the secondary reactions. This method has been confirmed to gives more accurate values of E_R and E_{OX} due to its technique to detect the electrochemical reactions before they are followed by the secondary reactions [126,127].

Although the values of I and EA of sensitizing dyes in solid state have been determined by several working groups with solid-state physics techniques such as ultraviolet photoelectron spectroscopy (UPS) [128–145], they cannot be related to the molecular structures and photographic capabilities of the dyes. Nielsen critically reviewed the theory and experiments for determining the sensitizing dyes' electronic energy levels [146]. Seki, Tani, and others systematically determined the ionization energy of cyanine and merocyanine dyes by UPS with respect to the Fermi level of the substrate (i.e., Ag) for the first time, and confirmed the close relation of ionization energy with molecular structure of sensitizing dyes and with their E_{OX} and ϵ_{HO} [147–149,127]. Then, determining the Fermi level of an evaporated Ag substrate was followed by determining the height of the top of the valence band of the AgBr layer evaporated on the Ag substrate and the HOMO of merocyanine dyes evaporated on the AgBr layer with respect to the Fermi level of the Ag substrate under ultra-high vacuum conditions throughout the experiments.

Figure 5.13 presents the result of the above-stated experiments indicating the electronic structures of merocyanine dyes with varying polymethine chain length and heterocyclic ring on evaporated AgBr layers with respect to the Fermi level of the Ag substrate [111,149,150]. The HOMO electronic energy levels of the dyes were directly determined by UPS. The LUMO levels were evaluated by taking into account their HOMO–LUMO transition energies, which can be obtained from the absorption spectra. The evaluated LUMO levels were situated above the bottom of the AgBr conduction band for dyes A, B, and E, and below that for dyes C and D. This result indicates that dyes A, B, and E should be good sensitizers for the dye sensitization of AgBr grains in emulsion layers, whereas dyes C and D should not. This was the first observation with UPS to indicate that the electronic structure of dyes with respect to that of AgX is consistent with the spectrally sensitizing abilities of the dyes for the AgX grains. It was noted that the vacuum level of each dye had shifted considerably with respect to that of AgBr at their interface.

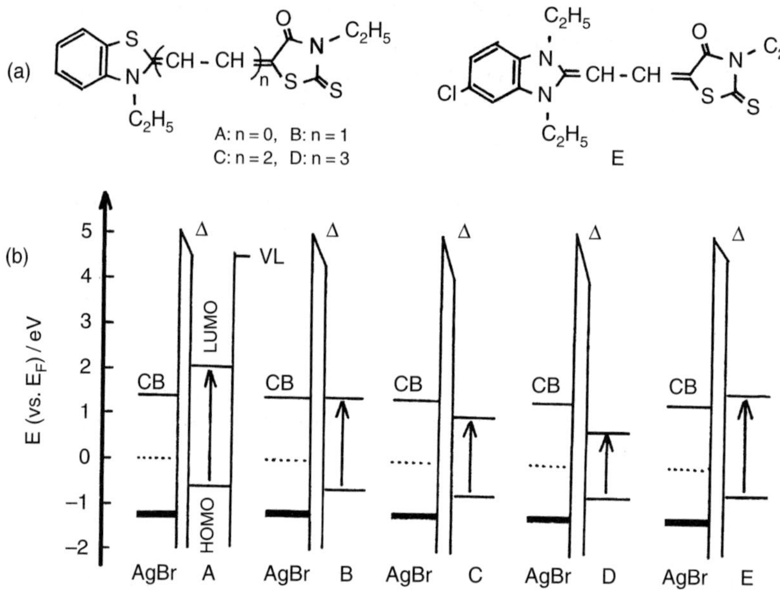

Figure 5.13 Electronic energy level diagram with respect to the Fermi level of bulk silver (E_F) for the interfaces between AgBr and merocyanine dyes as determined by ultraviolet photoelectron spectrometer, where VL is the vacuum level of materials, CB and VB are the bottom of the conduction band and the top of the valence band of AgBr, respectively, and Δ is the vacuum level shift at the interface between AgBr and a dye layer [150,151].

The electronic structure of the interface between dyes and silver halides supports the possibility of the electron transfer mechanism by both the quantum-chemical and electrochemical method and the photoelectron-spectroscopic method. Namely, the LUMO levels of sensitizing dyes are above the bottom of the AgXconduction band, while those of non-sensitizing dyes are below it. The relation between the electronic structures and the mechanism of dye sensitization is described in Section 5.9.

As seen in Figure 5.13, it was observed that the vacuum level of any of the dyes studied was lower than that of AgBr. This was the first observation of the shift of the vacuum level between a substrate and an organic layer, and indicated the formation of a layer of dipoles at the interface between them. The vacuum level shift could be observed by measuring the ionization energies of a thin AgBr layer with Ag as its substrate and then a thin dye layer on the AgBr layer with Ag as their substrate always with respect to the Fermi level of the Ag substrate, and gave the reason for the difference between the result by Seki et al. and those reported before that [128–145]. Seki et al. extended this observation to the interfaces between metal substrates and organic layers of wide variety and observed that the vacuum level of the latter was lower than that of the former in most cases [151,152]. The vacuum level shift is important, not only for the dye sensitization of photographic films, but also for organic devices, and is described in Sections 5.9 and 10.5.

As stated above, the heights of HOMO and LUMO of dyes are important and can be estimated by means of oxidation and reduction potentials, molecular orbital calculation, and photoelectron spectroscopy. From a practical viewpoint, the oxidation and reduction potentials as given by phase-selective second harmonic voltammetry are convenient and appropriate for estimating the heights of HOMO and LUMO for a wide variety of dyes, although the measurements are made on dye molecules in monomeric state in solutions.

Since J-aggregates of dyes are important for photographic films and differ in property and photographic performance from their monomers, it is important to know the difference in electronic structure between monomers and J-aggregates of dyes. Although the difference for a particular dye has been revealed [153] by a sophisticated method [153–155], it is still difficult to determine from those results the general rule for the displacement of the heights of HOMO and LUMO of J-aggregated dye molecules from those of monomeric ones.

5.6 Efficiency and Characterization of Light-Induced Electron Transfer

In the dye sensitization of photographic films, the electron transfer takes place from an excited dye molecule to an AgX grain, leaving a positive hole on the dye molecule (i.e., dye positive hole). The efficiency of the electron transfer is given by the relative quantum yield of dye sensitization (RQE; ϕ_r), which is the ratio of the quantum yield of the latent image formation on the excitation of sensitizing dyes to that on the excitation of AgX grains (i.e., the ratio of the number of quanta absorbed only by the AgX necessary to produce a specified optical density to the number absorbed quanta only by the dye at a wavelength λ within its absorption band required to produce the same density) [156]. It is noted that RQE gives the efficiency of the electron transfer regardless of the degree of the recombination between the transferred electrons and dye positive holes, and was therefore used widely for the analysis and design of the dye sensitization of photographic films. The validity of RQE in dye-sensitized solar cells is suggested in Section 9.5. The experimental procedure and examples of the measurements of RQE are described in detail [156,157,3, 4] with examination of the condition where RQE is available [157,4].

It has often been confirmed that the efficiency of the electron transfer is independent of the wavelength of exciting light within the absorption band of the dye molecules in the same aggregate state on the grains [156,158,92]. It has, however, been pointed out that the efficiency differed for light absorbed by the same J-aggregated dye with different size [159]. These results indicate that the electron transfer from an excited dye molecule to an AgX grain takes place, not from a 'hot' vibrational state, but from the equilibrium state in the excited state of the dye. It is also noted that cyanine dyes on AgX grains in color films form J-aggregates with extraordinary small Stokes shift [160]. Therefore, it is not probable that 'hot' electrons significantly contribute to the electron transfer in the dye sensitization for color films.

Figure 5.11 shows the electronic energy levels of a sensitizing dye in the excited state with respect to those of AgX and the potential energy profile for the electron transfer in dye sensitization. The energy gap for the electron transfer is the difference in energy between an excited electron in the dye molecule and an electron at the bottom of the conduction band of AgX as shown by ΔE in this figure, and is related to the reduction potential (E_R) of the dye studied. In the potential energy profile, curves a and b indicate the state before and after the electron transfer, and λ is the rearrangement energy of solvents in the Marcus theory [172,173]. In order to characterize the electron transfer process in dye sensitization, it is important to study the energy gap dependence of the rate and efficiency of the process. In the case of the dye sensitization of AgX grains in photographic films, consideration was needed on several points, which included the measurements of the rate and efficiency of the electron transfer process, the evaluation of the energy gap, the aggregate state of sensitizing dye molecules, the distribution of the energy of an excited electron in the dye molecule, and the influence of oxygen.

The energy gap dependence of dye sensitization was studied by several working groups [4,9,123,161–170]. In most studies, the efficiency of the electron transfer process instead of the rate was used to study the energy gap dependence, since the former was more accessible and reliable than the latter. Reduction potentials of dyes by phase-selective second-harmonic voltammetry [126] were much more precise than those by conventional methods and were used to study the energy gap dependence [168,169]. Muenter et al. used the experimental condition under which dye molecules acted as monomers [169]. However, cyanine dye molecules usually work in aggregated states on the surface of AgX grains. A sophisticated emulsion was proposed in order to avoid the influence of oxygen [166], and used to study the energy gap dependence [166,167,169,170]. The Marcus theory was applied to the energy gap dependence [167–170]. The distribution of the electronic energy levels of dyes was considered [168,4] and is described below.

In order to characterize the electron transfer in the dye sensitization of photographic films, the present author and others have used many cyanine dyes with similar molecular structure and varied electronic structure to determine common characteristics of the electron transfer from these dyes in the excited states to AgX grains, and studied the efficiency of the electron transfer process. Since cyanine dye molecules easily form aggregates, it is difficult to prepare the condition of AgX emulsions, under which cyanine dye molecules were monomeric on the surfaces of AgX grains. Taking this into consideration, as shown below, the present author used these dyes in aggregated states to determine the characteristics of the electron transfer common to cyanine dyes on AgX grains. Before the electron transfer, only excitons (i.e., excited dye molecules, each of which confines an excited electron and a positive hole) are present in single-crystalline aggregates of cyanine dyes, and are not likely to be substantially influenced by oxygen as described in Chapter 10. As seen in Figure 5.14, ϕ_r sharply depended upon E_R of the dyes, being in accord with the electron transfer mechanism. It is noted that ϕ_r could be unity for dyes with large ΔE and decreased steeply with decreasing ΔE.

The value of E_R is related to the height of LUMO of dyes in monomeric state in solutions, while dye molecules formed J-aggregates or H-aggregates on the grain surface. It does not seem that the difference in the type of aggregates apparently influenced the correlation between ϕ_r and E_R in Figure 5.14. As illustrated in Figure 5.8, aggregated dye molecules in the excited state have split electronic energy levels. It is considered that an electron is photo-excited from HOMO to one of the two excited electronic levels, and transferred to an AgX grain from the lower lying excited electronic level. It is considered that the deviation of the lower lying excited electronic level of an aggregate from the excited electronic level of a monomer did not differ too much among the dyes with similar molecular structure regardless of the difference in the type of aggregate.

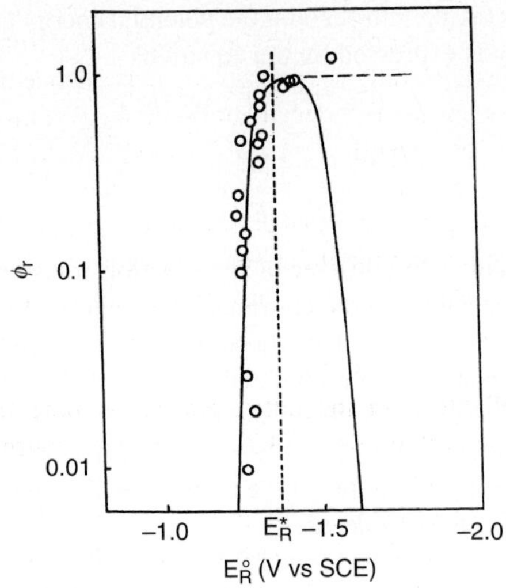

Figure 5.14 Energy gap dependence of ϕ_r of cubic AgBr grains by cyanine dyes with similar molecular structure and varied electronic energy levels. The energy gap was expressed by E_R of the dyes, which was measured by means of phase-selective second harmonic voltammetry. The solid curve represents the energy gap dependence derived from the Marcus theory with Eqs. (5.7)–(5.9), where $k_s{}^o$ and λ are 10^{11} s^{-1} and 0.05 eV, respectively. Reproduced with permission from Reference [168], copyright 1990 American Chemical Society.

The steep threshold of the energy gap dependence of ϕ_r in Figure 5.14 is in favor of electronic energy levels of dyes defined sharply rather than distributed widely. Nelson et al. proposed that quantities such as the ionization energy and the electron affinity of adsorbed dye layers were not sharply defined but were distributed over a band of energy values much greater than that associated with thermal fluctuations on the basis of the external photoemission from adsorbed dye layers on the substrates other than AgX in vacuum. The spread of ionization energies and electron affinities was attributed to perturbing interactions of adsorbed dye molecules with randomly distributed positively and negatively charged crystal defects and impurities on the surface of the substrate [140,143,144]. Nelson's proposal was followed by Sturmer et al. [161,162], Penner and Gilman [171], and James [124]. It is considered that the distribution of the electronic energy levels of cyanine dyes on AgX grains in emulsions is inclined to be sharper than on other substrates in vacuum. Namely, the positive charge of the chromophore in a cyanine dye molecule is surrounded by negative ions on the surface of AgX, since the dye molecules could smooth the surface of AgX by repelling silver ions at kink sites into interstitial positions especially when they formed their aggregates, as described in Section 5.4, and are surrounded by gelatin containing water and ions with high concentration. The observation that all the molecules in J-aggregated mono-molecular dye layers with size of several micrometers [74–76] were aligned in the same direction supported the idea that all the dye molecules are nearly under the same condition.

Marcus theory [172,173], which is well accepted as the explanation for the electron transfer phenomena in solutions, was applied to the electron transfer in photographic emulsions [167–170] by taking into account the potential energy profile in Figure 5.11 and Eq. (5.5). Marcus theory is expressed by the equations

$$k_s = k_s^o \exp(-\Delta G^{=} kT) \tag{5.7}$$

$$\Delta G^{=} = (\Delta E + \lambda)^2/4\lambda \tag{5.8}$$

$$\Delta E = E_R - E_R^*, \tag{5.9}$$

where k_s^o is the maximum rate constant of the electron transfer, $\Delta G^{=}$ is the activation energy for the electron transfer, λ is the rearrangement energy of the electron transfer, and E_R^* is the reduction potential of the dye whose ΔE is zero. Namely, the electron transfer from an excited dye molecule to the conduction band of AgX takes place through the tunnel effect at the crossing points between curve a (the potential energy of the state before the electron transfer) and curve b (that of the state after the electron transfer), and ΔG is needed to raise the potential energy of state a to the crossing point through the rearrangement of the surroundings of the molecule.

The solid curve in Figure 5.14 is derived from the above-stated equations with k_s^o of 10^{11} s^{-1} and a small rearrangement energy ($\lambda = 0.05$ eV), indicating that the region where ϕ_r steeply increases with ΔE is followed by the region where ϕ_r steeply decreases with ΔE (i.e., the inverted region). The experimental points were plotted along the curve in the first region, and apart from it in the inverted region. Namely, the energy gap dependence of ϕ_r in the present system was characterized by a small value of λ and the absence of the inverted region.

The electron transfer processes in solutions were generally associated with the inverted region and large values of λ, which are more than 0.3 eV [174,175]. On the basis of the temperature dependence of the rate of the electron transfer from excited dye molecules to AgX, it was observed that λ was 0.02 eV for J-aggreged dye molecules, while it was 0.3 eV for monomeric ones [176,177]. It is considered that the reason for the small value of λ is the absence of solvent molecules and/or the rigid structure of J-aggregates, being in accord with the fact that the value of λ for the electron transfer process with monomeric dyes in the dye sensitization of AgX grains [167,169] was larger than the above-stated value with aggregated cyanine dyes [168].

The appearance for the inverted region is explained by the potential energy profiles before and after the electron transfer (curves a and b, respectively) with variation of ΔE as shown in Figure 5.15. With increasing ΔE, the rate constant of the electron transfer increases uphill, reaches the maximum when the crossing point between curves a and b is at the bottom of curve a, and then decreases downhill, since the crossing point is leaving far behind the bottom of curve a. With increasing ΔE, the activation energy of the electron transfer decreases, becomes zero at the point with the maximum rate constant, and then increases.

In the case of electron transfer in the present system, the electron acceptor is the conduction band of AgX, which is composed of widely and densely distributed electronic energy levels. Thus, the state after the electron transfer could be represented by many curves as shown in Figure 5.15. Even in the downhill region, there is always some curve in the state after the electron transfer that crosses at the bottom of curve a. This is considered to be the reason for the absence of the inverted region in the electron transfer in dye sensitization of the photographic process. In this case, the absence of the inverted region of the Marcus theory was observed for photo-induced electron transfer from dyes to AgBr

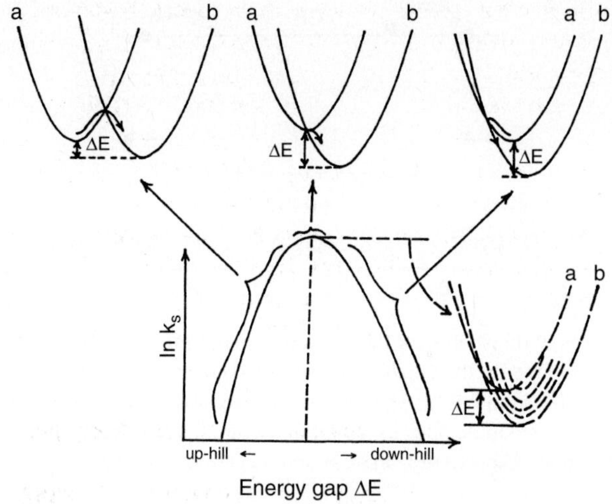

Figure 5.15 Relation between k_s and ΔE and corresponding potential energy profiles for the initial (a) and final (b) states of the electron transfer with variation of ΔE. The solid curves are for the electron transfer processes between molecules in solutions, and the broken curves for the electron transfer from excited dyes to the conduction band of AgBr.

grains with varying kind of dyes. The absence of the inverted region was also observed for the photo-induced electron transfer from a particular dye to metal oxide powders with varying electronic structures [178,179].

As expected from the situation shown in Figures 5.11, 5.14, 5.15, the increase in ΔE increased k_s and decreased its temperature dependence, which could finally disappear [180,181]. In accord with the fact that supersensitization increases ΔE, as illustrated in Figure 5.18, it also extinguishes the temperature dependence of the efficiency of dye sensitization, as observed by Frieser et al. [182]. Tani, Sano, and Saito confirmed through Dember photovoltage measurement as well as sensitometric results that the dye–supersensitizer combination could not only increase the efficiency of the dye sensitization, but also extinguish its temperature dependence [183].

As seen in Figure 5.16, there are two kinds of electron transfer processes associated with dye sensitization. One of them is the electron transfer from the excited dye molecule (Dye^*) to the conduction band of AgX to bring about dye sensitization. The other is the transfer of a free electron in the conduction band of AgX to the LUMO level of a dye molecule in the ground state (Dye) to give the dye in one-electron-surplus state (Dye^-), bringing about the decrease in the intrinsic sensitivity of the AgX grains (i.e., electron-trapping desensitization) [3,4]. As shown in Figure 5.17, the degree of desensitization by dyes exhibited steep energy gap dependence in AgBr emulsions [168]. The threshold of the energy gap dependence of the electron transfer from AgBr to Dye was 0.17 eV less negative in E_R than that of the energy gap dependence of the electron transfer from Dye^* to AgBr. This difference in the threshold of the energy gap dependence indicates that the energy level of an electron excited to LUMO in Dye^* was lower by 0.17 eV than that of an electron injected to LUMO in Dye^-, as schematically shown in Figure 5.17, and that the binding energy of excitons in mono-molecular layers of cyanine dyes on AgBr grains is

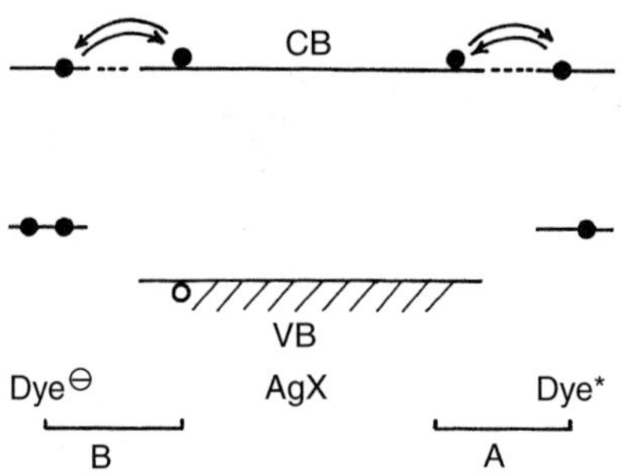

Figure 5.16 Two types of thresholds for the electron transfer processes in dye sensitization. Case A shows the transfer from Dye* to the conduction band (CB) of AgX to bring about dye sensitization. Case B shows the transfer from CB of AgX to Dye to bring about electron-trapping desensitization. Dye, Dye*, and Dye$^-$ are dye molecules in the ground state, the lowest excited singlet state, and the one-electron-surplus-state, respectively, and CB and VB are the conduction band and valence band of AgX.

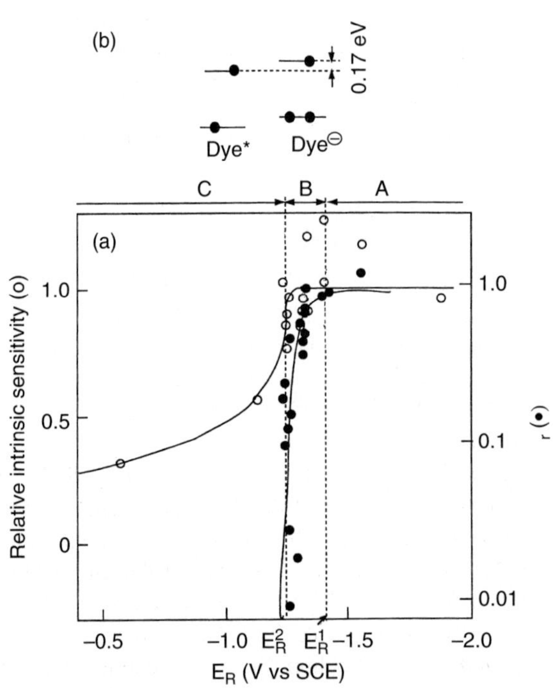

Figure 5.17 (a) Intrinsic sensitivity (○) of dyed emulsions with respect to that of an undyed emulsion and ϕ_r (●) as a function of E_R of dyes. The thresholds for dye sensitization ($E_R{}^1$) and for electron-trapping desensitization ($E_R{}^2$) were -1.42 and -1.25 eV, respectively, as shown here. Dyes are classified into three groups; A with ϕ_r of unity and without desensitization, B with ϕ_r smaller than unity and without desensitization, and C with ϕ_r much smaller than unity and with desensitization. (b) The electronic energy levels of Dye* and Dye$^-$ on the basis of the values of $E_R{}^1$ and $E_R{}^2$ and Eqs. (5.5) and (5.6), where Dye* and Dye$^-$ are dyes in the lowest excited singlet state and in one-electron-surplus-state. Reproduced with permission from Reference [168], copyright 1990 American Chemical Society.

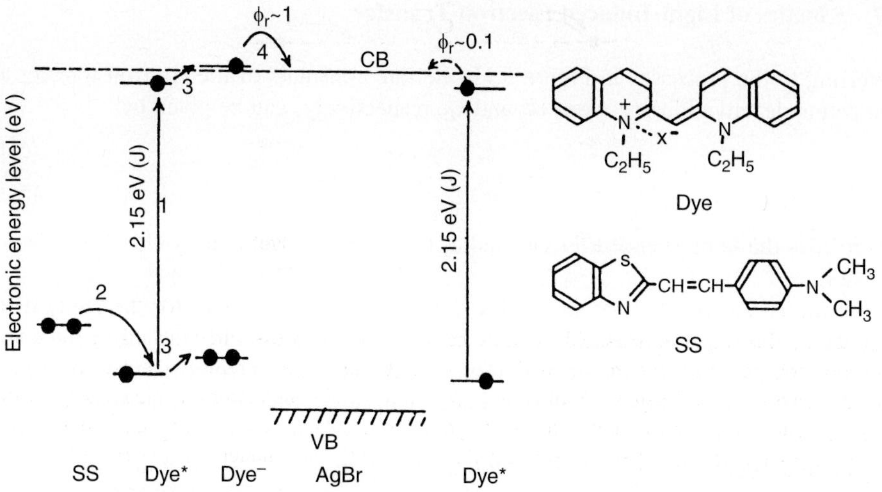

Figure 5.18 Electronic energy level diagram showing hole-trapping supersensitization of Dye 3 (Dye) by a supersensitizer (SS). Dye^* and Dye^- are Dye 3 in the lowest excited singlet state and in one-electron-surplus-state, respectively, CB and VB are the conduction band and the valence band of AgBr, respectively. The light excitation of Dye to give Dye* (1) is followed by the electron transfer fron SS to Dye* (2) to give Dye^- with electronic energy levels, which are elevated by the binding energy of an exciton in the dye (3). Thus, an electron is efficiently transferred from Dye^- to the conduction band of AgBr (4).

therefore 0.17 eV [184], being much smaller than the binding energies of excitons in organic layers, which is usually larger than 0.4 eV [185,186]. Since the binding energy of an exciton depends on the dielectric constant of its surroundings, it is considered that the small binding energy of an exciton in a mono-molecular dye layer on an AgBr grain resulted from the influence of AgBr with dielectric constant of 12.5 [36,184].

This explains the improvement of ϕ_r of the dyes with insufficient ΔE by hole-trapping supersensitization, as shown in Figure 5.18 for the case of 1,1'-diethyl-2,2'-quinocyanine (dye 3; Dye) with a supersensitizer (SS) [168]. Namely, k_s and ϕ_r of the dye alone were small, since the excited electron in Dye^* were situated below the bottom of the conduction band of AgBr. When Dye was excited to give Dye^* in the presence of SS, the electron transfer took place at first from SS to Dye^* to give Dye^-. Since the energy level of the excited electron in Dye^* was raised by 0.17 eV by the electron injection from SS, the electron could be situated above the bottom of the conduction band of AgBr, and transferred with high efficiency to AgBr [168]. The long lifetime of Dye^- as compared with Dye^* could be another cause for the efficient electron transfer from Dye^- to the conduction band of AgBr [187–189]. However, the fact that the temperature dependence of the efficiency of dye sensitization by dye 3 was extinguished by SS [180,181,183] could be explained, not in terms of the lifetime, but in terms of the electronic energy level of Dye^- as compared with Dye^*.

5.7 Kinetics of Light-Induced Electron Transfer

Referring to the processes in Figure 5.11, the rate constants of the electron transfer and competing deactivation processes (k_s and k_L, respectively) can be given by

$$k_s = \phi_r/\tau_F \tag{5.10}$$

$$k_L = (1 - \phi_r)/\tau_F, \tag{5.11}$$

where τ_F is the fluorescence lifetime, and ϕ_r is the relative quantum yield of the electron transfer.

The measurement of τ_F of sensitizing dyes is a key experiment for the study of the kinetics of the electron transfer process in dye sensitization and was made by several working groups with subnanosecond-, picosecond-, and femtosecond-spectroscopic methods [190–199,160]. For these measurements, each sample was exposed to an intense light pulse within a very short time period, or weak light pulses for many times. It should be taken into account when measuring accurate fluorescence lifetimes that the former condition is likely to bring about the singlet–singlet exciton annihilation [191–193],

$$S_1 + S_1 \rightarrow S_0 + S_n, \tag{5.12}$$

where S_0, S_1, and S_n are a dye in an aggregate in the ground state, and the first and *n*th excited singlet states. In the latter condition, it was a concern that exposed samples would be subjected to degradation during repeated exposure to many light pulses. Suzumoto and co-workers were able to depress the degradation of exposed samples by moving each sample to continually change its position, which was excited by light pulses for the measurement of the fluorescence lifetime [195].

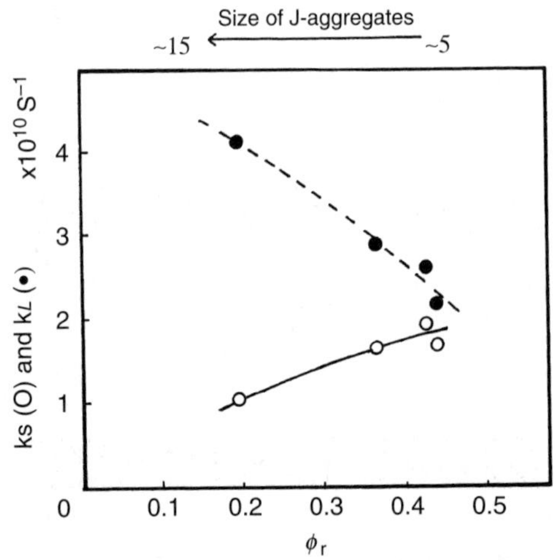

Figure 5.19 Relation of k_s and k_L with ϕ_r for the dye sensitization of octahedral AgBr grains by dye 1 with variation in aggregate size. The increase in agitation temperature of dyed emulsions resulted in the growth of J-aggregates of the dye and in the decrease in ϕ_r. In this figure, k_s, k_L, k_r, and k_{nr} are the rate constants of the electron transfer, deactivation channels, radiative and nonradiative recombination processes, respectively, as shown in Figure 5.11 and Eqs. (5.10) and (5.11). Reproduced with permission from Reference [195], copyright 1992 American Chemical Society.

Then, the picosecond spectroscopy was applied to the study of dye sensitization by J-aggregates of cyanine dyes with varying aggregate size [193–196]. As stated in Section 5.4, the growth of J-aggregates of dye 1 was enhanced by digesting the dyed emulsions at an elevated temperature [4]. Figure 5.19 shows k_s and k_L of dye 1 as a function of ϕ_r, which decreased with increasing aggregate size of the dye [193–195]. Namely, the decrease in ϕ_r resulted from the increase in k_L and the decrease in k_s.

The value of k_L is the sum of the rate constants of radiative and nonradiative recombination processes (i.e., k_r and k_{nr}, respectively). The value of k_{nr} tends to decrease with increasing aggregate size [200]. On the other hand, k_r increases with increasing aggregate size [55,199–203], as expressed by the equation

$$k_r(\mathrm{J}) = Nk_r(\mathrm{M})\cos^2\alpha, \tag{5.13}$$

where $k_r(\mathrm{J})$ and $k_r(\mathrm{M})$ are the rate constants of the radiative recombination processes of dyes in J-aggregate and monomeric states, respectively, N is the number of dye molecules in a J-aggregate, and α is the angle between the long axes of a J-aggregate and of a dye molecule in it. It is therefore considered that the increase in k_r with increasing aggregate size contributed to the increase in k_L, which caused the decrease in ϕ_r with increasing aggregate size.

The decrease in k_s with increasing aggregate size [193–195] is consistent with the mechanism of aggregate-partitioning supersensitization proposed by West and Carroll (Figure 5.20) [204,205], according to which the dye molecules at the ends of a J-aggregate are different in energy from others and act as traps for an exciton, providing suitable places

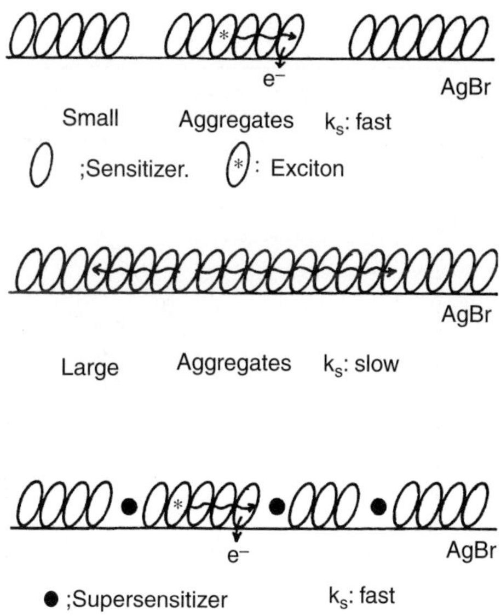

Figure 5.20 The performance of excitons in aggregates and aggregate-partitioning supersensitization. Molecules at the ends of an aggregate are different from others and act as traps for excitons in the aggregate, enhancing their dissociation to inject an electron to the conduction band of silver halides. Thus, the rate of the electron injection decreases with the size of aggregates, and increases by the partition of an aggregate by supersensitizer molecules [204,205].

for it to achieve the charge separation to inject an electron into the conduction band of an AgX grain.

Muenter et al. investigated the effect of aggregate size on the kinetics of dye sensitization by dye 1, varying the average physical size of the aggregate by diluting it with a close structure analog [196]. Among processes that might compete with dye sensitization, they point out the importance of the energy transfer to defects in the aggregates, which was enhanced by the increase in aggregate size. They recognize the increase in k_s with increasing aggregate size. It is anticipated that the method for controlling aggregate size, which differed considerably between Tani's group and Muenter's, brought about the difference in aggregate size dependence of k_s between them.

Picosecond spectroscopy was also applied to the analysis of hole-trapping supersensitization for dye 3 [190,192,193,195], as shown in Figure 5.18. Figure 5.21 shows the relation of k_s and k_L as estimated by Eqs. (5.10) and (5.11) with ϕ_r for the dye with and without a supersensitizer as well as for the dye with varying adsorption temperature. It is noted that k_s increased with increasing ϕ_r, while k_L was independent of ϕ_r. Namely, the increase in ϕ_r, which was brought about by hole-trapping supersensitization, was solely caused by the increase in k_s, being in accord with the mechanism of hole-trapping supersensitization as described in Figure 5.18.

It is also noted from these results that the rate constant of the electron transfer from an excited dye molecule to the conduction band of an AgX grain is on the order of 10^{-10}–10^{-11}s, and does not indicate the involvement of electrons at "hot" vibrational states in excited dye molecules in the electron transfer for the dye sensitization of AgX grains. This result and consideration are in accord with the fact that the efficiency of the electron transfer

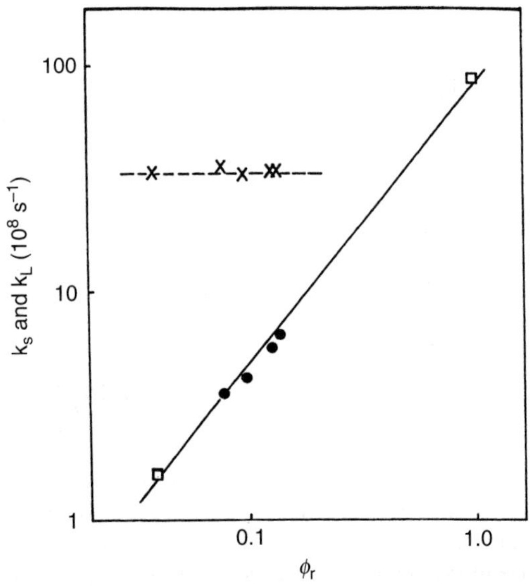

Figure 5.21 Relation of k_s and k_L with ϕ_r for dye sensitization of dye 3. The experimental points are k_s for the dyes on octahedral AgBr grains of varying aggregate size (●), k_s for the dye on cubic AgBr grains with and without (□) a supersensitizer, and k_L (×) for the dyes in the corresponding state.

from excited dye molecules to AgX grains was independent of the wavelength of exciting light within the absorption band of the dye molecules under the same aggregate condition [156,158,92]. As described in Section 9.6, it is considered that "hot" electrons could take part in the dye sensitization of TiO_2 on the basis of the results with femtosecond spectroscopy [206,207] and wavelength dependence of the efficiency of the electron transfer from excited sensitizing dye molecules to TiO_2 [208].

5.8 Back Electron Transfer of Dye Sensitization

An electron transfer process is likely to be followed by a back electron transfer. The electron transfer from an excited dye molecule to an AgX grain leaves a positive hole in the dye phase, which is called a dye positive hole (Dye^+). In photographic films, dye positive holes cause back electron transfer processes, expressed as

$$\mathrm{Dye}^+ + e^- \to \mathrm{Dye} \tag{5.14}$$

$$\mathrm{Dye}^+ \to \mathrm{Dye} + \mathrm{h}^+ \to \mathrm{Dye} + X_2/2 \tag{5.15}$$

$$Ag_n + X_2/2 \to Ag_{n-1} + AgX, \tag{5.16}$$

where h^+ is a positive hole, X_2 is a photolitically formed halogen molecule, and Ag_n is a latent image center.

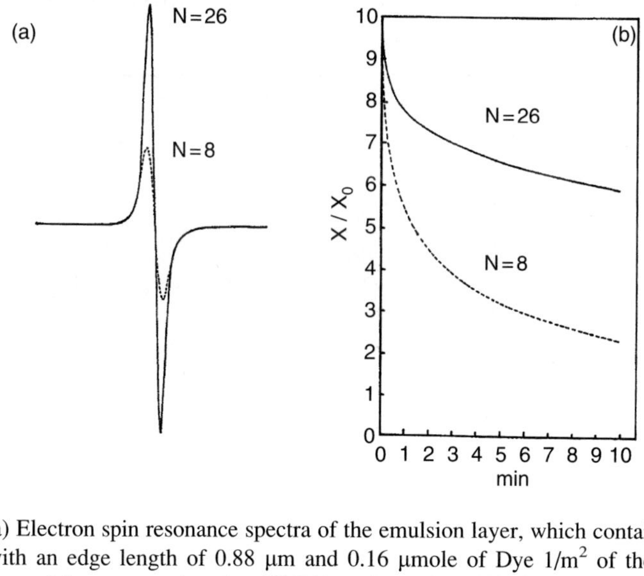

Figure 5.22 (a) Electron spin resonance spectra of the emulsion layer, which contained octahedral AgBr grains with an edge length of 0.88 μm and 0.16 μmole of Dye 1/m^2 of the grain surface [210–213]. The emulsions were agitated at 40°C (dotted curve) and 70°C (solid curve) for 60 min for the adsorption of the dye to the grains. The size of aggregates in terms of the number of the dye molecules per aggregate (*N*) was evaluated on the theoretical relationship between the absorption spectrum of the dye with *N*. Only the dye was excited during the ESR measurement. (b) Decay curves of light-induced ESR signals of the above-stated emulsion layers subjected to minus-blue exposure and ceased at $t = 0$.

Following the result with cyanine dyes on gelatin-free AgBr grains at liquid nitrogen temperature by Needler, Griffith, and West [209], Tani found that dye positive holes on octahedral and cubic AgBr grains in photographic films were paramagnetic and gave rise to an electron spin resonance (ESR) signal at room temperature [210–215]. Thus, the light-induced ESR signal arising from dye positive holes in photographic films is useful for analyzing the recombination through dye positive holes in dye sensitization. Lenhard and co-workers spectroscopically detected and analyzed dye positive holes photochemically and electrochemically formed on the surface of AgX grains in emulsions and in solutions [216,217].

The light-induced ESR signal is structureless and sharp as seen in Figure 5.22. The half-width decreased with increasing aggregate size as seen in this figure and with increasing temperature for the ESR measurement [210–212]. These results indicate that some kind of motional narrowing by the movement of positive holes in J-aggregates gave rise to such a sharp and structureless signal.

The light-induced ESR signal as stated above could be observed only with the dyes whose HOMO were situated significantly higher (by more than 0.3 eV) than the top of the valence band of octahedral AgBr grains. The signal decayed slowly with second-order kinetics. The rate constant of the decay increased with decreasing height of HOMO. These results indicate that a positive hole was trapped by the HOMO of a dye for producing Dye^+, which gave rise to the ESR signal.

By using dyes that cannot trap photoelectrons in their ground state, it was found that the desensitization (i.e., the decrease in the intrinsic sensitivity of AgX grains by dyes) was significant only for those dyes that showed the light-induced ESR signal in the emulsions studied [210–215]. Figure 5.23 shows the result of the time-resolved sensitometry of

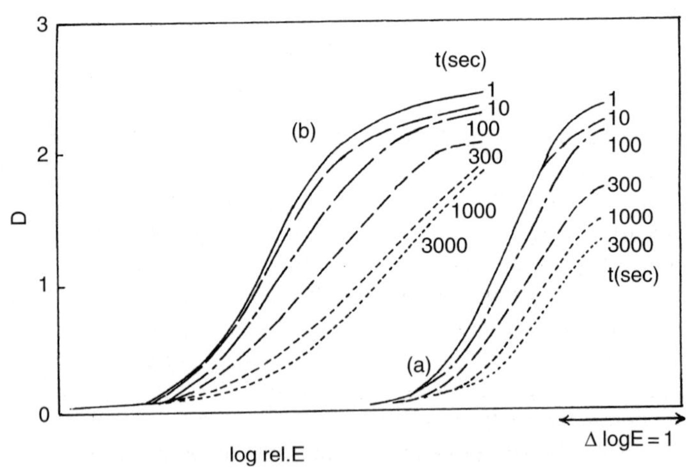

Figure 5.23 Time-dependent D-log *E* curves showing the decay of latent image centers in photographic films that contained octahedral AgBr grains (0.24 μm) and dye 1. The amount of the dye was 5.5×10^{-4} mole/mole AgBr. An emulsion layer was exposed for 10^{-3} s to (a) blue and (b) minus-blue light, kept for *t* s, immersed in water to terminate the decay of latent image centers, and developed by a MAA-1 surface developer. Reproduced with permission from Reference [213], copyright 1987 American Institute of Physics.

octahedral AgBr emulsion grains with J-aggregated dye 1 to indicate that the latent image centers also decayed in the presence of the dyes, which gave rise to the light-induced ESR signal [210–215] The decay of the latent image centers obeyed second-order kinetics. The rate constant and activation energy of the decay of the latent image centers in the presence of dyes was comparable to those of the decay of the light-induced ESR signal of the dye in the emulsions [214,215].

The above-stated result indicates that the back electron transfer through Processes (5.15) and (5.16) took place slowly, and decreased photographic sensitivity gradually, as seen in Figure 5.23. On the other hand, the back electron transfer through Process (5.14) and the decrease in sensitivity caused by it should take place rapidly [214,215]. The ESR measurement and time-resolved sensitometry were applied to J-aggregates of dyes of varying size on AgBr grains in photographic films. The decay rate of dye positive holes decreased with increasing size of the J-aggregates, while its temperature dependence was independent of the size of the J-aggregates. This result indicates that dye positive holes are more stable and mobile as the size of the J-aggregates increases. In accord with this result, the back electron transfer through Process (5.14) is enhanced by the increase in size of the J-aggregates, while that through Processes (5.15) and (5.16) is depressed.

As stated above, the lifetimes of dye positive holes in typical sensitizing dyes on AgX grains in photographic films are so long that they can be measured by means of conventional ESR apparatuses. However, conventional ESR apparatuses for measuring dye positive holes with short lifetimes are not available. A time-resolved ESR apparatus, which is usually available for detecting paramagnetic species with short lifetimes, was applied to photographic films without success, since electrons, which appeared in the conduction band of AgBr grains as a result of their dye sensitization, absorbed the microwave for the ESR measurement and thus disturbed it.

Hosoi et al. developed a new method for measuring the lifetimes of dye positive holes, which were as short as several microseconds [218]. They used a microwave photoconductivity apparatus with a time constant of several nanoseconds to measure the change in concentration of electrons in AgX grains created by two synchronized light pulses. The first light pulse excited dye molecules to form excitons, which were then dissociated to transfer electrons to the grains and to leave dye positive holes in the dye layers. The transferred electrons, detected by the microwave photoconductivity apparatus, rapidly decayed within 1 μs, while dye positive holes slowly decayed. After a certain time period (t) during which the transferred electrons had already disappeared, the second light pulse excited dye layers to form excitons. The formed excitons were deactivated by resultant dye positive holes that had been created by the first light pulse. Thus, the deviation in concentration of transferred electrons by the second light pulse from that by the first gave the concentration of resultant dye positive holes as a function of t and the lifetimes of the positive holes. As shown in Figure 5.24, the lifetime of dye positive holes increased with increasing height of the HOMO of dyes, and ranged from several microseconds to nearly infinity.

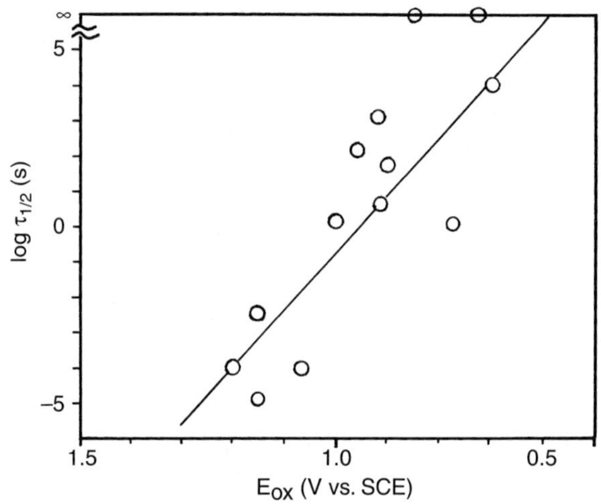

Figure 5.24. Half-life of dye positive holes $\tau_{1/2}$ as a function of oxidation potential E_{OX} of dyes [218]. The emulsions used were composed of AgBr grains with an average size of 0.88 μm and surface coverage of the grains of 8.5%. The values of $\tau_{1/2}$ were estimated from the data of the microwave photoconductivity technique with two synchronized light pulses. The solid line was expressed by $1/\tau = A\exp(-\Delta E/kT)$, and adjusted to meet the observed data by means of the least-square method.

5.9 Mechanisms of Dye Sensitization and Related Phenomena

5.9.1 Mechanism of Dye Sensitization

As stated above, the electron transfer mechanism is now accepted as the explanation of dye sensitization [4], as a result of intensive studies in light of various fundamental sciences and technologies for many years [3,4,92]. While various kinds of organic devises have attracted the keen interest of many investigators, color films, being composed of J-aggregated dye layers as light-sensitive organic semiconductors, as described in detail in Chapter 10, and tabular AgX grains as substrates, have been regarded among them [184,219]. It is therefore meaningful to review the progress of the mechanism of dye sensitization of photographic films in light of organic semiconductors and photochemistry.

Figure 5.25 illustrates the electronic energy levels of an organic semiconductor molecule, indicating the ionization energy and electron affinity of the molecule in vacuum and in solid as well as the polarization energies of the molecule by its surroundings on its electron-withdrawing and electron-accepting. This result, along with the results described in this chapter, reveal the following points to be considered for clarification of the mechanism of dye sensitization.

Point (1): The ionization energy of an organic molecule in solid is smaller by 1–2 eV than that of the molecule in vacuum owing to the polarization of the ionized molecule by its surroundings. Similarly, the electron affinity of an organic molecule in solid is larger by 1–2 eV than that of the molecule in vacuum owing to the polarization of the ionized molecule by its surroundings.

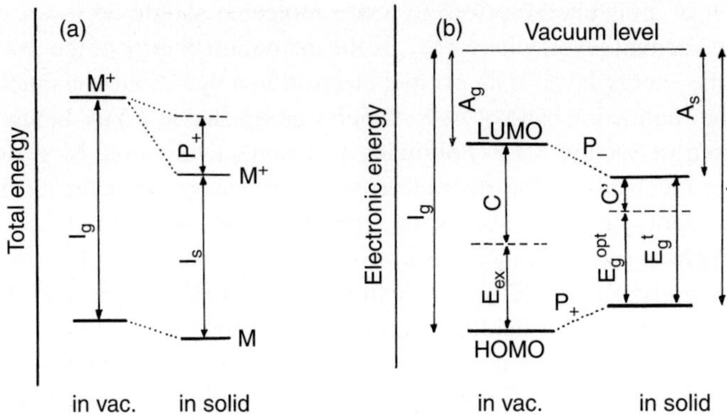

Figure 5.25 (a) Total energy levels for the ionization of an organic molecule, where M and M^+ are a molecule and an ionized molecule, respectively, I_g and I_s are the ionization energy of a molecule in vacuum and in solid, respectively, and P_+ and W_{vdW} are the polarization energy of the ionized molecule by its surroundings and the van der Waals energy among the molecules, respectively. (b) Electronic energy levels of an organic molecule showing its ionization energy and electron affinity in vacuum and in solid, where A_g and A_s are the electron affinity of a molecule in vacuum and in solid, respectively, P- is the polarization energy of the anion of the molecule by its surroundings, E_{ex} and C are the excitation energy and the binding energy of an exciton in the molecule, in vacuum respectively, and E_g^{opt}, E_g^{t}, and C' are the optical energy gap (i.e., the excitation energy to create an exciton in solid) and the total energy gap for the creation of electronic charge carriers, respectively.

Point (2): In many cases, the vacuum level of an organic layer is shifted downward with respect to that of its substrate when they are in contact with each other, especially in vacuum. The Fermi level exists and works in an organic layer only when it contains a substantial amount of impurities, and/or dopants, and is not always responsible for the shift in vacuum level.

Point (3): The energy transfer from an excited dye molecule to a molecule or a site with certain transition probability takes place through Foerster's resonance transfer process for a distance of more than several nanometers, while the electron transfer takes place for a distance of less than 1 nm.

Point (4): The binding energy of an exciton (i.e., an excited dye molecule) in an organic layer is so large (more than 0.4 eV) [185,186] that it is difficult for an exciton to dissociate at room temperature without supply of energy for its dissociation. On the other hand, the binding energies of excitons in silver halides are small enough for their dissociation at room temperature.

Point (5): The diffusion length of excitons in organic layers is usually as small as several nanometers.

In the past, several mechanisms for dye sensitization have been proposed. Major ones include the electron transfer mechanism proposed by Gurney and Mott [5] and the energy transfer one proposed by Mott [6], as described in Section 5.1. Gurney and Mott proposed the electron transfer mechanism under the assumption that the energy of an excited electron in a dye molecule is higher than the bottom of the conduction band of AgX. Then, Coulson applied the molecular orbital theory to sensitizing dyes and evaluated that the energy level of an excited electron in a dye molecule should be lower by about 7 - $h\nu_{\mathrm{max}}$ below the vacuum level, where $h\nu_{\mathrm{max}}$ is the excitation energy of the dye. It was thus evaluated that the energy level of an excited electron in a dye should be much lower than the bottom of the conduction band of AgX, which was situated at 3.5 eV below the vacuum level as described in Section 3.1.3. Following Coulson's evaluation, Mott dismissed the electron transfer mechanism, and instead proposed the energy transfer mechanism. According to the energy transfer mechanism, the energy in the excited dye molecule is transferred to AgX to create an electron in its conduction band by a Foerster-type long-range resonance transfer [220–222] or a Dexter-type short-range energy exchange [223].

This author entered into the field of photographic science in 1963, and found that the mechanism of dye sensitization of photographic films was left unsolved, while the mechanism of latent image formation was almost clarified. He considered it one of the most important subjects for photographic science at that time to make clear the possibility of electron transfer from excited dye molecules to the conduction band of AgX by evaluating the electronic energy levels of dye molecules on the surface of AgX. There were two methods for this subject. One was based on the property of molecules of dyes as characterized by quantum chemistry (i.e., molecular orbital theory) and organic electrochemistry. The other was based on the physical property of dye layers as characterized by UPS. The adequacy of these methods should be examined if they could explain the difference in sensitizing capability among dyes.

Then, the present author systematically evaluated the HOMO and LUMO electronic energy levels of dyes by means of the Hueckel-approximation molecular orbital method and was able to correlate the results well with their oxidation and reduction potentials, respectively [120–122,36,4]. The above-stated correlations were confirmed for cyanine dyes with varying heterocyclic rings and polymethine chain length, and extended for aromatic hydrocarbons with known ionization energy in vacuum and in crystal in order to evaluate the absolute electronic energy levels of dye molecules. It was considered that each dye molecule was isolated and had its positive charge compensated. As a result, he could confirm Coulson's estimation that the ionization energies of isolated cyanine dye molecules available for dye sensitization in the visible wavelength region were around 7 eV in vacuum. This estimation was also confirmed by measuring the ionization energy of dyes in vacuum by UPS [224].

This author extended his evaluation by taking into account Point (1), which Coulson and Mott failed to consider but was necessary for evaluating the electronic energy levels of dye molecules on the surface of AgX grains in photographic films [122,36,4]. Although Lyon pointed out that the ionization energy of an organic molecule in vacuum (I_{G}) is much larger than that of its crystal (I_{C}) due to the polarization of the molecule by its surrounding molecules at the crystal lattice at the instance of its ionization [225], he did not make it clear whether the ionization of an isolated organic molecule in solid (e.g., an isolated molecule on the surface of AgX) should be much different from that of the molecule in vacuum. Indeed, Terenin and Alimov measured the ionization energy of isolated dye molecules in vacuum by UPS to be about 7 eV. They considered that the ionization energy of an isolated dye molecule on an AgX grain should be close, not to that of the molecule in its crystal, but to that of the molecule in vacuum, and applied the measured value to the ionization energy of isolated dye molecules on the surface of AgX [224].

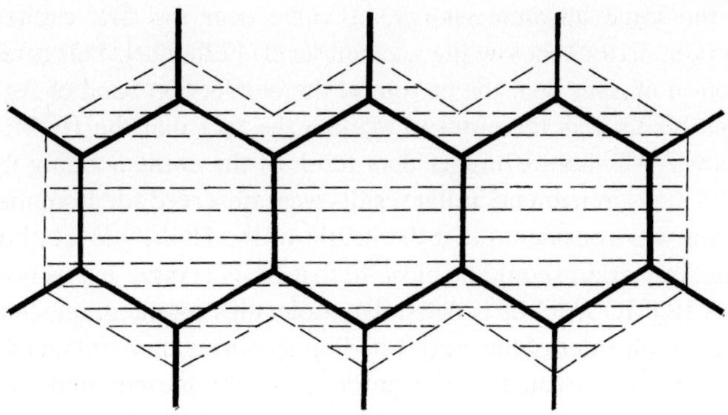

Figure 5.26 The projective area of the volume in an anthracene molecule, in which a charge is assumed to be distributed when it is ionized. The thickness of the area is assumed to be 3.0 Å, which was used to evaluate the volume V of the anthracene molecule where a charge is distributed when it is ionized. The equivalent radius r of the volume was given by $4\pi r^3/3$.

This author evaluated the polarization energy P as $(1\text{-}1/D)e^2/2r$, where r is the equivalent radius of the volume of the π-electron conjugated system in an organic molecule as exemplified in Figure 5.26 and D is the optical dielectric constant of its surroundings, and successfully applied it to explaining the difference in ionization energy between aromatic hydrocarbon molecules in vacuum and those in crystal. By using this method, the present author evaluated the ionization energies of isolated cyanine dye molecules on AgX grains covered with a gelatin layer [122,36,4]. As a result, he found that the ionization energies of cyanine dyes available for dye sensitization in the visible wavelength region were 5–5.5 eV, and that the electron transfer from these dyes in the excited state to silver halides was energetically possible.

It is noted that the electronic energy levels of cyanine dyes evaluated by molecular orbital methods and those by organic electrochemistry on the basis of quantum chemistry could be correlated well with the structures and photographic behaviors of dyes, distinguishing sensitizing dyes and desensitizing dyes clearly [4]. The oxidation potential and reduction potential as measured by means of the phase-selective second-harmonic voltammetry [126,127] became widely used for evaluating the HOMO and LUMO electronic energy levels of dyes of photographic interest [4].

Ultraviolet photoelectron spectroscopy is a direct method for measuring the HOMO electronic energy level of a dye molecule in vacuum and in crystalline form. There were two problems to be solved. As stated above, Terenin and Akimov measured the ionization energy of isolated dye molecules in vacuum as about 7 eV by means of UPS. On the other hand, UPS was applied to a evaporated layer of 1,1'-diamyl-4,4'-cyanine iodide in crystalline form by Fleischman to give its HOMO level at 5.5 eV below the vacuum level [226]. This result placed the excited level of the dye in the layer on a par with the bottom of the conduction band of AgBr. Similar results were reported for solid layers of sensitizing and other dyes [128,129,131,135,227]. Scheibe et al. observed that the first excited electronic levels of many organic molecules including dyes in solids are of approximately the same absolute energy, at value near the first excited level of the hydrogen atom (i.e., 3.19 eV below the vacuum level) [228,229]. This result also brought the excited electron of dyes near the bottom of the conduction band of AgBr.

One of the problems was therefore to explain the fact that the HOMO levels of the crystalline dyes was 1.5–2.0 eV higher than those of the corresponding dyes in vacuum owing to the UPS measurements. Similar results were observed for aromatic hydrocarbons [225]. Although it was considered that the height of the HOMO level of aggregated dye molecules on an AgX grain should be close to that of its crystal, it was not clear whether the height of the HOMO level of isolated dye molecules on the grain should be close to that of the isolated molecules of the corresponding dye in vacuum or that of crystalline dye molecules [92]. The above-stated idea as proposed by the present author on Point (1) for explaining the difference in ionization energy between organic molecules in solid and in vacuum could solve this problem and made it possible to evaluate the heights of both aggregated and isolated dye molecules on the grain [122,36].

Failing to take into account Point (1), Terenin and Akimov applied the ionization energies of sensitizing dyes in vacuum as measured by UPS to those of corresponding dyes on silver halides, denied the possibility of electron transfer from excited sensitizing dyes to silver halides, and supported the energy transfer mechanism [132]. Taking into account Point (3), Kuhn and co-workers observed and analyzed the dye sensitization of AgX as a function of the dye–AgX separation from van der Waals contact to greater distances in steps of 2.7 nm by use of a Langmuir–Blodgett membrane with fatty acids having long chain length [230–235]. After several discussions of the validity of their results [236–238], they provided evidence that under some conditions the energy transfer mechanism operated in dye sensitization of AgX.

Since it was proved that both the electron transfer and the energy transfer mechanisms were possible, the mechanism actually taking place in dye sensitization of silver halides was to be determined by examining whether it could quantitatively explain the phenomena related to dye sensitization of silver halides. It turned out that many phenomena could be quantitatively explained by the electron transfer mechanism [4,122,168], and that the rate of the energy transfer mechanism was considerably smaller than that of the electron transfer mechanism for those dyes in contact with AgX [235].

The other problem was that the electronic energy levels of dyes obtained by UPS measurements did not correlate with their molecular structures or photographic performance. Namely, UPS data could not distinguish between sensitizing and desensitizing dyes [92]. It seems that there were two points to be considered. One was that many dyes are composed of positively charged dye moieties and negatively charged counter ions, and that it was not necessarily clear whether photoelectrons came from dye moieties or from counter ions in the crystals of those dyes, especially in the case of desensitizing dyes, whose HOMO levels were much lower than those of sensitizing dyes. The other was related to Point (2), i.e., the shift of the vacuum levels of dyes and AgX when they are in contact with each other. In collaboration with Seki et al., the present author solved these problems by applying careful experiments to electrically neutral merocyanine dyes on AgBr layers, and found good correlation between their excited levels and dye-sensitizing abilities [150]. One of the reasons for the absence of correlation in the past was the neglect of the vacuum level shift. Actually, the vacuum level of all the dyes studied was determined to be considerably lower that that of AgBr. The inclusion of the vacuum level shift in the electronic structure of the interface between dyes and AgBr realized for the first time the correlation between the electronic structure of dyes as given by UPS measurement and their photographic behavior. This was also the first observation of Point (2), i.e., the vacuum level shift between substrates and organic layers, and was extended by Seki et al. to common systems composed of substrate/organic layer with a wide variety of both substrates and organic compounds [151,152]. Additional considerations on the vacuum level shift are described in Section 10.5.

As the result of the above-stated discussion, for many years the electron transfer mechanism was accepted for dye sensitization of silver halides. Although the energy transfer mechanism takes place much slower than the electron transfer mechanism for dye sensitization of silver halides, it is probable that it actually takes place in some cases. For example, it was reported that the photoelectric effect of an inorganic semiconductor was spectrally sensitized by the plasmons of metal nanoclusters [239,240], in which free electrons were collectively vibrated. Since a plasmon does not contain any optically excited electron, it is probable that the photoelectric effect of an inorganic semiconductor is spectrally sensitized by the plasmons of metal nanoclusters according to the energy transfer mechanism. Additional considerations on this subject are given in Section 8.4.2.

It was observed that the dye sensitization of the photoelectric effect of a semiconductor could be realized by a thick layer of a sensitizing dye, although its efficiency reached the maximum with less than a mono-molecular layer of the dye [241]. It was thus postulated that electronic charge carriers should be created in dye layers on their exposure to light. Although the dye sensitization of AgX grains in photographic films is brought about by mono-molecular layers of sensitizing dyes, it was reported that the dye sensitization of the grains could be realized by multi-molecular layers of sensitizing dyes [88,90]. Although the dye sensitization of the grains by the dye molecules in contact with the grains takes place through the electron transfer mechanism, the mechanism of the dye sensitization by the dye molecules in their layers, which are not in contact with the grains, was not clear. Gilman et al. proposed the formation of the dye multilayer structure on AgX grains, and the occurrence of the electron transfer from the excited dye molecules to the grains [88], although the dye multilayer structure could not be confirmed by careful measurement of the amount of adsorbed dye molecules [89]. Parton et al. proposed the excited energy transfer from dye molecules not in contact with the grains to those in contact with the grains [90]. Now, in the light of Point (4) [185,186], it is obvious that the excited energy in a dye molecule not in contact with AgX in a dye layer should be transferred to a dye molecule in contact with AgX for its contribution to dye sensitization.

Several groups of investigators including Meier [241–243] and Levy [244–246] proposed the semiconductor junction model for dye sensitization. According to this model, a macropotential difference across the junction between a p-type dye and n-type AgBr is developed by electron flow from AgBr to the dye in the dark because of an equalization of Fermi levels. On exposure of the dye to light, excited electrons readily move across the junction from the dye to AgBr. Since Fermi level actually exists in organic layers containing substantial amount of impurities or dopants in the light of Point (2) [247], it is not probable that a p–n junction forms between dye layers and AgX grains in photographic films due to Fermi level equalization, since mono-molecular layers of sensitizing dye molecules on the grains are regarded as single crystals [74,75],which should not contain impurities to bring about the above-stated Fermi level equalization, as described in Section 5.3.

It was observed that positive hole transfer [123,248–250] and electron transfer took place from dyes in the excited state to AgX. Dye-sensitizing phenomena with electron transfer and positive hole transfer were explained by the modified electron transfer mechanism for dye sensitization, according to which the HOMO and LUMO middle point of a dye was defined as its quasi-Fermi level, and used as the Fermi level of an excited layer of the dye, which should determine the net flow of electronic carriers across the interface between the dye layer and AgX [119,251–256].

5.9.2 Mechanisms of Supersensitization

Supersensitization is of great importance from the viewpoints of photographic science and engineering [3,4], and is defined in this chapter as phenomena in which ϕ_r is increased on adding a second substance. There are two major mechanisms for supersensitization in photography discussed in this chapter. One is the hole-trapping supersensitization, as described in Figure 5.18, and the other is the aggregate-partitioning supersensitization, as illustrated in Figure 5.20. The mechanisms of these two kinds of supersensitizations are described in Sections 5.6 and 5.7. In contrast, antisensitization is defined as a phenomenon in which ϕ_r is decreased on adding a second substance (i.e., an antisensitizer) [3,4].

The aggregate-partitioning supersensitization is based on the proposal of exciton-trapping by the dye molecules at the ends of J-aggregates, as proposed by West and Carroll [204,205] and described in Figure 5.20. Bird et al. proposed another explanation [257–260]. They considered that ϕ_r of larger J-aggregates of a sensitizing dye was smaller, since the larger aggregates were likely to incorporate more anti-sensitizers, and that aggregate-partitioning supersensitizers isolate such antisensitizers by dividing larger aggregates into smaller ones.

The hole-trapping supersensitization was used to improve the inefficiency of dye sensitization due to the fact that an excited electron in a sensitizing dye is situated below the bottom of the conduction band of an AgX grain, and is therefore likely to return to the dye's HOMO level without being transferred to the grain. Figure 5.18 illustrates the scheme of the hole-trapping supersensitization for dye 3. By observing that the HOMO electronic energy level of a supersensitizer should be about 0.2 eV higher than that of a sensitizing dye [187], Gilman et al. proposed that supersensitization took place due to the electron transfer from a supersensitizer to an excited sensitizing dye, preventing the return of an excited electron to the HOMO level of the dye and enhancing the transfer of an excited electron to the grain [187–189]. They considered that an excited electron in a sensitizing dye is still situated below the bottom of an AgX grain's conduction band after an electron was transferred from a supersensitizer to an excited sensitizing dye molecule. However, an electron below the bottom of an AgX grain's conduction band is likely to be attacked by oxygen and water in air, although it could be transferred to the grain and contribute to latent image formation in vacuum, as demonstrated by James et al. [261–263].

The present author introduced Point (4) into the mechanism of hole-trapping supersensitization to solve the above-stated problem. Namely, an excited dye molecule in a dye layer is regarded as an exciton. When an electron is transferred from a supersensitizer to an excited sensitizing dye molecule, the energy level of an excited electron in the dye molecule should be raised by the binding energy of cyanine dye excitons, which has been evaluated to be 0.17 eV [168,4]. Thus, the height of an excited electron in a sensitizing dye molecule should be raised by 0.17 eV and situated above the bottom of an AgX grain's conduction band due to the electron transfer from a supersensitizer to an excited sensitizing dye molecule. The difference in electronic energy level between the HOMO of a supersensitizer and that of a sensitizing dye, which should be kept for effective supersensitization, was more than about 0.2 eV, and attributed to the binding energy of an exciton in a sensitizing dye molecule.

The characteristics of hole-trapping supersensitization, i.e., the increase in efficiency of dye sensitization and elimination of the temperature dependence of the efficiency, can be confirmed by photoelectric effects as well as photographic sensitometry [183,264,4]. The transfer of a positive hole from excited dye molecules to supersensitizer molecules has been confirmed by ESR according to the following procedure [183]. Namely, ESR was applied to the supersensitizing system in Figure 5.18, where positive holes at supersensitizer molecules on octahedral AgBr grains gave rise to a light-induced ESR signal, whereas those at sensitizing dye molecules did not. It was observed that the action spectrum for the ESR signal coincided with the absorption spectrum of the dye on the grains.

5.9.3. Mechanisms of Desensitization

Desensitization is the decrease in intrinsic sensitivity caused by dyes and often provides severe problems for the development of AgX photographic films and papers. There are three major mechanisms for desensitization: (1) electron-trapping desensitization, (2) hole-trapping desensitization, and (3) desensitization due to dispersion of image centers caused by dyes, as described in Chapter 5 in this author's previous book [4] and other work [265,266]. The electron-trapping desensitization is described in Figures 5.16 and 5.17, and the hole-trapping desensitization in Section 5.8.

The binding energy of exitons in mono-molecular layers of cyanine dyes on silver halides was evaluated to be 0.17 eV in light of Point (4) [168,4]. Thus, even the dyes, whose excited electrons are situated below an AgX grain's conduction band, did not exhibit electron-trapping desensitization when the difference between them was less than 0.17 eV, and could be efficient sensitizing dyes with the aid of hole-trapping supersensitization.

References

1. H. Vogel, *Ber.*, 6, 1302 (1873).
2. T. Tani, S. Kikuchi, Report of the Institute of Industrial Science, University of Tokyo, 18, 51 (1968).
3. W. West, P. B. Gilman, in T. H. James (ed.), *The Theory of Photographic Sensitivity*, 4th edn (New York: Macmillan, 1977), Chapter 10.
4. T. Tani, *Photographic Sensitivity: Theory and Mechanisms* (New York/Oxford: Oxford University Press, 1995), Chapter 5.
5. R. W. Gurney, N. F. Mott, *Proc. R. Soc. London, Ser. A* 146, 151 (1938).

6. N. F. Mott, *Photogr. J.*, 88B, 119 (1948).
7. T. Tani, *J. Imaging Sci. Technol.*, 51, 110 (2007).
8. T. Tani, *Imaging Sci. J.*, 55, 65 (2007).
9. D. M. Sturmer, D. W. Heseltine, in James (ed.), *The Theory of the Photographic Process*, Chapter 8.
10. S. Daehne, *Science*, 199, 1163 (1978).
11. N. Tyutyulkov, J. Fabian, A. Mehlhorn, F. Dietz, A. Tadjer, *Polymethine Dyes* (Sofia: St. Kliment Ohridski University Press, 1991).
12. Y. Oishi, *J. Soc. Photogr. Sci. Technol. Jpn.*, 70, 334 (2007).
13. L. G. S. Brooker, R. H. Sprague, C. P. Smyth, G. L. Lewis, *J. Am. Chem. Soc.*, 63, 1116 (1940).
14. L. G. S. Brooker, G. H. Keyes, R. H. Sprague, R. H. Van Dyke, E. Van Lare, G. Van Zandt, F. L. White, H. W. J. Cressman, S. G. Dent, Jr., *J.Am. Chem. Soc.*, 73, 5332 (1951).
15. M. V. Deichmeister, I. I. Levkov, E. B. Lifshits, *Zh. Obshch. Khim.*, 23, 1529 (1953).
16. K. W. Hausser, R. Kuhn, G. Seitz, *Z. Physik. Chem. B*, 29, 391 (1935).
17. P. J. Wheatley, *J. Chem. Soc.*, 3245 (1959).
18. P. J. Wheatley, *J. Chem. Soc.*, 4096 (1959).
19. D. L. Smith, *Photogr. Sci. Eng.*, 18, 309 (1974).
20. K. Nakatsu, H. Yoshioka, S. Nishigaki, *Kwansei Gakuin Univ. Annual Studies*, 29, 213 (1980).
21. K. Nakatsu, H. Yoshioka, S. Nishigaki, *J. Soc. Photogr. Sci. Technol. Jpn.*, 46, 89 (1983).
22. A. P. Marchetti, C. D. Salzberg, E. I. P. Walker, *Photogr. Sci. Eng.*, 20, 107 (1976).
23. J. Delaney, M. Morrow, C. J. Eckhardt, *Chem. Phys. Lett.*, 122, 347 (1985).
24. J. Tanaka, M. Tanaka, M. Hayakawa, *Bull. Chem. Soc. Jpn.*, 53, 3109 (1980).
25. N. Sano, J. Tanaka, *Bull. Chem. Soc. Jpn.*, 59, 843 (1986).
26. J. Effinger, G. Germain, J. Meunier, J. Vanderauwera, M. Van Meerssche, *Bull. Soc. Chim. Belg.*, 69, 387 (1960).
27. B. Dammenier, W. Hoppe, *Acta Crystallogr.*, *Sect. B*, 27, 2364 (1971).
28. I. Kawada, R. Allmann, *Angew. Chem. Int. Ed. Engl.*, 7, 69 (1968).
29. K. Nakatsu, H. Yoshioka, T. Aoki, *Chem. Lett.*, 339 (1972).
30. H. Asanuma, K. Ogawa, H. Fukunaga, T. Tani, J. Tanaka, *Proceedings of International Congress on Imaging Science, September 6–12, 1998, Antwerp*, Vol.1, p. 178.
31. H. Kuhn, *J. Chem. Phys.*, 17, 1198 (1949).
32. H. Kuhn, W. Huber, G. Handsching, H. Martin, F. Schaefer, F. Baer, *J. Chem. Phys.*, 32, 467 (1960).
33. F. Baer, W. Huber, G. Handsching, H. Martin, H. Kuhn, *J. Chem. Phys.*, 32, 470 (1960).
34. S. Olszewski, *J. Chem. Phys.*, 26, 1205 (1957).
35. R. A. Berg, G. Moyano, R. A. Pierce, *J. Phys. Chem.*, 71, 3352 (1967).
36. T. Tani, S. Kikuchi, *Photogr. Sci. Eng.*, 11, 129 (1967).
37. D. Leupold, *Z. Phys. Chem. (Leipzig)*, 223, 405 (1963).
38. T. Tani, *J. Photogr. Sci.*, 19, 161 (1971).
39. T. Tani, *Photogr. Sci. Eng.*, 16, 258 (1972).
40. T. Tani, *J. Electrochem. Soc.*, 120, 254 (1973).
41. D. M. Sturmer, W. S. Gaugh, *Photogr. Sci. Eng.*, 17, 146 (1973).
42. J. K. Elwood, *J. Org. Chem.*, 38, 2430 (1973).
43. E. G. MacRae, *J. Phys. Chem.*, 61, 562 (1957).
44. W. West, A. L. Geddes, *J. Phys. Chem.*, 68, 837 (1964).
45. E. B. Lifshitz, *Dokl. Akad. Nayk. SSSR*, 179, 596 (1968).
46. T. Tani, S. Kikuchi, *Bull. Soc. Sci. Photogr. Jpn.*, 17, 1 (1967).
47. E. E. Jelly, *Nature*, 138, 1009 (1936).
48. G. Scheibe, *Angew. Chem.*, 49, 563 (1936).
49. G. Scheibe, A. Mareis, R. Schiffmann, *Z. Phys. Chem.*, *B*, 49, 324 (1941).
50. B. Zuckerman, H. Mingace, *J. Chem. Phys.*, 50, 3432 (1969).

51. S. de Boer, K. J. Vink, D. A. Wiersma, *Chem. Phys. Lett.*, 137, 99 (1989).
52. E. W. Knapp, *Chem. Phys.*, 85, 73 (1984).
53. L. G. S. Brooker, F. L. White, D. W. Heseltine, G. H. Keyes, S. G. Dent, Jr., E. J. Van Lare, *J. Photogr. Sci.*, 1, 173 (1953).
54. D. J. Fry, in *Dye Sensitization: Symposium Bressanone, 1967* (London: Focal Press, 1970), 44.
55. E. G. McRae, M. Kasha, *J. Chem. Phys.*, 28, 721 (1958).
56. E. G. McRae, M. Kasha, in L. Augenstein, R. Mason, B. Rosenberg (eds), *Physical Processes in Radiation Biology* (New York: Academic Press, 1964), 23.
57. M. Kasha, H. R. Rawls, M. A. Elbayoumi, *Pure Appl. Phys.*, 11, 371 (1965).
58. G. S. Levinson, W. T. Simpson, W. Curtis, *J. Am. Chem. Soc.*, 79, 4314 (1957).
59. K. Norland, A. Ames, T. Taylor, *Photogr. Sci. Eng.*, 14, 2951 (1970).
60. A. E. Rosenoff, V. K. Walworth, G. R. Bird, *Photographic Sci. Eng.*, 14, 328 (1970).
61. A. S. Davidov, *Theory of Molecular Excitons*, trans. M. Kasha, M. Oppenheimer, Jr. (New York: McGraw-Hill, 1962).
62. D. S. McClue, *Electronic Spectra of Molecules and Ions in Crystals* (New York: Academic Press, 1959).
63. J. M. Simson, *Photogr. Sci. Eng.*, 18, 302 (1974).
64. C. Reich, *Photogr. Sci. Eng.*, 18, 335 (1974).
65. S. L. Scrutton, *J. Photogr. Sci.*, 22, 69 (1974).
66. J. D. Wright, *Molecular Crystals* (Cambridge: Cambridge University Press, 1987).
67. V. Czikkery, H. D. Foersterling, H. Kuhn, *Chem. Phys. Lett.*, 6, 11 (1970).
68. H. Buecher, H. Kuhn, *Chem. Phys. Lett.*, 6, 183 (1970).
69. H. Asanuma, T. Tani, *J. Phys. Chem.*, 101, 2149 (1997).
70. G. E. Ficken, *J. Photogr. Sci.*, 21, 11 (1973).
71. D. F. O'Brien, *Photogr. Sci. Eng.*, 18, 16 (1974).
72. A. H. Herz, *Photogr. Sci. Eng.*, 18, 323 (1974).
73. L. Daehne, K. Kamiya, J. Tanaka, *Bull. Chem. Soc. Jpn.*, 65, 2328 (1992).
74. J. E. Maskasky, *Langmuir*, 7, 407 (1991).
75. J. E. Maskasky, *J. Imaging Sci.*, 35, 29 (1991).
76. J. C. Hansen, J. E. Maskasky, K. W. Williams, The preprint of *International Symposium on Silver Halide Technology, Montreal, Canada, September 10–14, 2000.*
77. Mitsuo Kawasaki and Hironori Ishii, *J. Imaging Sci. Tech.*, 39, 210 (1995).
78. G. Jannssens, F. Touhari, J. W. Gerritsen, H. van Kempen, P. Callant, G. Deroover, D. Vanderbroucke, *Chem. Phys. Lett.*, 344, 1 (2001).
79. He Jiang, Xin-Min Ren, Chung-Li Bai, *Chinese Sci. Bull.*, 38, 48 (1993).
80. W. West, B. H. Carroll, D. Whitcomb, *J. Phys. Chem.*, 56, 1054 (1952).
81. W. West, B. H. Carroll, D. Whitcomb, *J. Photogr. Sci.*, 1, 145 (1953).
82. A. H. Herz, in James (ed.), *The Theory of the Photographic Process*, Chapter 9.
83. A. H. Herz, R. Danner, G. Janusonis, in *Adsorption from Aqueous Solution* (Washington, DC: American Chemical Society, 1968), 173.
84. T. Tani, T. Suzumoto, *J. Appl. Phys.*, 70, 3626 (1991).
85. T. Tani, S. Kikuchi, *J. Photogr. Sci.*, 17, 33 (1969).
86. T. Hoffend, Jr., H. Price, Y. T. Tan, P. R. Baukema, A. J. Hopfinger, *J. Colloid Interface Sci.*, 130, 568 (1989).
87. Y. Suda, S. Matsuzaka, *J. Soc. Photogr. Sci. Technol. Jpn.*, 50, 281 (1987).
88. T. I. Penner, P. B. Gilman, *Photogr. Sci. Eng.*, 20, 97 (1976).
89. K. Yamashita, H. Takei, T. Ikeda, T. Tani, in *Proceedings Intern. Cong. Imaging Sci., May, Tokyo*, 93.
90. R. L. Parton, T. L. Penner, W. J. Harrison, J. C. Deaton, A. Muenter, in *AgX 2004: International Symposium on Silver Halide Technology* (Springfield, VA: IS&T, 2004), 161.

91. T. Suzumoto, K. Yamashita, S. Aiba, K. Kobayashi, T. Hioki, in the preprint of *Annual Conf. Soc. Photogr. Sci. Tech. Japan, May 2005*, p. 38.
92. W. West, B. H. Carroll, in C. E. K. Mees and T. H. James (eds), *The Theory of the Photographic Process*, 3rd edn (New York: Macmillan, 1966), Chapter 12.
93. S. E. Sheppard, H. Crouch, *J. Phys. Chem.*, 32, 751 (1928).
94. S. E. Sheppard, R. Lambert, R. Walker, *J. Chem. Phys.*, 7, 265 (1939).
95. R. Berendsen, H. Borginon, *J. Photogr. Sci.*, 16, 194 (1968).
96. T. Tani, *J. Imaging Sci.*, 29, 165 (1985).
97. S. Boyer, in James (ed.), *The Theory of the Photographic Process*, 244.
98. T. Tani, *Photogr. Sci. Eng.*, 16, 163 (1972).
99. S. Takada, T. Tani, *J. Appl. Phys.*, 45, 4767 (1974).
100. T. Tani, S. Takada, *Photogr. Sci. Eng.*, 18, 620 (1974).
101. L. E. Oppenheimer, T. H. James, A. Herz, in A. L. Smith (ed.), *Particle Growth in Suspensions* (London: Academic Press, 1973), 159.
102. A. Herz, J. Helling, *J. Colloid Interface Sci.*, 22, 391 (1966).
103. A. Herz, J. Helling, *J. Colloid Sci.*, 17, 293 (1962).
104. E. Gunther, E. Moisar, *J. Photogr. Sci.*, 13, 280 (1965).
105. R. Matejec, *Dye Sensitization, Bressanone, Symposium* (London: Focal Press, 1970), 26, 57.
106. S. E. Sheppard, R. Lambert, R. Keenan, *J. Phys. Chem.*, 36, 174 (1932).
107. J. Leermakers, B. H. Carroll, C. J. Staud, *J. Chem. Phys.*, 5, 878 (1937).
108. S. V. Natanson, N. I. Sennikova, *Trans. Conf. Theory and Technology of Photogr. Process, Zakopane, Poland, September 1968.*
109. T. Tani, S. Kikuchi, *Bull. Soc. Sci. Photogr. Jpn.*, 18, 1 (1968).
110. T. Arai, K. Seki, S. Narioka, Y. Tanaka, T. Yokoyama, T. Ohta, S. Watanabe, T. Tani, *J. Appl. Phys.*, 32 (Suppl.), 434 (1993).
111. K. Seki, T. Araki, E. Ito, K. Oich, S. Narioka, T. Yokoyama, T. Ohta, S. Watanabe, A. Matsunaga, T. Tani, *J. Imaging Sci. Technol.*, 37, 589 (1993).
112. S. E. Sheppard, R. H. Lambert, R. D. Walker, *J. Phys. Chem.*, 50, 210 (1946).
113. A. Stanienda, *Naturwissenschaften* 47, 353 (1960).
114. A. Stanienda, *Z. Phys. Chem., N.F.*, 32, 238 (1962).
115. A. Stanienda, *Z. Physik. Chem., N.F.*, 33, 170 (1962).
116. A. Stanienda, *Z. wiss. Photogr.*, 59, 76 (1965).
117. S. Daehne, *Z. wiss. Photogr.*, 59, 113 (1966).
118. M. Tamura, H. Hada, *Photogr. Sci. Eng.*, 11, 82 (1967).
119. T. Tani, S. Kikuchi, K. Honda, *Photogr. Sci. Eng.*, 12, 80 (1968).
120. T. Tani, S. Kikuchi, *Kogyo Kagaku Zasshi*, 69, 2049 (1966).
121. T. Tani, S. Kikuchi, *Kogyo Kagaku Zasshi*, 69, 2053 (1966).
122. T. Tani, S. Kikuchi, *Kogyo Kagaku Zasshi*, 69, 2237 (1966).
123. R. W. Berriman, P. B. Gilman, Jr., *Photogr. Sci. Eng.*, 17, 235 (1973).
124. T. H. James, *Photogr. Sci. Eng.*, 16, 120 (1972).
125. R. F. Large, in R. J. Cox (ed.), *Photographic Sensitivity* (London/New York: Academic Press, 1973), 241.
126. J. Lenhard, *J. Imaging Sci.*, 30, 27 (1986).
127. T. Tani, K. Ohzeki, K. Seki, *J. Electrochem. Soc.*, 138, 1411 (1991).
128. F. I. Vilesov, *Dokl. Akad. Nauk SSSR*, 132, 632 (1960).
129. F. I. Vilesov, A. N. Terenin, *Akad. Nauk. SSSR*, 133, 1060 (1960).
130. I. A. Akimov, V. A. Bentsa, F. I. Vilesov, A. N. Terenin, *Phys. Status Solidi*, 20, 771 (1967).
131. A. N. Terenin, I. A. Akimov, *Z. Phys. Chem.,(Leiptiz)*, 217, 307 (1961).
132. A. Terenin, I. Akimov, *J. Phys. Chem.*, 69, 730 (1965).
133. R. C. Nelson, *J. Phys. Chem.*, 71, 2517 (1967).

134. R. C. Nelson, *J. Opt. Soc. Am.*, 46, 1016 (1956).
135. R. C. Nelson, *J. Opt. Soc. Am.*, 51, 1186 (1961).
136. R. C. Nelson, *J. Opt. Soc. Am.*, 54, 897 (1965).
137. R. C. Nelson, *J. Mol. Spectroscopy*, 23, 213 (1967).
138. R. G. Selsby, R. C. Nelson, *J. Mol. Spectroscopy*, 33, 1 (1970).
139. R. C. Nelson, R. G. Selsby, *Photogr. Sci. Eng.*, 14, 342 (1970).
140. S. S. Choi, R. C. Nelson, *Photogr. Sci. Eng.*, 16, 341 (1972).
141. J. W. Trusty, R. C. Nelson, *Photogr. Sci. Eng.*, 16, 421 (1972).
142. R. C. Nelson, *J. Photogr. Sci.*, 20, 221 (1972).
143. P. Yanoulis, R. C. Nelson, *Photogr. Sci. Eng.*, 18, 94 (1974).
144. R. C. Nelson, P. Yanoulis, *J. Photogr. Sci.*, 22, 17 (1974).
145. R. C. Nelson, *J. Chem Phys.*, 47, 4451 (1967).
146. P. Nielsen, *Photogr. Sci. Eng.*, 18, 186 (1974).
147. T. B. Tang, K. Seki, H. Inokuchi, T. Tani, *J. Appl. Phys.*, 59, 5 (1986).
148. T. B. Tang, H. Yamamoto, K. Imaeda, H. Inokuchi, K. Seki, M. Okazaki, T. Tani, *J. Phys. Chem.*, 93, 3970 (1989).
149. K. Seki, T. Tani, *Oyo Butsuri*, 60, 1018 (1991).
150. K. Seki, H. Yanagi, Y. Kobayashi, T. Ohta, T. Tani, *Phys. Rev. B*, 49, 2760 (1994).
151. K. Seki, T. Tani, H. Ishii, *Thin Solid Films*, 273, 20 (1996).
152. H. Ishii, K. Sugiyama, E. Ito, K. Seki, *Adv. Mater.*, 11, 605 (1999).
153. J. R. Lenhard, B. R. Hein, *J. Phys. Chem.*, 10, 17287 (1996).
154. M. Kawasaki, T. Sato, *J. Phys. Chem. B*, 105, 796 (2001).
155. M. Kawasaki, D. Yoshidome, T. Sato, M. Iwasaki, *J. Electroanal. Chem.*, 543, 1 (2003).
156. J. Spence, B. H. Carroll, *J. Phys. Colloid Chem.*, 52, 1090 (1948).
157. T. Suzumoto, T. Tani, *J. Imaging Sci. Tech.*, 40, 56 (1996).
158. J. A. Leermakers, *J. Chem. Phys.*, 5, 889 (1937).
159. B. Zuckerman, *Photogr. Sci. Eng.*, 14, 336 (1970).
160. I. V. Rubtsov, K. Ebina, F. Satou, J. W. Oh, S. Kumazaki, T. Suzumoto, T. Tani, K. Yoshihara, *J. Phys. Chem.*, 106, 2795 (2002).
161. D. M. Sturmer, W. S. Gaugh, B. J. Bruschi, *Photogr. Sci. Eng.*, 18, 49 (1974).
162. D. M. Sturmer, W. S. Gaugh, B. J. Brushi, *Photogr. Sci. Eng.*, 18, 56 (1974).
163. I. H. Leubner, *Photogr. Sci. Eng.*, 20, 61 (1976).
164. I. H. Leubner, *Photogr. Sci. Eng.*, 22, 270 (1978).
165. I. H. Leubner, *Photogr. Sci. Eng.*, 24, 138 (1980).
166. R. K. Hailstone, *J. Photogr. Sci.*, 32, 25 (1984).
167. A. A. Muenter, P. B. Gilman, Jr., J. R. Lenhard, T. L. Penner, in *Fall Conf. Soc. Photogr. Sci. Technol. Japan, November 1985, Kyoto, Japan.*
168. T. Tani, T. Suzumoto, K. Ohzeki, *J. Phys. Chem.*, 94, 1298 (1990).
169. R. S. Eachus, A. P. Marchetti, A. A. Muenter, *Annu. Rev. Phys. Chem.*, 50, 117 (1999).
170. J. R. Fyson, P. J. Twist, I. R. Gould, in V. Balzani (ed.), *Electron Transfer in Chemistry*, 5th edn (Weinheim, Germany: Wiley-VCH , 2001), 285.
171. T. L. Penner, P. B. Gilman, Jr., *Photogr. Sci. Eng.*, 19, 102 (1975).
172. R. A. Marcus, *Ann. Rev. Phys. Chem.*, 15, 155 (1964).
173. V. G. Levich, in H. Eyring, D. Henderson, and W. Jost (eds), *Physical Chemistry*, vol. 9B (New York: Academic Press, 1970), 986.
174. J. R. Miller, L. T. Calcaterra, G. L. Closs, *J. Am. Chem. Soc.*, 106, 3047 (1984).
175. L. Eberson, *Adv. Phys. Org. Chem.*, 18, 79 (1982).
176. M. T Spitler, A. Ehert, R. Kietzmann, F. Willig, *J. Phys. Chem. B*, 101, 2552 (1997).
177. F. Willg, M. T. Spitler, *J. Imaging Sci. Tech.*, 41, 272 (1997).
178. K. Hashimoto, M. Hiramoto, A. B. P. Lever, T. Sakata, *J. Phys. Chem.*, 92, 1016 (1988).

179. K. Hashimoto, M. Hiramoto, T. Sakata, *Chem. Phys. Lett.*, 148, 215 (1988).
180. T. Tani, *Photogr. Sci. Eng.*, 28, 150 (1984).
181. T. Tani, *J. Imaging Sci.*, 33, 17 (1989).
182. H. Frieser, A. Graf, D. Eschrich, *Z. Elektrochem.*, 65, 870 (1961).
183. T. Tani, Y. Sano, M. Saito, *Photogr. Sci. Eng.*, 23, 240 (1979).
184. T. Tani, IEICE: Technical Report 109, ED2009–12, 49 (2009).
185. G. Hill, A. Kahn, Z. G. Soos, R. A. Pascal, Jr., *Chem. Phys. Lett.*, 327, 181 (2000).
186. V. I. Arkhipov, H. Boessler, in W. Bruetting (ed.), *Physics of Organic Semiconductors* (Weinheim, Germany: Wiley-VCH, Verlag, 2005), 183.
187. P. B. Gilman, Jr., *Photogr. Sci. Eng.*, 11, 222 (1967).
188. P. B. Gilman, *Photogr. Sci. Eng.*, 12, 230 (1968).
189. P. B. Gilman, Jr., *Photogr. Sci. Eng.*, 18, 418 (1974).
190. A. A. Muenter, *J. Phys. Chem.*, 80, 2178 (1976).
191. S. Daehne, F. Fink, E. Klose, K. Tauchner, S. Bach, J. V. Grossmann, *J. Signal AM*, 6, 105 (1978).
192. D. V. Brumbaugh, A. A. Muenter, W. Knox, G. Mourou, *J. Luminescence*, 31&32, 783 (1984).
193. K. Takahashi, K. Obi, I. Tanaka, T. Tani, *Chem. Phys. Lett.*, 154, 223 (1989).
194. K. Kemnitz, K. Yoshihara, T. Tani, *J. Phys. Chem.*, 94, 3099 (1990).
195. T. Tani, T. Suzumoto, K. Kemnitz, K. Yoshihara, *J. Phys. Chem.*, 00396, 2778 (1992).
196. A. A. Muenter, D. V. Brumbaugh, J. Apolito, L. A. Horn, F. C. Spano, S. Mukamel, *J. Phys. Chem.*, 96, 2783 (1992).
197. J. M. Lanzafame, A. A. Muenter, D. V. Brumbaugh, *Chem. Phys.*, 210, 79 (1996).
198. J. Moll, W. J. Harrison, D. V. Brumbaugh, A. A. Muenter, *J. Phys. Chem. A*, 104, 847 (2000).
199. J. W. Oh, S. Kumazaki, I. V. Rubtsov, T. Suzumoto, T. Tani, K. Yoshihara, *Chem. Phys. Lett.*, 352, 357 (2002).
200. B. Scharf, U. Dinur, *Chem Phys. Lett.*, 105, 78 (1984).
201. V. Sundstrom, T. Gillbro, R. A. Gadonas, A. J. Piskarkas, *Chem. Phys.*, 89, 2754 (1988).
202. J. Grad, G. Hernandez, S. Mukamel, *Phys. Rev. A*, 37, 3835 (1988).
203. F. C. Spano, S. Mukamel, *J. Chem. Phys.*, 91, 683 (1989).
204. W. West, B. H. Carroll, *J. Chem. Phys.*, 15, 529 (1947).
205. W. West, B. H. Carroll, *J. Chem. Phys.*, 19, 417 (1951).
206. T. Hannapel, B. Burfeidt, W. Storek, F. Willig, *J. Chem. Phys. B*, 101, 6799 (1997).
207. J. B. Asbury, R. J. Ellington, N. H. Chosh, S. Ferrere, A. Nozik, T. Lian, *J. Phys. Chem. B*, 103, 3110 (1999).
208. J. E. Moser, M. Graetzel, *Chimia*, 52, 160 (1998).
209. W. C. Needler, R. L. Griffith, W. West, *Nature*, 191, 902 (1961).
210. T. Tani, *Photogr. Sci. Eng.*, 19, 356 (1975).
211. T. Tani, Y. Sano, *J. Photogr. Sci.*, 27, 231 (1979).
212. T. Tani, in M. Hair, M. D. Croucher (eds). *Colloids and Surfaces in Reprographic Technology*, ACS Symposium Series 200 (Washington, DC: Am. Chem. Soc., 1982), 71.
213. T. Tani, *J. Appl. Phys.*, 62, 2456 (1987).
214. T. Tani, Y. Sano, *J. Appl. Phys.*, 69, 4391 (1991).
215. T. Tani, Y. Sano, in B. Levy (ed.), *The Proceedings of Symposium on Electronic and Ionic Properties of Silver Halides-Common Trends with Photocatalysis* (Springfield, VA: IS&T, 1991), 109.
216. J. R. Lenhard, A. A. Muenter, *Proc. Electrochem. Soc.*, 88, 97 (1088).
217. J. R. Lenhard, A. D. Cameron, *J. Phys. Chem.*, 97, 4916 (1993).
218. K. Hosoi, A. Hirano, T. Tani, *J. Appl. Phys.*, 90, 6197 (2001).
219. T. Tani, K. Seki, K. Yoshihara, J. Hanna, *Intern. J. Photoenergy*, 1 (2006).
220. Th. Foerster, *Ann, Phys.*, 6, 2, 55 (1948).
221. Th. Foerster, *Discuss. Faraday Soc.*, 27, 7 (1954).

222. Th. Foerster, in O. Sinanoglue (ed.), *Modern Quantum Chemistry*, Part III (New York: Academic Press, 1965), 93.
223. D. l. Dexter, *J. Chem. Phys.*, 21, 836 (1953).
224. A. Terenin, I. Akimov, *J. Phys. Chem.*, 69, 730 (1965).
225. L. E. Lyons, *J. Chem. Soc.*, 5001 (1957).
226. R. Fleischmann, *Ann. Physik.*, (5)5, 73 (1930).
227. R. C. Nelson, *J. Opt. Soc. Am.*, 50, 1029 (1960).
228. G. Scheibe, D. Brueck, F. Doerr, *Chem. Ber.*, 85, 867 (1952).
229. G. Scheibe, F. Doerr, in H. Sauvenier (ed.), *Scientific Photography (Proc. Intern. Colloq. Liege, 1959)* (New York: Pergamon Press, 1962), 512.
230. H. Buecher, H. Kuhn, B. Mann, D. Moebius, L. von Szentpaly, P. Tillman, *Photogr. Sci. Eng.*, 11, 233 (1967).
231. L. von Szentpaly, D. Moebius, H. Kuhn, *J. Chem. Phys.*, 52, 4618 (1970).
232. H. Kuhn, in W. F. Berg, U. Mazzucato, H. Meier, G. Semerano (eds), *Dye Sensitization (Proceedings Symposium, Bresanone, 1967)* (London/New York: Focal Press, 1970), 199.
233. H. Kuhn, D. Moebius, *Angew. Chem. Intern. Edn.*, 10, 620 (1971).
234. D. Mobius, *Photogr. Sci. Eng.*, 18, 413 (1974).
235. R. Steiger, H. Hediger, P. Junod, H. Kuhn, D. Mobius, *Photogr. Sci. Eng.*, 24, 185 (1980).
236. D. F. O'Brien, *Photogr. Sci. Eng.*, 17, 226 (1973).
237. K. H. Drexhage, *Photogr. Sci Eng.*, 18, 627 (1974).
238. V. I. Saunders, *Photogr. Sci. Eng.*, 19, 71 (1975).
239. K. Yu, N. Sakai, T. Tatsuma, *Electrochemistry*, 76, 161 (2008).
240. N. Sakai, Y. Fujiwara, Y. Takahashi, T. Tatsuma, *Chem. Phys. Chem.*, 10, 766 (2009).
241. H. Meier, *Spectral Sensitization* (London/New York: Focal Press, 1968).
242. H. Meier, *Photochem. Photobiol.*, 16, 219 (1972).
243. H. Meier, W. Albrecht, U. Tschirwitz, *Photogr. Sci. Eng.*, 18, 276 (1974).
244. B. Levy, M. Lindsey, *Photogr. Sci. Eng.*, 16, 389 (1972).
245. B. Levy, M. Lindsey, *Photogr. Sci. Eng.*, 17, 135 (1973).
246. B. Levy, M. Lindsey, *Photogr. Sci. Eng.*, 17, 423 (1973).
247. N. Hayashi, E. Ito, H. Ishii, Y. Ouchi, K. Seki, *Synthetic Metals*, 121, 1717 (2001).
248. H. E. Spencer, R. E. Atwell, *J. Opt. Soc. Am.*, 58, 1131 (1968).
249. T. Tani, *Photogr. Sci. Eng.*, 15, 21 (1971).
250. V. I. Saunders, *Photogr. Sci. Eng.*, 21, 163 (1977).
251. T. Tani, *Photogr. Sci. Eng.*, 26, 213 (1982).
252. T. Tani, *Photogr. Sci. Eng.*, 13,231 (1969).
253. T. Tani, *Photogr. Sci. Eng.*, 14, 72 (1970).
254. T. Tani, *Photogr. Sci. Eng.*, 14, 237 (1970).
255. T. Tani, *Photogr. Sci. Eng.*, 15, 21 (1971).
256. T. Tani, *Photogr. Sci. Eng.*, 18, 165 (1974).
257. B. Zuckerman, *Photogr. Sci. Eng.*, 14, 326 (1970).
258. A. E. Rosenoff, K. S. Norland, A. E. Ames, V. K. Walworth, G. R. Bird, *Photogr. Sci. Eng.*, 12, 184 (1968).
259. G. R. Bird, K. S. Norland, A. E. Rosenoff, H. B. Michaud, *Photogr. Sci. Eng.*, 12, 196 (1968).
260. G. R. Bird, *Photogr. Sci. Eng.*, 18, 429 (1974).
261. T. H. James, *J. Photogr. Sci.*, 20, 182 (1972).
262. T. H. James, *Photogr. Sci. Eng.*, 18, 100 (1974).

263. P. M. Fergson, T. A. Babcock, T. H. James, *Photogr. Sci. Eng.*, 19, 266 (1975).
264. S. S. Collier, *Photogr. Sci. Eng.*, 20, 249 (1976).
265. M. Ihama, T. Tani, *J. Imaging Sci.*, 31, 157 (1987).
266. T. Tani, M. Ihama, in F. Granzer, E. Moisar (eds), *Progress in Basic Principles of Imaging Systems, Proceedings of the Intern. Cong. Photogr. Sci., Cologne, 1986* (Braunschweig/Wiesbaden: Friedr. Vieweg & Soh, 1987), 335.

Part II

Application of Photographic Science to New Areas

6

Analog and Digital Photography

6.1 Introduction

Since the invention of silver halide (AgX) photography in 1839 by Daguerre [1], AgX photographic materials have made remarkable progress in achievements such as huge speed increases by as large as 400,000 times [2] and full color photography. In the preceding chapters, fundamental knowledge on which AgX photographic films and papers are based is described in detail. Since the pictures taken with digital still cameras have become of good enough quality for amateur consumers as those taken with color film, many people have come to use convenient digital still cameras. However, there are substantial differences in image quality and performance between the two based on differences in technology, and which explain why people will continue to use color film in addition to digital still cameras in the future. In this chapter, AgX photographic films are characterized by the knowledge discussed in the preceding chapters. Taking into account the fact that photographic films represent analog photography and have a long history with fruitful experiences, the difference between analog photography and digital photography, as represented by digital still cameras, is analyzed to serve as references to the understanding and development of digital photography, which has been recently developed and has a long future.

6.2 Technologies in Color Films

6.2.1 *Structure, Constituents, and Production*

A color film is characterized by the structure of its photosensitive layers. An example of color film, a section of which is shown in Figure 1.1, is composed of 14 different functional layers with thickness of about 20 μm in total and coated on a triacetylcellulose (TAC) film base about 100 μm thick. In addition to photosensitive layers, TAC film bases had been refined for many years [3,4]. White lines and spots are the sections of AgX grains. As seen in Figure 1.1, a color film is a three-dimensional sensor with many functional layers in pile. There are three major layers, each of which is sensitive to one of the three primary colors, i.e., blue, green, and red. In each major layer, there are two or three sub-layers with high, medium, and low sensitivities. A color film is thus designed to capture an image in full color and large dynamic range by taking advantage of its three-dimensional structure.

However, it was theoretically predicted that the red-sensitive layer should have considerable negative sensitivity in the blue–green region in order to achieve faithful color reproduction by a color film with layers sensitive to three primary colors only. In order to achieve this prediction, a fourth layer was introduced into color films, as seen in

Figure 1.1, to retard the development in the red-sensitive layer upon exposure to blue–green light [5,6].

As described in Chapter 1, a light image is captured by AgX grains covered with sensitizing dyes, recorded by forming on the grains latent image centers composed of Ag clusters, and displayed by dyes formed during color development, as exemplified in Figure 1.5. During the development, only the grains with latent image centers on the surfaces are reduced to form the dyes. It is generally accepted that the smallest latent image center is composed of four silver atoms.

Since AgX grains used for color films are now very thin, as shown in Figure 1.2, they can absorb blue light only weakly. In blue-, green-, and red-sensitive layers, AgX grains are covered with mono-molecular layers of J-aggregated sensitizing dye molecules, which absorb blue, green, and red lights, respectively, and then inject electrons into the conduction band of AgX grains, as illustrated in Figure 1.4 and described in Chapter 5.

Color films contain more than 100 functional organic compounds and tremendous number of dyed AgX grains [4], which are in concert with each other in the above-stated structure, and are produced in large scale and with high precision. As described in Section 2.2.2, fine AgX grains are produced with specially designed apparatus under precisely controlled conditions [7]. As described in Sections 6.3.1 and 10.8, the above-stated many functional layers are coated continuously for a very long time on an above-stated TAC film at large scale and high speed.

Figure 6.1 shows the scanning electron micrograph of a section of color negative film. As shown, photosensitive layers, as stated above, are coated on a TAC, which is 100 μm thick. As stated above, the structure and performance of photosensitive layers are highly optimized and should be sensitive to the structure and properties of the film base on which

Figure 6.1 A scanning electron micrograph of a section of color negative film composed of photosensitive layers (~20 μm thick in total) with AgX grains as white lines and spots suspended in gelatin layers on a TAC film base (~100 μm thick). Reproduced with permission from Reference [10], copyright 2007 IS&T, The Society for Imaging Science and Technology.

they are coated. The TAC films, which have been developed for color films, are optically and physically very uniform, and have turned out to be highly useful for liquid crystal displays [3,4].

6.2.2 *Factor Analysis: Light Absorption*

As illustrated in Figure 1.3, it is obvious that the factors influencing photographic sensitivity are the light absorption of dyed AgX grains in photographic emulsion layers, the efficiency of latent image formation, and the size of the smallest latent image center [8–10]. The efficiency of latent image formation depends on the quantum yield of the formation of free electrons, the formation of appropriate sites for latent image formation, and the prevention of recombination between photoelectrons and positive holes. The present state of color films with respect to the ultimate limit, the room for improvement, and the technologies expected to contribute to the improvement are analyzed from the viewpoints of the above-stated factors, and summarized in Table 6.1.

As described in Section 4.3.2, the present state of light absorbance is about one-half, leaving room for the light absorption to improve by two times. The present state of latent image formation efficiency is about one-third, leaving room for the efficiency to improve by three times. It is judged that the smallest latent image center in current color films is composed of three atoms. This condition will not be improved upon for the time being. In total, it is expected that the sensitivity of color films has the potential to be increased several times without deteriorating image quality [8–10].

According to Marchant, replacing conventional AgX grains with thin tabular ones developed at that time [11,12] can increase the light absorbance of a color film up to about one-half [13]. At the present, thin tabular AgX grains covered with mono-molecular layer of J-aggregated sensitizing dyes are used in color films. It is expected that the light absorbance of a color film will be increased by two times through the improvement of tabular AgX grains and the achievement of multi-molecular layer of sensitizing dyes on each grain. Electron micrographs in Figure 6.2 show recent progress in tabular AgX grains in color films, as reflected in the decrease in their thickness with time.

Table 6.1 Factors Influencing Photographic Sensitivity and Technologies for Improving Them

Factors	Present state	Theoretical limit	Room for improvement
Light absorption	$\sim$1/2	1	Two times
	Thin tabular AgX grains Antenna dye sensitization		
Efficiency of latent image (LI) formation	$\sim$1/3	1	Three times
(1) Yield of free electrons	Two-electron sensitization		
(2) Site for LI formation	Chalcogen sensitization (S $\rightarrow$ Se, Te)		
(3) Prevention of recombination	Dislocations to locate electrons Two-electron sensitization		
The size of the smallest LI center	3 atoms	3 atoms	No room

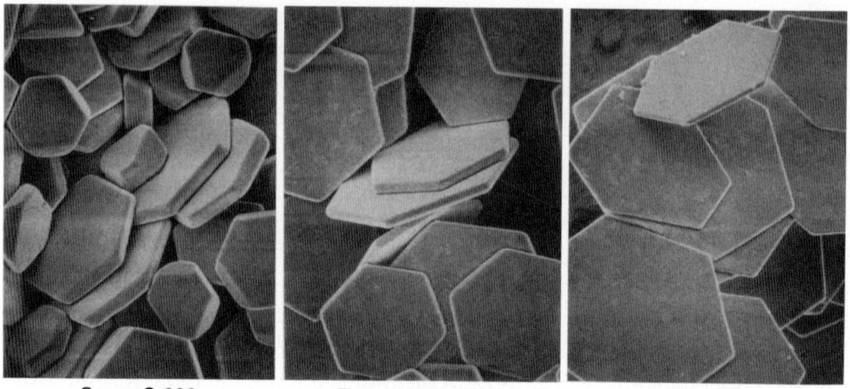

Figure 6.2 Scanning electron micrographs of tabular AgX grains developed in 1993, 2000, and 2003 for color negative films. Reproduced with permission from Reference [10], copyright 2007 IS&T, The Society for Imaging Science and Technology.

In order to use sensitizing dyes in amounts larger that that for their mono-molecular layer on the grain surface, several proposals have been made, including dye multilayer by Gilman et al. [14], junction of dye and AgX crystals by House [15], and antenna dye system with energy transfer among dye molecules by Bird et al. [16]. Although the formation of the dye multilayer could not be confirmed as it was through careful measurement of the amount of dye adsorbed on AgX grains [17], an antenna dye sensitization with bi-molecular layers of sensitizing dyes on the grain surface was realized through the development of the molecular design of sensitizing dyes specified for enhancing the attractive interaction between the first and the second layers [18,19], and for use with color films [18]. Such molecular designs include the introduction of hydrophobic effect and the enhancement of attractive Coulombic interaction between the first and the second layers.

6.2.3 Factor Analysis: Efficiency of Latent Image Formation

As described in Section 4.3, the efficiency of latent image formation is expressed by the quantum sensitivity, which is defined as the number of absorbed photons per grain required for the formation of latent image centers on half of the grains present. The ultimate limit has been evaluated as 1.7 absorbed photons per grain [20]. As described in Section 4.4.1, the smallest latent image center formed in color films is composed of three atoms, while one absorbed photon creates at most two electrons. Based on this data, Hailstone evaluated the ultimate limit of quantum sensitivity as 1.7 by taking into account that the number of absorbed photons per grain fluctuates according to Poisson distribution. Although the observed quantum sensitivity could be less than three for fine AgBr grains, the highest quantum sensitivity of AgX grains in a current color film has been reported as five [21]. Based on these results, the efficiency of latent image formation in current color films was determined to be one-third, leaving room for the sensitivity of color films to increase by three times.

The efficiency was thus predicted to increase by three times as a result of enhancing the yield of free electrons, optimizing the sites for latent image formation, and preventing the recombination of electrons and positive holes [10]. The quantum yield of free electrons would be improved through refinement of the two-electron sensitization technology, according to which one absorbed photon creates two free electrons, as has already been achieved in part by reduction sensitization [22–24]. This led to the proposals of new types of two-electron sensitization with formate [25] and fragmentable electron donors [26,27] as sensitizers. In the latter case, an excited dye injects an electron into the conduction band of AgX and leaves a dye positive hole. The oxidation of a fragmentable electron donor by a dye positive hole gives a radical with an activated electron whose energy level is higher than the bottom of the AgX conduction band.

Among the sites for latent image formation, a sulfur sensitization center is the most popular and has been used since it was discovered by Sheppard in 1925 [28]. As described in Section 4.5.2, it is believed that a sulfur sensitization center is composed of a dimer of two substitutional sulfur ions on an AgX grain, which are associated with two interstitial silver ions for the compensation of excess negative charges on the sulfur ions [29–31]. While each interstitial silver ion is associated with a hydrogen-like orbital, to which an electron is loosely bound, the two hydrogen-like orbitals interact with each other to give bonding and anti-bonding orbitals. An electron is captured at the bonding orbital.

As shown in Figure 4.4, Hailstone theoretically estimated quantum sensitivity in terms of the number of photons absorbed per grain for latent image formation as functions of the depth and number of sites for latent image formation, showing that the sensitivity has a tendency to increase with decreasing trap depth [32].

According to the above-stated structure of a sulfur sensitization center, it is expected that its trap depth can be decreased by increasing the sizes of the ions in the center, since this would decrease the degree of interaction between the two hydrogen-like orbitals associated with the two interstitial silver ions in the center. One way to achieve this is to replace one of the interstitial silver ions with a gold ion to give a sulfur-plus-gold sensitization center [31], as described in Section 4.5.3. In accord with this prediction, it was already confirmed that the trap depth of a sulfur-plus-gold sensitization center is shallower than that of a sulfur sensitization center [33,34]. Sulfur sensitization and sulfur-plus-gold sensitization have been put to practical use for many years.

Another way to decrease the trap depth of a sulfur sensitization center is to replace sulfur ions in a sulfur sensitization center with selenium and tellurium ions to form selenium and tellurium sensitization centers, respectively. Morimura et al. comparatively characterized all the possible chalcogen sensitizations and found that the trap depths were 0.28 eV (sulfur sensitization center), 0.17 eV (selenium one), and 0.12 eV (tellurium one) [35]. Increasing the amount of sulfur-sensitizer brought about a significant decrease in sensitivity upon exposure to high-intensity light. This phenomenon, which is known as high-intensity reciprocity law failure (HIRF), results from the difficulty of growing image centers, bringing about the dispersion of latent image formation. Selenium sensitization is superior to sulfur sensitization in that the degree of HIRF is smaller in selenium-sensitized emulsions than in sulfur-sensitized ones. Tellurium sensitization is superior to sulfur and selenium sensitizations in that HIRF was nearly absent. On the other hand, the ability to enhance the nucleation of latent image formation and retard the low-intensity reciprocity law failure was in the order of sulfur sensitization, selenium sensitization, and tellurium sensitization. While tellurium sensitization was realized [35], all the above-stated chalcogen sensitizations have been studied for practical use in order to meet the necessary condition for new products.

One way to hinder the recombination of electrons and positive holes is obviously the two-electron sensitization technology. Another way is to introduce dislocations along the edges of a thin tabular AgX grain covered with sensitizing dye molecules as shown in Figure 2.10. Electrons are thus localized by dislocations along the edges, and spatially separated from positive holes, which are trapped by sensitizing dyes on the main surfaces of the grain [36,37,10].

6.3 Comparison between Color Films and Digital Still Cameras

6.3.1 Device Structure and Alignment of Photosensitive Elements

A digital still camera is equipped with a charge-coupled device (CCD) [38] or complementary metal-oxide-semiconductor (CMOS), in which pixels are placed side by side to form a two-dimensional image-capturing device. Figure 6.3 illustrates the cross-section of a pixel in a current CCD. Micro lenses in each pixel focus incident photons on a diode, where absorbed photons generate electrons in proportion to the number of absorbed photons. Electrons generated in a diode are accumulated in a potential well in each pixel, and transferred to give a signal composed of the number of electrons in each pixel in series. As illustrated in Figure 6.4, a CCD thus operates with the generation, accumulation, transfer, and detection of carriers. In this section, comparison in performances between a color film and a CCD are made from various viewpoints to make clear the differences in technology between them and to serve as references for imaging science and technology.

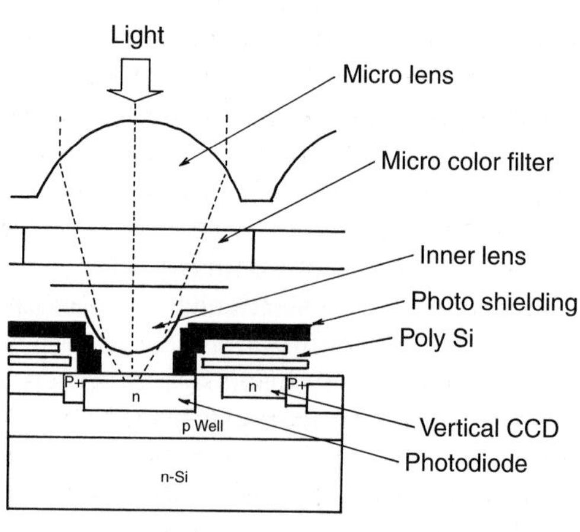

Figure 6.3 The cross-section of a pixel of a CCD.

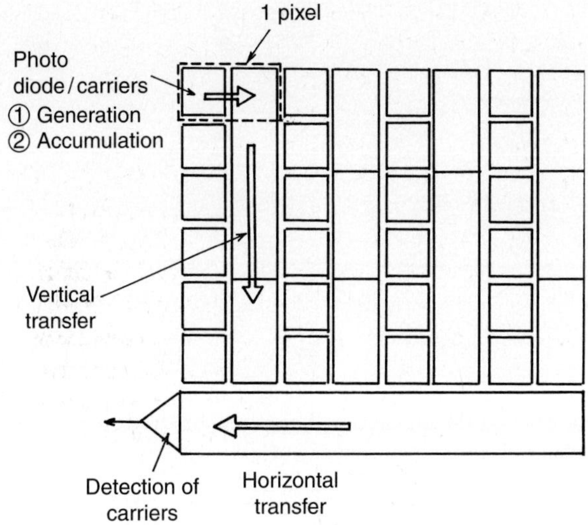

Figure 6.4 Structure and operation of an interline transfer CCD with generation, accumulation, transfer, and detection of carriers.

One of the most important differences between a color film and a CCD is that the structure of the former is three-dimensional, while that of the latter is two-dimensional. In order to use a CCD for color photography, it is necessary to put in each pixel a filter that transmits one of the three primary colors and absorbs the other two. Namely, it has been noted that two-thirds of incident photons are absorbed by a filter in each pixel and are not used for image formation. It has also been pointed out that the three-dimensional structure of a color film contributes to the increase in dynamic range and the number of pixels in a color film. An idea to introduce onto a CCD or CMOS a three-dimensional sensor composed of three organic semiconductor layers that absorb the light of one of the three primary colors and transmit the other two was proposed [39–43].

One of other important differences between a color film and a CCD is in that photosensitive elements in a color film (i.e., AgX grains) are randomly arrayed, while those in a CCD (i.e., pixels) are regularly arrayed. In general, it is easier to prepare wide and uniform layers with randomly arrayed materials than with regularly arrayed materials. Despite the fact that a color negative film as shown in Figure 1.1 is composed of 14 layers with different functions, they are uniformly coated at once with very high speed on a very wide TAC film base by means of a multilayer coating method, as described in Section 10.8 [44,45]. In contrast, CCD and CMOS are produced through more than one hundred sophisticated batch processes in small scale.

Many efforts have been made to regularly array AgX grains in an AgX photographic film for practical use. Figure 6.5 shows the electron micrographs of regularly arrayed mono-grain layers of cubic and octahedral AgBr grains prepared by Heki [46]. It seems that some of the reasons for the unsuccessful result can be found in the performance of a CCD as described below.

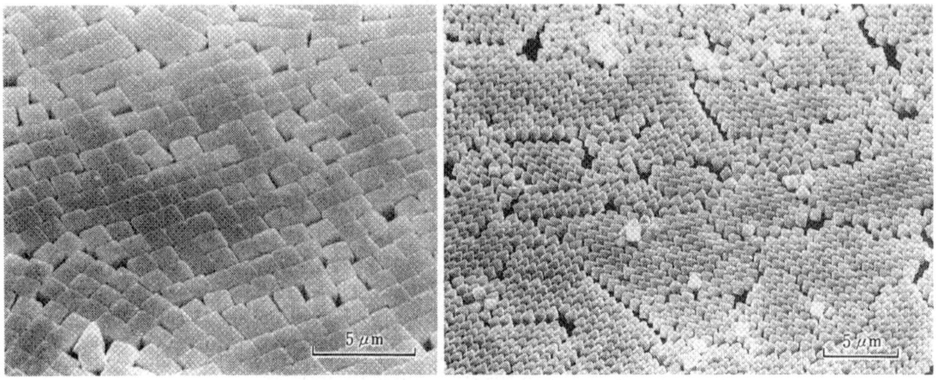

Figure 6.5 Scanning electron micrographs of Langmuir–Blodgett membranes of cubic (left) and octahedral AgBr grains (right) [46] (courtesy of T. Heki).

The frame area is also an important factor that characterizes a color film and a CCD. Although a device with small frame area is convenient, the number of pixels in a device increases in proportion to its frame area [47]. The frame area of a color film for a compact camera (i.e., 135 format) is about 30 times larger than that of a CCD for a compact camera (i.e., 1/2 inch). In contrast to color films, it is difficult to develop CCD with a large frame area because Si used for CCD and CMOS is expensive, and the pixels in a CCD are regularly arrayed and connected with each other in series. Namely, the function of a CCD is very sensitive to dust and defects in it, since it is probable that one particle of dust or defect can deteriorate the whole function of a CCD [48,49]. Actually, the productive yield of a CCD decreases with increasing frame area. In addition, a deteriorated pixel among regularly arrayed ones is noticeable. This problem was much reduced by the development of the technology called interpolation, by which the response at a deteriorated pixel is interpolated by those of its adjacent pixels [50–55]. It became possible to produce CCDs with advanced photo system (APS) and 135 formats available for single lens reflex cameras and other purposes. This is considered to be one of the most important reasons why digital still cameras have made more extensive progress than the present author had predicted previously [8–10,56,57].

Since the signal of a corresponding pixel in a color film is given by the accumulation of the signals from several hundred AgX grains, which works digitally and independently, dust and defects in small concentration hardly deteriorate the function of a color film regardless of the frame size. Since AgX grains are randomly arrayed as seen in Figure 1.1, a dust or a deteriorated grain is not noticeable in the photosensitive layers in a color film. Since a picture taken with photographic film with a very large frame area can be of very high quality and precision, photographic film is used for precise reproductions of famous paintings and old important documents as exemplified by the Polaroid Museumreplica Collection. For many years, Benrido in Kyoto has produced many precise reproductions of Japanese traditional paintings and documents by taking full-size pictures with large films and a big camera of height, width, and depth of 2.5, 1.7, and 3.0 m, respectively, as shown in Figure 6.6, and applied them to color type printing[58].

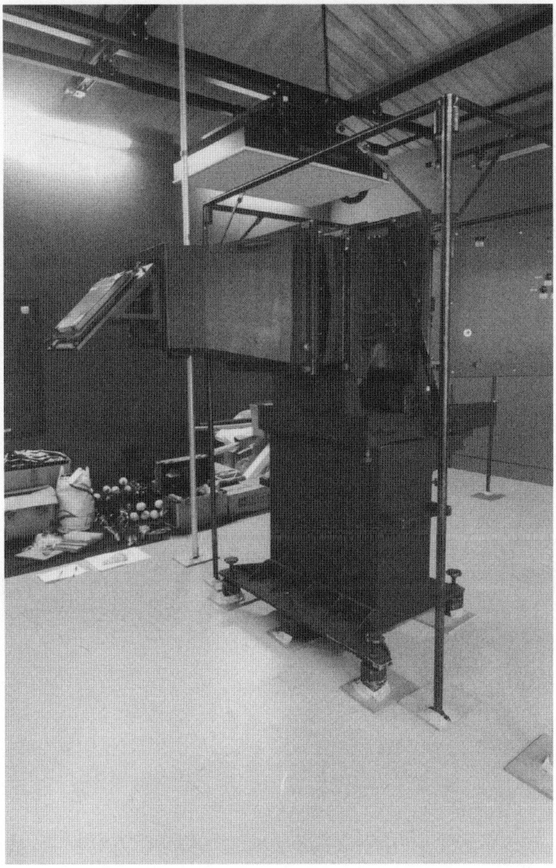

Figure 6.6 A large camera with height, width, and depth of 2.5, 1.7, and 3.0 m, respectively, for precise reproduction of valuable paintings and documents [58] (courtesy of S. Nishimura of Benrido).

6.3.2 Responses, Image Processing, and Image Performance

As described above and in Figure 6.3, a photosensitive element in a CCD, i.e., a pixel, acts as an analog sensor, creating photoelectrons in proportion to the number of absorbed photons. On the other hand, a photosensitive element of a color film, i.e., a tabular AgX grain, acts as a digital sensor, being developed only when the number of photons absorbed by it is larger than its quantum sensitivity (i.e., the number of absorbed photons per AgX grain for rendering half of the grains developable). Figure 6.7 shows the characteristic curves of a CCD and a photographic film with AgX grains. A light image captured by a CCD as an analog signal with a characteristic curve, as illustrated in Figure 6.7 (top), is converted into a digital signal by image processing. Although an AgX grain acts as a digital sensor with a characteristic curve, as illustrated by the broken line in Figure 6.7 (bottom), a photographic film with many AgX grainss acts as an analog sensor with a

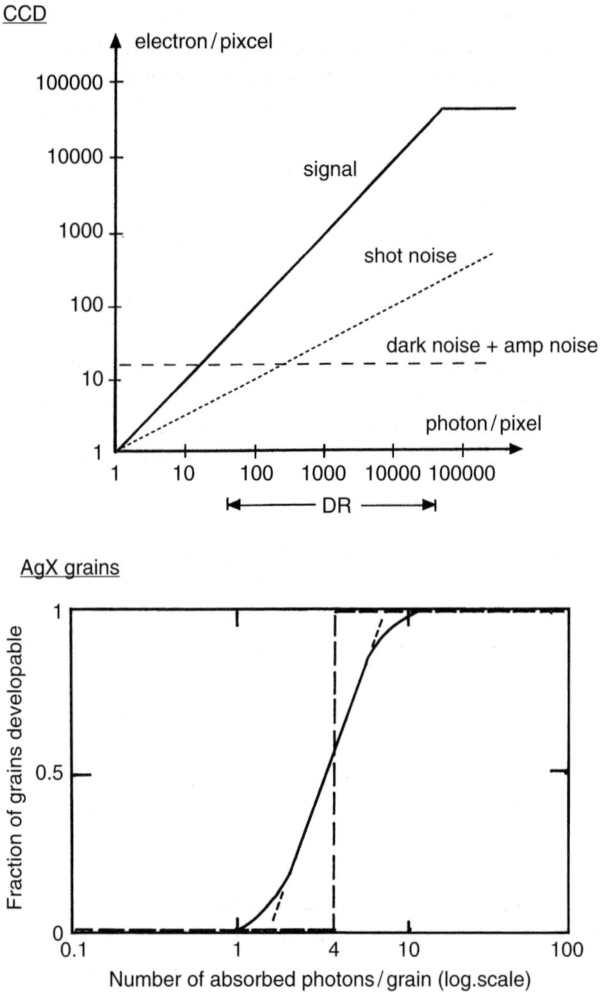

Figure 6.7 Typical characteristic curves of a CCD with the logarithm of the number of electrons per pixel as a function of the logarithm of the number of absorbed photons (top) and of a photographic film with the fraction of developed grains as a function of the logarithm of the number of photons absorbed per grain (bottom).

characteristic curve, as illustrated by the solid line in Figure 6.7 (bottom), since the number of photons absorbed in each grain fluctuates according to the Poisson distribution.

The output signal versus input signal should be linear in an emissive display such as a cathode ray tube. On the other hand, the output signal versus log(input signal) should be linear in a non-emissive display such as a color print, since the intensity of light in a non-emissive display obeys the Lambert–Beer law, according to which the optical density of a layer used for the display (i.e., the fraction of developed grains in the case of color films) is proportional to the logarithm of the intensity of the light passing through the layer. The characteristic curve of a CCD, which is linear as illustrated in Figure 6.7 (top), is suitable, as it is, for an emissive display, and should be treated for a non-emissive display. The characteristic curve of a color film, in which output signal versus log(input signal) is linear for the most part, is suitable, as it is, for a non-emissive display, and should be treated for an emissive display.

The latitude is the exposure range, within which incident photons are detected, and is one of the most important characteristics for a color film and a CCD. In a CCD, the minimum detectable exposure is given by its dark noise, while the maximum detectable exposure is given by the number of photoelectrons that fill a potential well in each pixel. The latitude of a CCD can be expanded by decreasing the minimum detectable exposure and/or by increasing the maximum detectable exposure. Even when the latitude of a photographic layer is limited, the latitude of a color negative film can be expanded by multilayering photographic emulsions with varied sensitivity, as shown in Figure 1.1. As illustrated in Figure 6.7 (top), the performance of a CCD is associated with shot noise, which results from the number of absorbed photons fluctuating in each pixel. In a photographic film, the fluctuating number of absorbed photons by each AgX grain is used to provide a photographic film with expanded latitude. The graininess originating from AgX grains should be the main source of noise for a color film, and increases with increase in sensitivity (i.e., increase in the size of the grains). When the image quality is judged to be the same, a color film is superior to a digital still camera in the number of pixels per frame by four times, while a digital still camera is superior to a color film in granularity [59].

As illustrated in Figure 6.8, the resolution of a light sensor is usually evaluated by its modulation transfer function (MTF), which is given by the ratio of the amplitude of the output signal to that of an input light signal with a sine curve as a function of its frequency [60,61]. As shown in Figure 6.9, the MTF of a color film gradually decreases with frequency, maintaining a positive value until the considerably high frequency region. While the MTF of a CCD also decreases with frequency, it becomes zero at the Nyquist frequency, which is given by $1/2p$, where the pixel pitch is p, and even becomes negative with further increases in frequency. This phenomenon is known as "aliasing" in a one-dimensional image, as exemplified in Figure 6.10, and as "moiré" in a two-dimensional

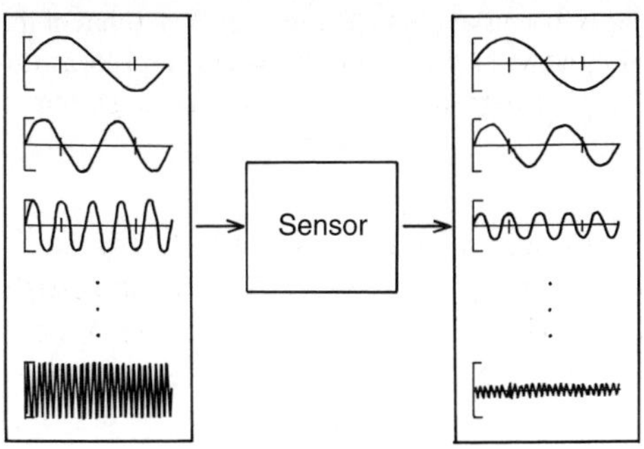

Figure 6.8 The procedure for measuring the modulation transfer function (MTF) of a sensor by giving the input signals of sine waves with a fixed amplitude and varied frequency to it and obtaining the output signals to evaluate their amplitude of the waves as a function of the frequency.

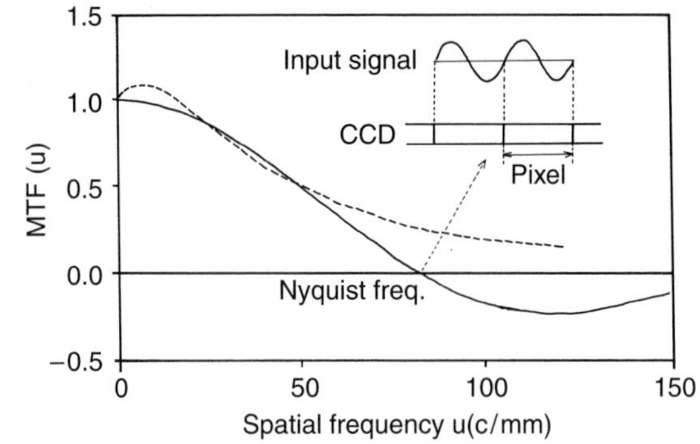

Figure 6.9 Typical MTF values of a color film (dotted line) and a CCD ((solid line) as a function of frequency. The inset indicates the relation between the Nyquist frequency and the period of pixels in a CCD.

one. An image taken by a CCD is thus processed by a low-pass filter in a digital still camera in order to eliminate its response in the frequency region, which is near and larger than the Nyquist frequency. Although this processing results in a decrease in the resolving power of a digital still camera, the actual effect of this treatment on the image quality of pictures taken by a digitial still camera has been recently diminishing due to the decrease in size of pixel in CCDs and CMOS.

As stated above, image processing is a powerful tool and has been successfully applied to digital photography with CCDs and CMOS as photo-sensors. Image processing was chemically applied to color films with success, and can serve as reference to digital photography with CCD and CMOS. Some of them are described below.

One such technology is known as colored coupler. In color negative films, the dye-forming couplers in the green- and red-sensitive layers are colored yellow and magenta

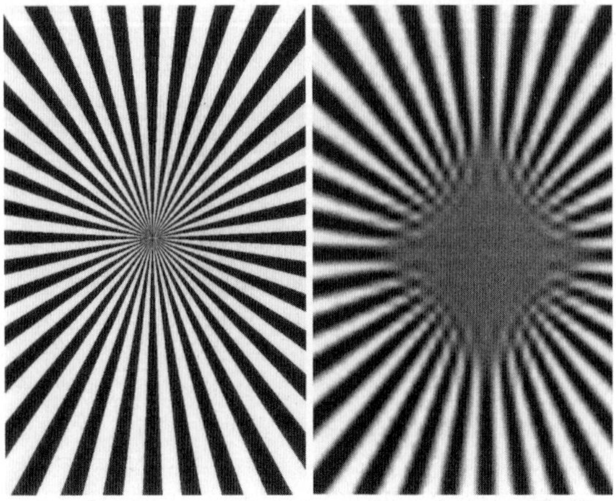

Figure 6.10 Typical input (left) and output (right) signal of a CCD showing the aliasing taking place in it.

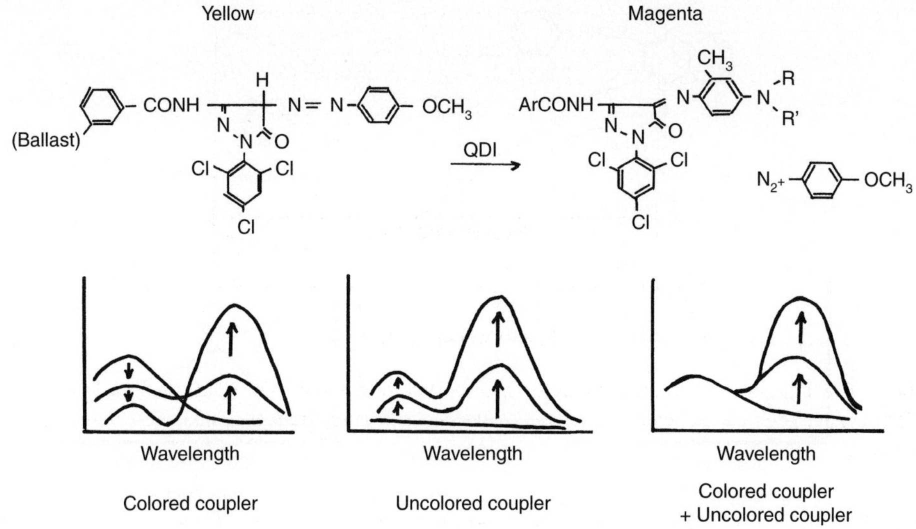

Figure 6.11 The effect of a colored coupler with changes in absorption spectra of a colored coupler and an uncolored one owing to the progress of coupling reactions as indicated by arrows.

due to the attachment of yellow- and magenta-imparting groups to the coupling positions of the respective couplers. As illustrated in Figure 6.11, these groups split off during color development and are replaced by the oxidized developing agent. Thus, colored couplers that remain in the green- and red-sensitive layers form positive color mask images that neutralize the unwanted secondary absorptions of the magenta and cyan image dyes, respectively [62–64].

An inter-image effect, which takes place during color development and is used for controlling image reproduction, is enhanced by development-inhibitor-releasing (DIR) couplers [65–67]. As illustrated in Figure 6.12, fragments that inhibit silver development such as phenylmercaptotetrazoles and benzotriazoles are released from the coupling position of DIR couplers as a result of its color coupling reaction. The magnitude and range of action of the inhibitor can be controlled by its inhibiting strength and diffusivity, as adjusted by the molecular design of the couplers. Thus, a DIR coupler in a layer suppresses development, not only within that layer, but in adjacent layers depending on the diffusivity of its inhibitor. The inhibitor brings about the reduction in size of developed silver and therefore in the size of dye clouds formed by the color coupling reaction. In addition, the inhibitor promotes the fringe effect, which enhances image sharpness and reduces image contrast in all affected layers. The inter-image effect caused by DIR couplers enhances the difference in color contrast produced by white light.

One of the most important achievements in color films and papers is image permanence over a long period. It was predicted that pictures taken by the latest color films will exist without noticeable deterioration in image quality and be available for more than 100 years [68]. It is not clear at the present whether we can keep pictures recorded in electronic media under currently available conditions over such a long period. It is also not certain how long electronic recording materials will be able to be stored before degradation sets

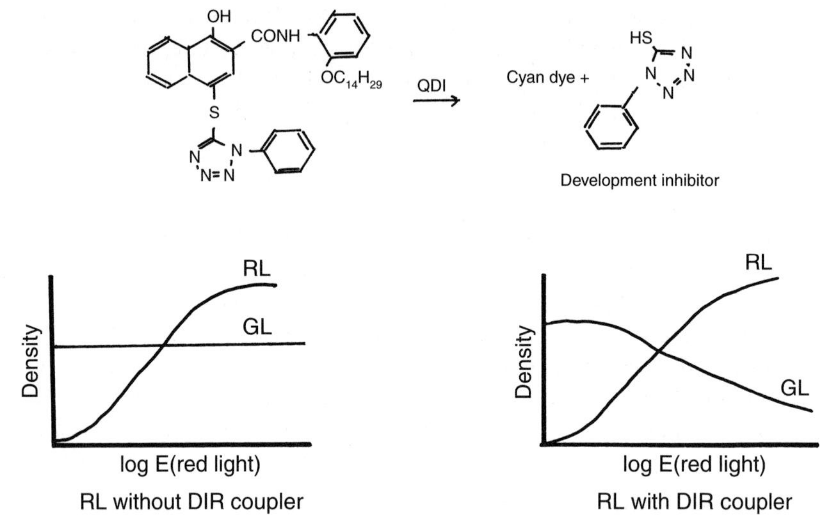

Figure 6.12 The reaction of a DIR coupler during color development (top) and the changes in optical densities of a green-sensitive layer (GL) and a red-sensitive layer (RL) in the absence (left) and presence (right) of a DIR coupler as a function of exposure (bottom).

in. In addition, electronic recording media have often been replaced rapidly by new ones of different systems and thus become obsolete.

6.4 New Trends in CCD and CMOS in Relation to Color Films

The increase in the number of pixels per frame is achieved by increasing the area of frame, decreasing the size of each pixel, and introducing three-dimensional structure into CCD and CMOS sensors. However, increasing the frame area is difficult from the viewpoint of cost. Decreasing pixel size also faces difficulty because Si should be thick enough to efficiently capture incident visible light, since the light absorption of Si in the visible region is attributed to indirect transition and therefore is weak. It is noted that the transitions corresponding to the light absorptions of AgBr and AgCl in the longer wavelength region are also indirect and weak. By use of this property, the three-dimensional structure of a color film was applied to a CMOS. Namely, the images of three primary colors were captured by silicon at different depths due to the difference in penetration among the lights of the three primary colors in silicon [69,70]. This system is free from color moiré problems without aid of any low-pass filter. However, it seems that the color separation of this system was not good enough, since the light absorption of Si with respect to wavelength is not selective enough for it.

In contrast to inorganic semiconductors such as Si, the light absorption of organic semiconductors is usually strong and selective with respect to wavelength. Aihara et al. proposed novel image sensors using organic photoconductive layers, which are around 1 μm thick and sensitive to blue, green, and red lights. Selectivities of the spectral responses of these layers were good enough to divide incident light into three primary color components [39–41]. At ICIS'06, Takada and others proposed a three-dimensional sensor with organic thin layers as photosensitive materials [42,43]. This was followed by Ihama et al.'s proposal to develop image sensors with organic semiconductor layers by taking advantage of their strong light absorption [71].

Following the idea that the dynamic range of a color film is expanded by use of two or three emulsion layers with different sensitivities in pile, the dynamic range of a CCD has been expanded by such a method that each pixel is equipped with two diodes with different sensitivities side by side [72,73]. The fourth layer technology for improving color reproduction in color films [5,6] was applied to the design of a CCD by using one of two green-sensitive pixels in the Bayer arrangement to reduce the response of the red-sensitive pixel upon exposure to blue-green light [74].

While digital still cameras have become very popular due to satisfactory image quality and overwhelming convenience, color films are characterized by several strong points as stated above, which cause us to believe that people will continue to use color films along with digital still cameras in the future.

References

1. Reese Jenkins, in Eugene Ostroff (ed.), *Pioneers of Photography: Their Achievements in Science and Technology* (Springfield, VA: SPSE, 1987).
2. E. Kaprelian, *Photographic Engineering*, 1, 42 (1950).
3. SPSTJ (ed.), *Principles of Photographic Science and Technology*, rev edn (Tokyo: Corona, 1998), Section 3.7.
4. J. M. Sturge (ed.), *Neblette's Handbook of Photography and Reprography, Materials, Processes, and Systems*, 7th edn (New York: Van Nostrand Reinhold, 1977), Chapter 6.
5. N. Sasaki, K. Takahashi, H. Ikoma, *J. Soc. Photogr. Sci. Technol. Jpn.*, 52, 41 (1989).
6. N. Sasaki, *J. Soc. Photogr. Sci. Technol. Jpn.*, 67, 504 (2004).
7. H. Saito, *Kagaku Souchi*, 34, 61 (1992).
8. T. Tani, *J. Soc. Photogr. Sci. Technol. Jpn.*, 66, 2 (2003).
9. T. Tani, *J. Dispersion Sci. Technol.*, 25, 375 (2004).
10. T. Tani, *J. Imaging Sci. Technol.*, 51, 110 (2007).
11. Anonymous, *Res. Discl.* (Item 22534), 225, 20 (January 1983).
12. J. T. Kofron and R. E. Booms, *J. Soc. Photogr. Sci. Technol. Jpn.*, 49, 15 (1986).
13. J. C. Marchant, in *The International Congress of Photographic Science, Cologne, Germany, 1986*.
14. T. L. Penner and P. B. Gilman, *Photogr. Sci. Eng.*, 20, 97 (1976).
15. G. L. House, in *The International Congress of Photographic Science, Cologne, Germany, 1986*.
16. G. R. Bird, R. C. Jones, A. E. Ames, *Appl. Opt.*, 8, 2389 (1969).
17. K. Yamashita, H. Takei, T. Ikeda, T. Tani, in *Proceedings Intern. Cong. Imaging Sci., May 2002, Tokyo*, p. 93.
18. R. L. Parton, T. L. Penner, W. J. Harrison, J. C. Deaton, and A. Muenter, in *AgX 2004: International Symposium on Silver Halide Technology, Sponsored by IS&T and SPSTJ, September 2004, Ventura, CA*, p. 161.
19. T. Suzumoto, K. Yamashita, S. Aiba, K. Kobayashi, T. Hioki, in the preprint of *Annual Conf. Soc. Photogr. Sci. Tech. Jpn., May 2005, Tokyo*.

20. R. K. Hailstone, N. B. Leibert, M. Levy, R. T. McCleary, S. R. Gilolmo, D. L. Jeanmaire, C. R. Boda, *J. Imaging Sci.*, 32, 113 (1988).
21. J. D. Bologa, , in *Proceedings of the PICS Conference, May 1998, Portland, IS&T*, p. 299.
22. J. M. Harbison, H. E. Spencer, in T. H. James (ed.), *The Theory of the Photographic Process*, 4th edn (New York: Macmillan, 1977), 152.
23. T. Tani, *J. Imaging Sci.*, 30, 41 (1986).
24. R. K. Hailstone, N. B. Liebert, M. Levy, R. T. McCleary, S. R. Gilolmo, D. L. Jeanmaire, C. R. Boda, *J. Imaging Sci.*, 30, 41 (1986).
25. J. Belloni, M. Treguer, H. Remita, and R. De Keyzer, *Nature*, 402, 865 (1999).
26. I. R. Gould, J. R. Lenhard, A. A. Muenter, S. A. Godleski, and S. Y. Farid, *J. Am. Chem. Soc.* 122, 11934 (2000).
27. T. D. Pawlik, A. A. Muenter, R. S. Eachus, J. R. Lenhard, in *AgX 2004: International Symposium on Silver Halide Technology, Sponsored by IS&T and SPSTJ, September, 2004, Ventura, CA.*
28. S. E. Sheppard, *Photogr. J.*, 65, 380 (1925).
29. T. Tani, *J. Imaging Sci. Tech.*, 39, 386 (1995).
30. T. Tani, *J. Imaging Sci. Tech.*, 42, 135 (1998).
31. T. Tani, *J. Imaging Sci. Tech.*, 44, 243 (2000).
32. R. K. Hailstone, in Boris Levy (ed.), *The Proceedings of Symposium on Electronic and Ionic Properties of Silver Halides: Common Trends with Photocatalysis (Part of the 44th Annual Conference, May, 1991)* (Springfield, VA: IS&T, 1991), 102.
33. J. F. Hamilton, J. M. Harbison, D. L. Jeanmaire, *J. Imaging Sci.*, 32, 17 (1988).
34. L. M. Kellogg, J. Hodes, in the preprint of *40th Annual Conference of SPSE, Rochester, New York, May 1987.*
35. K. Morimura, H. Sasaki, T. Kojima, M. Ihama, S. Yamashita, T. Tani, H. Mifune, in *2005 Beijing International Conference on Imaging, May 2005, Beijing.*
36. K. Makino, S. Yamaryo, Y. Mimaki, K. Kawai, in *The 29th International Congress of Imaging Science, Tokyo, 2002.*
37. F. Ueda, M. Goto, S. Kuwashima, K. Morimura, N. Haraguchi, M. Toyoda, H. Fukuzawa, H. Ikeda, *ICIS'06: The 30th International Congress of Imaging Science, May 2006, Rochester, New York.*
38. S. M. Sze, *Physics of Semiconductor Devices* (New York: Wiley, 1981), 407.
39. S. Aihara, Y. Hirano, T. Tajima, K. Tanioka, M. Abe, N. Saito, *Appl. Phys. Lett.*, 82, 511 (2003).
40. S. Aihara, K. Miyakawa, Y. Ohkawa, T. Matsubara, T. Takahata, S. Suzuki, M. Kubota, K. Tanioka, N. Kamata, D. Terunuma, *Jpn. J. Appl. Phys.*, 44, 3743 (2005).
41. S. Aihara, *J. Soc. Photogr. Sci. Tech. Jpn.*, 69, 320 (2006).
42. S. Takada, M. Ihama, M. Inuiya, *Proc. SPIE*, 6068, 60680A (2006).
43. M. Ihama, M. Hayashi, Y. Maehara, T. Mitsui, S. Takada, *Proc. SPIE*, 6656, 66560A (2007).
44. T. A. Russell, R. M. Wilson, C. R. Sanford, 'Multiple Coating Apparatus', US Patent 2, 761, 417 (1956).
45. J. Hens, W. Van Abbenyen, in S. F. Kistler, P. M. Schweizer (eds), *Liquid Film Coating Scientific Scientific Principles and Their Technological Implications* (London: Chapman & Hall, 1997), Chapter 11d.
46. For example, T. Heki, N. Inoue, *Forma*, 4, 55–8 (1989).
47. M. A. Kriss, *J. Soc. Photogr. Sci. Tech. Jpn.*, 50, 357 (1987).
48. M. Tabei, Y. Mizobuchi, *J. Soc. Photogr. Sci. Technol. Jpn.*, 49, 125 (1986).
49. S. Ikenoue, M. Tabei, *J. Imaging Sci.*, 34, 187 (1990).
50. N. Sakaguchi, N. Ozawa, *J. Inst. Television Eng. Jpn.*, 50, 1218 (1996).
51. H. Sugiura, K. Asakawa, T. Yamada, T. Kuno, N. Kontani, J. Yoshida, J. Fujino, *J. Inst. Television Eng. Jpn.*, 48, 210 (1994).

52. Y. Asa, Y. Bai, K. Sato, *J. Inst. Image Inform. Television Eng.*, 54, 294 (2000).
53. M. Sato, C. Matsumoto, K. Sato, *J. Inst. Image Inform. Television Eng.*, 56, 463 (2002).
54. N. Ozawa, *J. Inst. Image Inform. Television Eng.*, 57, 1119 (2003).
55. T. Saito, T. Komatsu, *Proceedings of Electronic Imaging Science and Technology, Digital Photography II*, SPIE 6069, 606905 (2006).
56. T. Tani, *Photographic Sensitivity: Theory and Mechanisms* (New York/Oxford: Oxford University Press, 1995).
57. T. Tani, *Photogr. Sci. Photochem.*, 21, 357 (2003).
58. S. Nishimura, *J. Soc. Photogr. Sci. Tech. Jpn.*, 68, 253, 338 (2005).
59. T. Noguchi, H. Ikoma, *J. Soc. Photogr. Sci. Technol. Jpn.*, 61, 8 (1998).
60. M. A. Kriss, in James (ed.), *The Theory of the Photographic Process*, Chapter 21.
61. J. Nakamura, in J. Nakamura (ed.), *Image Sensors and Signal Processing for Digital Still Cameras* (New York/London: Taylor and Francis, 2006), Chapter 3.
62. W. T. Hanson, Jr., P. W. Vittum, *PSA J. B*, 95 (1947).
63. W. T. Hanson, Jr., *J. Opt. Soc. Am.*, 40, 171 (1950).
64. Peter Krause, in J. Sturge, V. Walworth, A. Shepp (eds), *Imaging Processes and Materials* (New York: Van Nostrand Reinhold, 1989), Chapter 4.
65. C. R. Barr, J. R. Thirtle, P. W. Vittum, *Photogr. Sci. Technol.*, 13, 94 (1969).
66. L. K. J. Tong, in James (ed.), *The Theory of the Photographic Process*, 339.
67. SPSTJ (ed.), *Principles of Photographic Science and Technology*, rev edn (Tokyo: Corona, 1998), Section 3.6.
68. F. Ueda, M. Goto, S. Kuwashima, K. Morimura, N. Haraguchi, M. Toyoda, H. Fukuzawa, H. Ikeda, in *ICIS'06: The 30th Intern. Cong. Imaging Sci., May 2006, Rochester, New York.*
69. R. F. Lyon, P. M. Hubel, in *IS&T/TSID 10th Color Imaging Conference Proceedings, Scottsdale, Arizona, USA; 2002*, 349.
70. P. M. Hubel, J. Liu, R. J. Gutosch, *Proc. SPIE*, 5301, 402 (2004).
71. M. Ihama, T. Mitsui, K. Nomura, Y. Maehara, H. Inomata, T. Gotou, Y. Takeuchi, in *Proc. 16th Intern. Display Workshops, Miyazaki, Japan, December 2009.*
72. Kazuya Oda, Hirokazu Kobayashi, Kazuhiko Takemura, Yutaka Takeuchi, and Tetsuo Yamada, Technical Report of Institute of Image Information and Television Engineerings, 27, 17 (2003).
73. T. Ashida, H. Yamashita, M. Yoshida, O. Saito, T. Nishimura, K. Iwabe, in M. M. Blouke, N. Sampat, R. J. Motta (eds), *Sensors and Camera Systems for Scientific, Industrial, and Digital Photography Application V* (Proc. of SPIE-IS&T Electronic Imaging, 2004).
74. Noboru Ohta, Summer Seminar Sponsored by SPSTJ, October, 2004, Fujiyoshida, Japan.

7

Silver Halide Nanocrystals and Nuclear Particle Detectors

Nuclear emulsions for the detection of neutrinos through the international project OPERA (Oscillation Project with Emultion-tRacking Apparatus) and dark matter (i.e., WIMPs: weakly interacting massive particles) are characterized in the overview of photographic emulsions of silver halide (AgX) grains with respect to the grain size dependence of their sensitivities. As stated in Section 4.3.1, the emulsions are classified into three groups. The grains in class I, which are larger than 0.5 μm in equivalent diameter, are highly sensitive due to the large light absorbance of each AgX grain and are used for photographic films with high sensitivity as exemplified by color films. The grains in class II, which are 0.1∼0.5 μm in equivalent diameter and include those for OPERA films, can be principally defect-free and therefore afford efficient latent image formation. The grains in class III, which are smaller than 0.1 μm in equivalent diameter and expected to be available for WIMPs detection, are still poorly understood and interesting to study from the viewpoints of the formation and characterization of quantum-sized AgX grains. Descriptions of the reset function mechanism brought about by the "refresh" treatment for OPERA films and its applicability to those for WIMPs detection have also been made.

7.1 Characterization of Photographic Emulsions

7.1.1 Silver Halide Grains and Photographic Sensitivity

Since Becquerel discovered radiation by use of nuclear emulsions, nuclear emulsions have a glorious history [1–4], and made a fresh start after Niwa Laboratory in Nagoya University developed an automatically scanning system for reading nuclear tracks in emulsion layers [5–7]. The attributes of nuclear emulsions are (1) high degree of uniformity of grain size, (2) virtual absence of chemically fogged grains, and (3) high proportion of AgX compared to gelatin. In particular, attribute (1) has encouraged a strong motivation in photographic scientists and engineers to develop mono-dispersed AgX grains.

Poly-dispersed AgX grains have been used for many years in photographic films for taking pictures. In that time, much progress has been made in the method and apparatus for preparing photographic emulsions, as described in Chapter 2. Now, aqueous solutions of $AgNO_3$ and potassium halide (KX), for example, are simultaneously added to a reaction vessel initially containing an aqueous gelatin solution under the precise adjustment of addition rates of the $AgNO_3$ and KX solutions, and temperature and silver ion concentration in a reaction solution [8,9] to determine the size and shape and rate of growth of AgX grains [10,11] under desirable conditions during the reaction. The solution in a vessel is always stirred from the beginning to the end of the reaction. As described in Section 2.2.2, many ideas have been proposed for the development of appropriate stirrers for this purpose. One has been designed to guide the aqueous solutions of $AgNO_3$ and KX and add them to a reaction solution in a vessel at positions close to each other in order that the silver and halide ions react with each other immediately. Another one was designed to dilute the solutions of $AgNO_3$ and KX immediately upon the addition of a reaction solution before they reacted with each other.

Since silver halides of photographic interest are face-centered cubic in crystal structure, they are typically cubic, octahedral [8,9], or tabular [12,13] in shape, as shown in Figure 1.2. Now, it is possible to prepare mono-dispersed grains in one of these shapes at any size having narrow distribution with the coefficient of variation being less than 10%.

Figure 4.1 illustrates the process for latent image formation. In each grain, there are many interstitial silver ions in the interior and many sensitization centers acting as shallow electron traps on the surface. A photon absorbed by a grain creates a conduction electron and a positive hole. A conduction electron migrates and is trapped by one of sensitization centers. A trapped electron is thermally detrapped and migrates again. During the multiple trapping and release processes, a trapped electron is attacked by an interstitial silver ion and reacts with it to form a silver atom. The repetition of these steps at the same site leads to the growth of a silver cluster and finally to the formation of a latent image center. Since an electron is randomly travelling within a grain, the place where a latent image center is formed on each grain is not certain and independent of the place where an incident photon is absorbed. The smallest latent image center is composed of three or four atoms in the presence or absence of gold sensitization [14–16] regardless of grain size. Accordingly, intrinsic photographic sensitivity is proportional to grain volume when the efficiency of latent image formation is the same among the grains compared [17,16]. In order to realize high sensitivity, appropriate technologies are necessary to prevent positive holes from recombining with conduction electrons.

Figure 4.15 illustrates the development process. A latent image center acts as a deep electron trap, and receives an electron from a developer composed of an aqueous solution of reducing agents. An interstitial silver ion is attracted by a received electron and reacts with it to form and add a silver atom to the center. The repetition of these steps converts wholely an AgX grain into an Ag grain. Namely, a latent image center grows to a spherical silver particle until its diameter reaches about 20 nm. Then, the particle continues to grow as a silver filament with thickness of about 20 nm, since the diameter of the circle where a freshly formed silver atom can migrate before it finds its stable site on the surface of silver is about 20 nm (Figure 8.6 [18]).

7.1.2 Enhancement of Latent Image Formation by Chemical Sensitization

The framework of sensitization centers is a sulfur sensitization, which was invented by Sheppard [19]. A sulfur sensitization center is composed of a dimer of silver sulfide [20,21], i.e., a dimer of substitutional sulfide ions associated with two interstitial silver ions for the compensation of excess negative charges on the sulfide ions [22–24]. The sensitization centers provide transient electron traps with depths of about 0.3 eV [25,26,21] and should decrease the diffusion length of electrons in an AgX grain. Figure 4.12 shows the electron micrograph of silver-decorated sulfur sensitization centers on AgBr grains with an average diameter of 0.2 μm [27]. There are 300–400 centers on each grain. The replacement of sulfide ions in a sulfur sensitization center by selenium and tellurium ions gives a selenium sensitization center and a tellurium one, respectively, both of which have become available [28]. The degree of high-intensity reciprocity law failure, which results from an inefficiency in latent image formation due to the appearance of too many electrons within a short time period, as described in Section 4.3.1 [29], was in the order of sulfur-, selenium-, and tellurium-sensitized AgBr emulsions in accord with the fact that the depths of the electron traps provided by these sensitization centers are in the order of sulfur, selenium, and tellurium [28].

The replacement of one of the two interstitial silver ions in a sulfur sensitization center by an aurous ion gives a sulfur-plus-gold sensitization center [30,31,24]. The sensitization centers provide transient electron traps with a depth of about 0.2 eV [25,26] and therefore tend to decrease the diffusion length of electrons in AgX grains. Since the electron traps provided by sulfur-plus-gold sensitization centers are shallower than those provided by sulfur sensitization centers, it is expected that the diffusion length of electrons in sulfur-plus-gold-sensitized emulsions is longer than that in corresponding sulfur-sensitized ones, although it is shorter than that in unsensitized ones. In addition, an aurous ion is reduced to give a gold atom, which is incorporated into a latent image center formed on a sulfur-plus-gold-sensitized grain and enhances the activity of the center to initiate development. Thus, sulfur-plus-gold-sensitized emulsions usually exhibit a sensitivity higher than that of corresponding sulfur-sensitized emulsions, and are commonly used for highly sensitive AgX photographic emulsions.

One of the most effective technologies for preventing the recombination of electrons and positive holes is reduction sensitization [32], including hydrogen hypersensitization [33], where AgX grains are treated by a reduction sensitizer or by hydrogen gas to form reduction sensitization centers composed of silver dimers [34,16]. Equations (4.3)–(4.7) describe the scheme of reduction sensitization [35,36,16].

The absorption of a photon by a grain creates a conduction electron and a positive hole. A positive hole reacts with a reduction sensitization center (Ag_2) to give Ag_2^+, which is unstable and dissociates to give a silver atom and an interstitial silver ion. A silver atom is unstable and dissociates to give a conduction electron and an interstitial silver ion [37,38]. Thus, one absorbed photon can create two conduction electrons in the presence of a reduction sensitization center. In addition, reduction sensitization prevents positive holes from recombining with conduction electrons. It is probable that reduction sensitization enhances fog formation during storage due to the creation of a conduction electron according to the processes (4.4), (4.5), and (4.6) after the oxidation of a reduction sensitization center by oxygen and water [39].

On the other hand, the scheme of latent image formation, as described in Figure 4.1, indicates that a latent sub-image center composed of a silver dimer captures a conduction electron in order to grow. It is therefore noted that there are two kinds of Ag dimers. One is an electron trap, while the other is a positive hole trap [40]. It has been revealed that the former is formed at a positively charged kink site, and the latter at a neutral kink site [41,42,43,16].

7.1.3 Sensitivity to Light and Particles as a Function of Grain Size

The observed fraction of grains developed as a function of the number of photons absorbed per grain for the AgBr grains in sulfur-plus-gold- and hydrogen-hypersensitized emulsion revealed that less than three photons absorbed per grain, i.e., less than three electron–hole pairs per grain, were needed for latent image formation on the grains [44,36]. This is the highest efficiency ever observed for latent image formation, and was achieved, not with large grains in class I (> 0.5 μm in equivalent diameter), but with small grains in class II (0.1∼0.5 μm in equivalent diameter). For this reason, the present author divided AgX grains into three groups, class I, II, and III (<0.1 μm in equivalent diameter).

Table 7.1 summarizes the number of electron–hole pairs per grain needed for latent image formation (N_L) in various emulsions. They are ∼5000 in an unsensitized emulsion, ∼30 in a reduction-sensitized and a sulfur-sensitized emulsion [45], ∼10 in a sulfur-plus-gold-sensitized emulsion [14,46], and ∼3 in a sulfur-plus-gold-sensitized and hydrogen-hypersensitized or a sulfur-plus-gold-plus-reduction-sensitized emulsion [44,36]. The values of N_L are nearly the same among sulfur-, selenium-, and tellurium-sensitized emulsions under normal exposure conditions, and should be in the order of sulfur-, selenium-, and tellurium-sensitized emulsions under exposure with high intensity for a short time period [28]. Appropriate technologies are needed to stabilize hydrogen-hyper-sensitized and reduction-sensitized emulsions by hindering the enhancement of fogging by hydrogen hypersensitization and reduction sensitization during storage.

Then, it is necessary to know the number of electron–hole pairs per grain created by a nuclear particle (N_p(P)). Spence [47] estimated the number of conduction electrons per grain created by a singly charged relativistic particle for AgX grains with varying size based on the energy loss of ∼500 keV/mm, Yamakawa's figure of 5.8 eV for the creation of one conduction electron [48], and a continuous rate of energy loss per grain. The present author took his value as N_p(P) and evaluated it as a function of grain size, as listed in Table 7.2. It has been judged that tracks can be formed by neutrinos or WIMPs when N_L is smaller than N_p(P).

On exposure of an AgX grain to light, each quantum adds to a growing Ag cluster a single Ag atom and two Ag atoms in the absence and presence of reduction sensitization centers, respectively. Thus, a growing silver cluster becomes a latent image center when it contains more than three atoms. On the other hand, a single high-energy quantum usually creates conduction electrons of sufficient number per grain to form a latent image center on each grain. As a result, the optical density of a developed film is principally in a linear relationship with exposure to high-energy radiation [49], while it is not with exposure to light quanta.

Table 7.1 Number of electron–hole pairs per grain needed to render half of AgX grains developable in various emulsions (N_L) and that created by a nuclear particle in an OPERA film [N_p(P)].

Sensitizing condition	N_L	N_p(P)	Stability for storage	Availability for OPERA film
Unsensitized	∼5000	18	Stable	No
R-sensitized	∼30	18	Unstable	No
S-sensitized	∼30	18	Stable	No
S+Au-sensitized	∼10	18	Stable	Yes
S+Au+R-sensitized	∼3	18	Unstable	Technology required

Table 7.2 Number of electron–hole pairs per grain needed for latent image formation in various emulsions (N_L) and that created by a nuclear particle and light (N_p(P) and N_p(L), respectively), and concentrations of the pairs created by a nuclear particle and light (C_p(P) and C_p(L), respectively) together with the numbers of pairs required for latent image formation (N_L(L))

Size (nm)	200 (Class II)	100 (Class II)	50 (Class III)	20 (Class III)
N_L(S+Au)	10	10	(10) [a]	(10) [a]
N_L(S+Au+R)	3	3	(3) [a]	(3) [a]
N_p(P)	18	9	4.5	1.8
C_p(P) (rel.)	1	4	16	100
N_p(L) (rel.)	18	2.3	0.28	0.018
C_p(L)	1	1	1	1

[a] The values expected under the condition that the efficiency of latent image formation is the same between the grains in classes II and III.

7.2 Large Silver Halide Grains for Color Films

As a representative of a photographic product containing AgX grains in class I, a color negative film is briefly characterized in this section. Figure 1.1 shows the scanning electron micrograph of a cross-section of its photosensitive layers, which are about 20 μm thick in total and composed of some 14 layers with different functions in pile [50]. There are three major layers, each of which is sensitive to the light of one of the three primary colors (i.e., blue, green, and red, respectively). In each major layer, there are sub-layers with high, medium, and low sensitivities. Namely, a color negative film is a three-dimensional sensor designed to capture a full-color image with wide dynamic range. White lines in this electron micrograph are the cross-sections of thin tabular AgX grains, as shown in Figure 1.2. They belong to class I, since the diameter of the sphere, whose volume is equal to the average volume of those tabular grains, is around 1 μm. Each tabular grain is covered with a mono-molecular layer of sensitizing dyes, which absorb incident light quanta and then inject electrons into the conduction band of an AgX grain.

As described in the previous section, photographic sensitivity depends on the absorbance of incident light by a grain and the efficiency of latent image formation in it. Although the grains in class I cannot exhibit image formation of the highest efficiency, they are used for color films due to their large light absorbance. Since incident light on each grain is mostly absorbed by a mono-molecular layer of sensitizing dyes, the amount of which should be proportional to the surface area of the grain, thin tabular grains are prepared and used for highly sensitive layers in a color film [50].

The reason why the highest efficiency of latent image formation has not been achieved with class I grains is the presence of transient electron traps, which deteriorate the transport of electrons. In the presence of the above-stated electron traps, the diffusion length of electrons in AgBr grains was estimated to be 0.4–0.5 μm [51,52] and smaller than the size of class I grains. Namely, a certain fraction of electrons cannot reach a growing image center, and does not contribute to latent image formation. Moreover, crystal defects, which provide traps for electrons, are likely to be formed in large class I grains. As a result, the formation of more than one latent image center on each grain is probable and results in the inefficiency of latent image formation.

7.3 Small Silver Halide Grains for OPERA Films

7.3.1 Sensitivity of OPERA Films

While the size of class I grains is larger than the diffusion length of electrons, the size of the grains in class II is 0.1–0.5 μm, and comparable to or smaller than the diffusion length. It is therefore expected that nearly all the electrons can reach a growing image center and contribute to latent image formation. This expectation is supported by the following results. The traps for electrons in silver halides are usually provided by crystal defects as described in Section 3.4.1. The agreement between grain size and crystallite size as provided by X-ray diffraction analysis with Scherrer's equation indicates that class II AgX grains are free from crystal defects, unless crystal defects are intentionally introduced into the grains [53]. The lifetime of photoelectrons in class II AgBr grains as measured by the microwave photoconductivity method [54,55] was proportional to their size [56]. Namely, the rate of electron trapping, which is the reciprocal of electron lifetime, was proportional to the surface-to-volume ratio of the grains, indicating that the traps for photoelectrons were located, not in the interior, but on the grain surfaces. In agreement with these results, only one latent image center is usually formed on each class II grain regardless of exposure [22]. This condition is ideal for efficient latent image formation, and indicates that the class II grains among those in the three classes are the best for the latent image center with the highest efficiency. The AgX grains used for OPERA films belong to class II.

The international project OPERA is currently focused on directly proving the oscillation from a μ neutrino to a τ neutrino (i.e., $\nu\mu \rightarrow \nu\tau$ chanel) by detecting the short-lived τ lepton produced by an oscillating ν_τ [57,7]. The track of a τ lepton is to be observed by the system with nuclear emulsion films (i.e., OPERA films) and an automatically scanning system. The former was newly developed and produced by Fujifilm [58], and the latter was extensively developed by Niwa Laboratory at Nagoya University [5,6]. The average diameter of the mono-dispersed AgX grains used for the OPERA films is 200 nm with coefficient of variation of less than 10%.

Figure 7.1 shows the electron micrograph of AgX grains in an emulsion used for OPERA films. As indicated by the scanning electron micrograph of the section of an OPERA film in Figure 7.2, this emulsion was coated twice on both sides of a TAC film with thickness of 200 μm. The thickness of each layer is 22 μm, resulting in a total thickness of emulsion layers of 88 μm. Figure 7.3 shows the scanning electron micrograph of the emulsion layers on one side of an OPERA film compared with that of the emulsion layers of a color negative film. It is noted that the emulsion layers in an OPERA film are thick and contain a high concentration of fine AgX grains, making it possible to detect incident nuclear particles with high sensitivity.

Table 7.1 lists the numbers of electron–hole pairs per grain needed for latent image formation (i.e., for rendering half of the AgX grains developable) in various emulsions (N_L). Since the size of the AgX grains used for OPERA films is 200 nm on average, the number of electron–hole pairs per grain created by a singly charged particle (N_p(P)) is

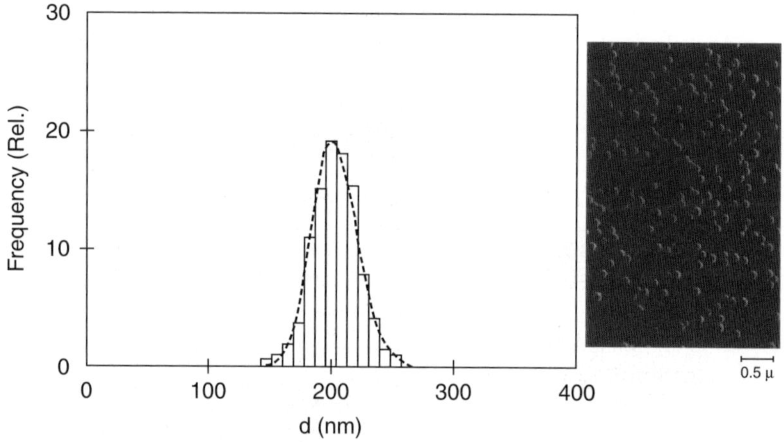

Figure 7.1 An electron micrograph of the carbon replicas of AgX grains used for an OPERA film and their size distribution. The length of the bar in this figure is 0.5 μm. Rerpoduced with permission from Reference [58], copyright 2004 The Society of Photographic Science and Technology of Japan.

estimated to be 18. Since N_L is smaller than $N_p(P)$ for sulfur-plus-gold-sensitized and sulfur-plus-gold-plus-reduction-sensitized emulsions, it is considered that they are sensitive enough to form the track of nuclear particles. Although a sulfur-plus-gold-plus-reduction-sensitized emulsion is more sensitive than a sulfur-plus-gold-sensitized one, it is not available unless the enhancement of fogging by reduction sensitization during storage is reduced by the appropriate technologies. Therefore, OPERA films are equipped with sulfur-plus-gold-sensitized emulsions. Furthermore, they were designed to be free of reduction sensitization centers, which can form otherwise during the preparation of AgX grains and other processes owing to the reducing ability of gelatin [59].

Figure 7.2 A scanning electron micrograph of a section of an OPERA film in comparison with that of a color film (left). White spots are the sections of AgX grains. Reproduced with permission from Reference [58], copyright 2004 The Society of Photographic Science and Technology of Japan.

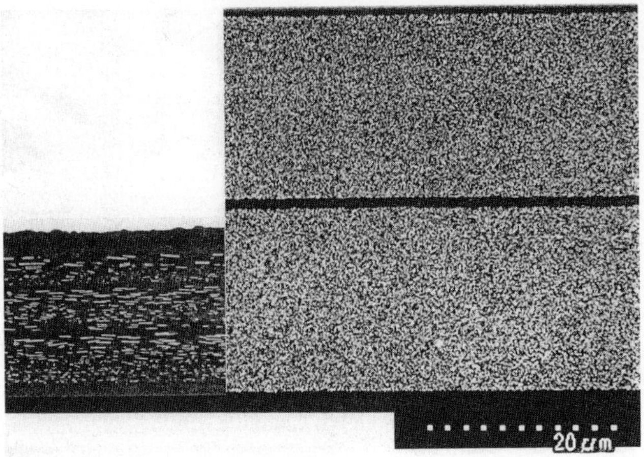

Figure 7.3 Scanning electron micrographs of the AgX emulsion layers of a color negative film (left) and those on one side of an OPERA film (right), where white spots are the sections of AgX grains. Reproduced with permission from Reference [58], copyright 2004 The Society of Photographic Science and Technology of Japan.

7.3.2 Reset Treatment of OPERA Films

One of the most outstanding features of OPERA films is the reset function [58], which is brought about by the "refresh" treatment. The detectors for neutrinos and WIMPs are required to be three-dimensional and have the reset function. Among various sensors including color films, CCD, and CMOS, only OPERA films can meet these two requirements.

The reset function is for deactivating the developability of latent image centers without deteriorating the activity of sensitization centers. Figure 7.4 shows the optical micrographs indicating the formation of the tracks by electron beams in an OPERA film and the disappearance of the tracks without deteriorating the activity of sensitization centers by the reset treatment. Latent image centers are Ag clusters, while sensitization centers are essentially dimers of Ag_2S. Figure 7.5 shows the amounts of silver and Ag_2S powder in gelatin layers as a function of the redox potential of a buffer solution in which the gelatin layers were immersed [60]. While the redox potential ranged from 400 to –200 mV (vs. saturated Ag/AgCl), the oxidation of silver powder was observed in the buffer solution, while that of the powder of Ag_2S was not. The oxidation potential of the Ag powder, which could be defined as the redox potential required to oxidize half of its amount, was around 100 mV. One can find an appropriate redox potential, say, 200 mV, at which silver powder was nearly completely oxidized, while the Ag_2S powder was unchanged.

The reset treatment is carried out by keeping OPERA films in high humidity at relatively high temperature in order to make it possible for the gelatin phase in the films to bring about redox reactions at an appropriate rate. Figure 4.14 shows the redox potential of an aqueous gelatin solution as a function of pH [59]. As seen in this figure, one could adjust the redox potential of an aqueous gelatin solution to be around 200 mV at reasonable pH value.

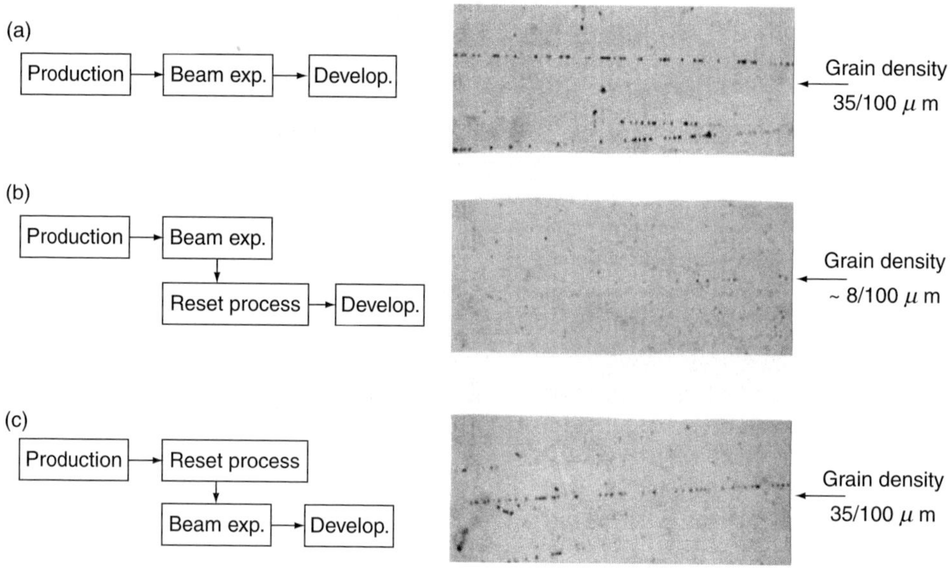

Figure 7.4 Optical micrographs of sections of emulsion layers in OPERA films showing the effect of the reset treatment. They were subjected to electron beam exposure and development (a), electron beam exposure, reset treatment, and development (b), and reset treatment, electron beam exposure, and development (c). Reproduced with permission from Reference [58], copyright 2004 The Society of Photographic Science and Technology of Japan.

Figure 7.6 shows the electronic energy level diagram, where the highest occupied electronic energy levels of latent image centers formed under various conditions and sensitization centers composed of Ag_2S clusters correspond to their observed oxidation potentials, while the Fermi level of an aqueous gelatin solution corresponds to its redox potential [60]. It is considered from the result in this figure that the reset function could be realized by adjusting the Fermi level of the wet gelatin phase in OPERA films to be somewhere between the highest occupied levels of latent image centers and sensitization centers, and that latent image centers, which were formed by nuclear particles and should exhibit their tracks when developed, were oxidized during the reset treatment to such an extent that their developability was reduced to be too small to exhibit the tracks.

The discrimination in developability between freshly formed latent image centers and those deactivated by the reset treatment could be enhanced by a functional compound such as methyl-benzotriazole [58]. As described in Section 4.5.5, this compound is an antifoggant, which strongly hinders the developability of development centers and is active only when it is ionized. It is known that this compound is ionized in an alkaline solution with pH larger than 8.5. An emulsion of OPERA films is weakly acidic, while a developer is alkaline. Namely, this compound is inactive in a dried emulsion layer with low pH and active in a developer with high pH, depressing photographic development [24].

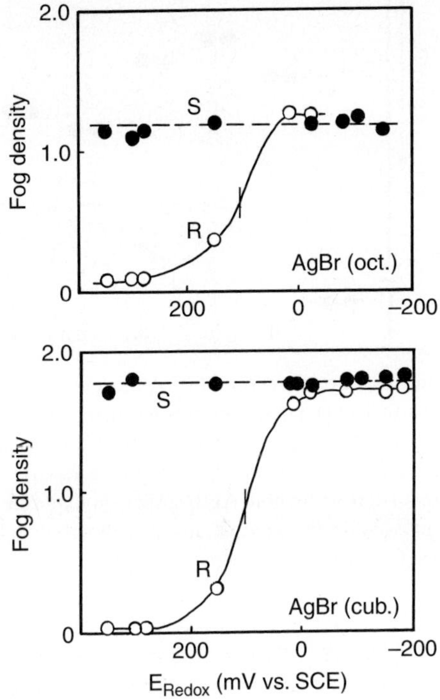

Figure 7.5 Optical densities after development of emulsion layers composed of fogged octahedral (top) and cubic (bottom) AgBr grains as a function of redox potential (mV vs SCE) of buffer solution, with which the fog centers were in equilibrium. The fog centers were composed of Ag_2S nanoparticles produced by excessive sulfur sensitization (S), and Ag nanoparticles produced by excessive reduction sensitization (R). Reproduced with permission from Reference [60], copyright 1983 IS&T, The Society for Imaging Science and Technology.

7.3.3 Advances in the OPERA Project

Neutrino oscillation, which is important for nuclear physics, was predicted nearly 50 years ago [61], and has been studied [62]. While the oscillations with the $\nu_\mu \rightarrow \nu_\tau$ and $\nu_\mu \rightarrow \nu_e$ channels (i.e., the oscillation from a μ neutrino to an electron neutrino) should be considered, the latter was experimentally excluded as the dominant process [63,64]. The $\nu_\mu \rightarrow \nu_\tau$ channel of the oscillation in disappearance mode (i.e., the disappearance of a μ neutrino) has already been confirmed experimentally [65,66]. However, direct observation of the transition through the detection of the τ lepton produced by an oscillating ν_τ has never been observed. The appearance of ν_τ will prove unambiguously that $\nu_\mu \rightarrow \nu_\tau$ oscillation is the dominant transition channel. This was the main goal of the OPERA experiment with high-energy long baseline neutrino beam from CERN to the underground Gran Sasso Laboratory (LNGS).

The OPERA neutrino detector in LNGS was designed to perform the first detection of neutrino oscillation in direct appearance mode through the study of the $\nu_\mu \rightarrow \nu_\tau$ channel. The challenge of the OPERA experiment was to detect the short-lived τ lepton ($c\tau$ = 87

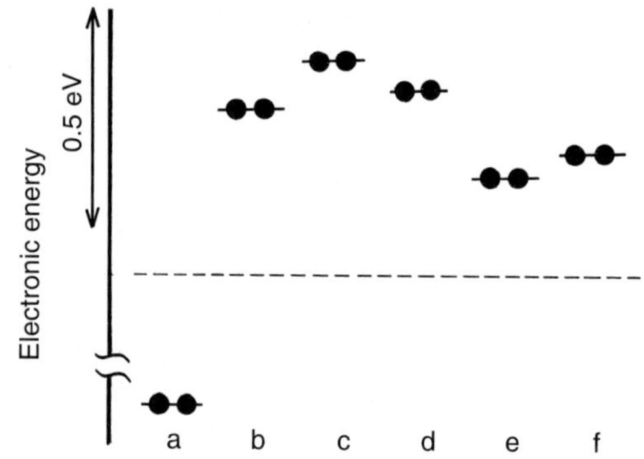

Figure 7.6 Explanation of the reset treatment for OPERA films with the highest occupied electronic energy levels of Ag_2S nanoparticles (a), latent sub-image centers formed in sulfur-sensitized emulsions with cubic and octahedral AgBr grains on exposure for 10^{-2} s (b), latent image center formed in an unsensitized emulsion with cubic AgBr grains on exposure for 10^{-5} s (c), latent image centers formed in an unsensitized emulsion with cubic AgBr grains on exposure for 10 s (d), latent image centers formed in a sulfur-plus-gold sensitized emulsion with cubic AgBr grains on exposure for 10 s (e), and developed silver (f). The highest occupied electronic energy levels of those centers were evaluated on the basis of their oxidation potentials. The horizontal broken line indicates the Fermi level, which corresponds to the redox potential of the gelatin phase in an OPERA film during its reset treatment. Reproduced with permission from Reference [60], copyright 1983 IS&T, The Society for Imaging Science and Technology.

μm) produced in the charged-current (CC) interaction of ν_τ. The hybrid apparatus consists of an emulsion/lead target complemented by electronic detectors and is placed in the CNGS beam 730 km away from the neutrino source. After runs with CNGS neutrinos were successfully carried out, the observation of the first candidate for ν_T CC interaction in the OPERA detector at LNGS was realized in 2010, and evaluated to have a significance of 2.36 σ of not being a background fluctuation [67]. The OPERA experiment would require the detection of a few more candidate events in order to firmly establish neutrino oscillation in direct appearance mode through identification of the final charged lepton.

7.4 Very Small Grains for Detecting Dark Matter

7.4.1 Number of Electron–Hole Pairs Created by Light and Particles

Because the solubility of AgX grains is relatively large and increases with decreasing grain size [68,69], it is difficult to prepare very small grains in class III and to keep their sizes unchanged during various processes. It is also difficult and challenging to measure precisely the size of very small AgX grains. Even in the case of class II grains, it has been pointed out that an electron-microscopic method cannot give accurate values due to the indistinct edges of grains in electron micrographs and has been replaced by an electrolytic method, according to which each AgX grain is reduced by a microelectrode to give a current pulse corresponding to the number of silver ions in it [36]. However, an electrolytic method is not available for class III grains, since they are too small to give detectable current pulses. An X-ray diffraction method with Scherrer's equation for measuring crystallite size is recommended for measuring the size of very small AgX grains, since their crystallite size is equal to their grain size due to the absence of any crystal defect in class III grains [53]. Direct observation of lattices in very small grains [70] might be available for their size evaluation.

Table 7.3 summarizes very small AgX grains prepared for various purposes [71–73]. The preparation of AgX grains with an average diameter of 13 nm by Berry in 1967 was a pioneering work in the field of nanotechnology. The grains prepared for basic research were $\leq$ 10 nm [74,75,70]. The grains produced for some photographic products are around 40 nm [76]. Since the range of a WIMP is estimated to be about 500 nm, Niwa [5,6] has suggested that AgX grains, which are much smaller than those for OPERA films, are needed for the detection of WIMPs. It is judged that AgX grains with 40 nm in diameter will be practically possible, and those with 20 nm in diameter might be the ultimate target.

Table 7.2 shows the grain size dependence of the number of electron–hole pairs per grain created by a singly charged particle $N_P(P)$ and the relative number of electron–hole pairs per grain created by light $N_P(L)$. The numbers of electron–hole pairs per grain needed for latent image formation (N_L) are 10 and 3 on sulfur-plus-gold-sensitized and sulfur-plus-gold-plus-reduction-sensitized AgBr grains, respectively. It is expected that $N_P(L)$, which should be proportional to the sensitivity to light, decreases very steeply with decreasing grain size, since the light absorbance of AgX grains is proportional to their volume (i.e., cube of grain size). On the other hand, it is expected that $N_P(P)$, which should be proportional to the sensitivity to nuclear particles, decreases much more slowly with decreasing grain size, since $N_P(P)$ is proportional to grain size and has wide distribution. Niwa already tried to use AgX grains of about 40 nm in diameter, and observed the sensitivity to nuclear particles [5,6], which should be much higher than that expected from its proportionality to grain volume. Although a sulfur-plus-gold-sensitized emulsion would be available, a sulfur-plus-gold-plus-reduction-sensitized emulsion would be more sensitive and useful than a sulfur-plus-gold-sensitized emulsion when appropriate technologies can be used to prevent the enhancement of fogging by reduction sensitization during storage.

Table 7.3 Very small AgX grains Belonging to Class III

Purpose	Size (nm)	Capability	References
Basic science	13	Quantum size effect and impurity exclusion	Berry (1967)
	8–22		Kanzaki et al. (1991)
	10		Marchetti et al. (1991)
Holography	8–90	>5000 lines/mm	Zacharovas et al. (2000)
IC plate	50	~3000 lines/mm	
Nuclear track formation	200	Neutrino detection	Niwa (2004); Kuwabara et al. (2004)
	(20) [a]–40	WIMPs detection	Niwa (2004)

[a] Expected as the ultimate limit.

7.4.2 Efficiency of Latent Image Formation and Detection of Dark Matter

Although the detection of WIMPs by use of nuclear emulsions with very small AgX grains is judged to be promising on the basis of the analysis on the number of electron–hole pairs per grain, it's just beginning to be used and leaves many challenging and interesting aspects unclear. They include quantum size effect, impurity exclusion, ballistic transport of conduction electrons, a high rate of recombination, and a very small fraction of fogged grains in very small AgX grains. While some of these would be favorable for WIMPs detection, others would not be, as judged on the basis of the following considerations.

Figure 3.2 illustrates the electronic structure of silver halides of photographic interest. As seen there, the transition between the top of the valence band and the bottom of the conduction band (i.e., the transition of an electron from the former to the latter resulting from light absorption and leaving a positive hole in the latter, and the recombination between an electron in the former and a positive hole in the latter resulting in free exciton emission at low temperature) is indirect and has small probability owing to the condition that it should be associated with change in momentum [77]. The intensity of the free exciton emission was very weak or absent in class I and II grains , and distinctly increased with decreasing size of very small class III AgBr grains , indicating the occurrence of the so-called quantum-size effect [74,75,70]. It is expected that the quantum-size effect enhances the indirect transition and increase $N_P(P)$ in very small grains available for the detection of WIMPs.

Iodide ions are residual impurities in AgBr. At low temperature, AgBr grains in classes I and II exhibit strong iodine–exciton emission, which results from the recombination of electrons with positive holes bound by iodide ions. In very small AgBr grains in class III, the iodine–exciton emission decreased distinctly with decreasing grain size [74,70]. This result indicates the impurity exclusion, by which the fraction of grains without any iodide ion increased with decreasing grain size. The impurity exclusion might be also valid for sensitization centers. As seen in Figure 4.12, there are 300–400 sensitization centers on a grain of 200 nm in diameter. It is therefore expected that there will be only 3–4 sensitization centers on each grain of 20 nm in diameter, and that the latent image formation under this condition is quite different from that under conventional condition.

The drift mobility of conduction electrons is reduced by their collision with phonons in class I and II grains. While the mean-free path of electrons is estimated to be 7.4 nm [78], it is expected that the fraction of conduction electrons, which reach a growing image center without any collision with phonons, increases with decreasing size for the grains in class III.

Table 7.2 shows the grain size dependence of the concentration of electron–hole pairs created by nuclear particles and light $C_p(P)$ and $C_p(L)$, as calculated on the basis of $N_p(P)$ and $N_p(L)$, respectively. It is noted that the concentration of conduction electrons created by particles, and therefore the concentration of positive holes created by particles, sharply increases with decreasing grain size, while the concentrations of electrons and positive holes created by light do not. Since the rate of recombination is proportional to the product of the concentrations of conduction electrons and positive holes, it is expected that the rate of recombination between them sharply increases with decreasing grain size when conduction electrons and positive holes are created by nuclear particles. For the detection of WIMPs, it is desirable to introduce into the emulsions with very small grains

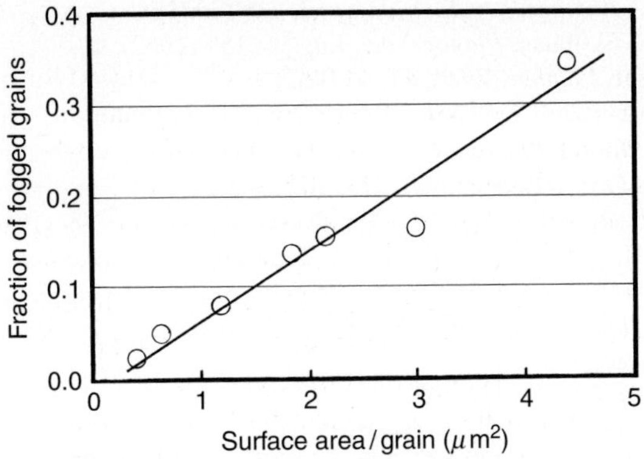

Figure 7.7 Fraction of fogged grains as a function of average surface area of sulfur-plus-gold-sensitized tabular AgBr grains (courtesy of K. Ohzeki).

appropriate technologies for hindering the recombination between conduction electrons and positive holes with high concentrations in the grains.

The inefficiency of latent image formation due to the appearance of electron–hole pairs with high concentration is called high-intensity reciprocity law failure (HIRF) in photography. The HIRF in sulfur-sensitized AgBr grains was attributed to the recombination between free positive holes and electrons trapped by sulfur-sensitization centers [29]. The HIRF and inefficiency of latent image formation with electron–hole pairs in high concentration in the grains for WIMPs detection might be depressed by such technologies as introducing shallow transient electron traps into the grains [79] and replacing sulfur-sensitization centers with selenium- and tellurium-sensitization centers, which are shallower in trap depth than sulfur-sensitization centers [28].

As seen in Figure 7.7, the fraction of fogged AgX grains in an emulsion layer decreases with decreasing surface area of the grains [45]. Since very small AgX grains for the detection of WIMPs have very small surface areas, it is expected that they have strong resistance to fogging, and can stand various emulsion technologies, which tend to bring about fogging.

References

1. P. Demers, *Ionographie Les Emulsions Nacleaires Principes et Applications* (Montreal: Les Presses Univ. Montreal, 1958).
2. C. F. Powell, P. H. Fowler, D. H. Perkins, *The Studies of Elementary Particles by the Photographic Method* (London: Pergamon Press, 1959).
3. W. H. Barkas, *Nuclear Research Emulsions*, Vol. 1: *Techniques and Theory* (New York: Academic Press, 1963).
4. W. H. Barkas, *Nuclear Research Emulsions*, Vol. 2: *Particle Behavior and Emulsion Applications* (New York: Academic Press, 1973).
5. K. Niwa, *J. Phys. Soc. Jpn.*, 59, 871 (2004).
6. K. Niwa, *J. Soc. Photogr. Sci. Tech. Jpn.*, 67, 561 (2004).

7. T. Toshito, *J. Soc. Photogr. Sci. Tech. Jpn.*, 67, 538 (2004).
8. C. R. Berry, D. C. Skillman, *Photogr. Sci. Eng.*, 6, 159 (1962).
9. E. Klein, E. Moisar, *Photogr. Wiss.*, 11, 3 (1962).
10. A. Hirata, S. Honishi, *Bull. Soc. Sci. Photogr. Jpn.*, 16, 1 (1966).
11. J. S. Wey, R. W. Strong, *Photogr. Sci. Eng.*, 21, 248 (1977).
12. Anonymous, Res. Discl. (Item 22534), 225, 20 (1883).
13. J. T. Kofron, R. E. Booms, *J. Soc. Photogr. Sci. Tech. Jpn.*, 49, 499 (1986)
14. R. K. Hailstone, N. B. Liebert, M. Levy, J. F. Hamilton, *J. Imaging Sci.*, 31, 185 (1987).
15. R. K. Hailstone, N. B. Liebert, M. Levy, *J. Imaging Sci.*, 33, 165 (1989).
16. T. Tani, *Photographic Sensitivity: Theory and Mechanisms* (New York: Oxford University Press, 1995), Chapter 4.
17. T. Tani, *J. Imaging Sci.*, 29, 93 (1985).
18. C. R. Berry, *Photogr. Sci. Eng.*, 13, (1969).
19. S. E. Sheppard, *Photogr. J.*, 65, 380 (1925).
20. J. E. Keevert, V. V. Gokhale, *J. Imaging Sci.*, 31, 243 (1987).
21. H. Kanzaki, Y. Tadakuma, *J. Phys. Chem. Solids*, 55, 631 (1994).
22. T. Tani, *J. Imaging Sci. Technol.*, 39, 386 (1995).
23. T. Tani, *J. Imaging Sci. Technol.*, 42, 135 (1998).
24. Tani, *Photographic Sensitivity: Theory and Mechanisms*, Chapter 6.
25. J. F. Hamilton, J. M. Harbison, D. L. Jeanmaire, *J. Imaging Sci.*, 32, 17 (1988).
26. L. M. Kellogg, J. Hodes, in the preprint of *40th Annual Conference of SPSE, Rochester, New York, 1987.*
27. H. Mifune, M. Mizuno, Y. Toyama, T. Shiozawa, J. Okuda, *J. Imaging Sci. Technol.*, 46, 262 (2002).
28. K. Morimura, H. Sasaki, T. Kojima, M. Ihama, S. Yamashita, T. Tani, H. Mifune, in the preprint of *Beijing International Conference: Technology and Application for 21st Century, May 2005, Beijing, China.*
29. R. H. Hailstone, N. B. Liebert, M. Levy, *J. Imaging Sci.*, 34, 169 (1990).
30. E. T. Larson, *PSA J. (Photogr. Sci. Tech.)*, 17B, 19 (1951).
31. T. Tani, Y. Yoshida, *J. Imaging Sci. Technol.*, 44, 242 (2000).
32. G. W. Lowe, J. E. Jones, H. E. Roberts, in J. W. Mitchell (ed.), *Fundamentals of Photographic Sensitivity* (Proc. Bristol Symposium, 1950) (London: Butterworths Science , 1951), 112.
33. T. A. Babcock, M. H. Sewell, W. C. Lewis, T. H. James, *Astron. J.*, 79, 1479 (1974).
34. T. Tani, *J. Appl. Phys.*, 91, 4595 (2002).
35. T. Tani, *J. Imaging Sci.*, 30,41 (1986).
36. R. H. Hailstone, N. B. Liebert, M. Levy, R. T. McCleary, S. R. Gilolmo, D. L. Jeanmaire, C. R. Boda, *J. Imaging Sci.*, 32, 113 (1988).
37. J. H. Web, *J. Opt. Soc. Am.*, 40, 3 (1950).
38. C. R. Berry, *J. Photogr. Sci.*, 18, 169 (1970).
39. R. P. Cliford, *J. Photogr. Sci.*, 25, 116 (1977).
40. H. E. Spencer, L. E. Brady, J. F. Hamilton, *J. Opt. Soc. Am.*, 57, 1020 (1967).
41. T. Tani, *Photogr. Sci. Eng.*, 15, 181 (1971).
42. J. F. Hamilton, R. C. Baetzold, *Photogr. Sci. Eng.*, 25, 189 (1981).
43. T. Tani, T. Tasaka, M. Murofushi, K. Hosoi, A. Hirano, *Imaging Sci. J.*, 47, 1 (1999).
44. T. A. Babcock, T. H. James, *J. Photogr. Sci.*, 24, 19 (1976).
45. K. Ohzeki, Doctoral Thesis to Tokyo Institute of Technology, 2008.
46. K. Ohzeki, *J. Imaging Sci. Technol.*, 40, 591 (1996).
47. J. Spence, in *The Theory of the Photographic Process*, 3rd edn (New York: Macmillan, 1966), Chapter 9.
48. K. A. Yamakawa, *Phys. Rev.*, 82, 522 (1951).

49. P. Broadhead, in T. H. James (ed.), *The Theory of the Photographic Process*, 4th edn (New York: Macmillan, 1977), Chapter 23.
50. T. Tani, *J. Imaging Sci. Technol.*, 51, 110 (2007).
51. G. C. Farnell, R. B. Flint, J. B. Chanter, *J. Photogr. Sci.*, 13, 25 (1965).
52. R. Matejec, E. Moisar, *Photogr. Sci. Eng.*, 12, 133 (1968).
53. K. Miyake, T. Tani, O. Nittono, *J. Soc. Photogr. Sci. Tech. Jpn.*, 60, 4 (2000).
54. L. M. Kellogg, *Photogr. Sci. Eng.*, 18, 378 (1974).
55. R. J. Deri, J. P. Spoonhower, *Appl. Phys. Lett.*, 43, 65 (1983).
56. T. Oikawa, N. Saeki, T. Kaneda, A. Hirano, T. Tani, *J. Imaging Sci. Technol.*, 39, 233 (1995).
57. Opera Collaboration, *An Appearance Experiment to Search for* $\nu_\mu \rightarrow \nu_\tau$ *Oscillation in the CNGS Beam*, Experimental Proposal.
58. K. Kuwabara, S. Nishiyama, *J. Soc. Photogr. Sci. Tech. Jpn.*, 67, 521 (2004).
59. H. Nakatsugawa, T. Tani, *J. Imaging Sci. Technol.*, 47, 78 (2003).
60. T. Tani, *Photogr. Sci. Eng.*, 27, 75 (1983).
61. B. Pontecorvo, *J. Exptl. Theoret. Phys.*, 33, 549 (1957).
62. SUPER-KAMIOKANDE collaboration, Y. Fukuda et al., *Phys. Rev. Lett.*, 81, 1562 (1998).
63. CHOOZ collaboration, M. Apollonio, et al., *Eur. Phys. J. C* 27, 331 (2003).
64. Palo Verde Collaboration, A. Piepke, *Prog. Part. Nucl. Phys.*, 48, 113 (2002).
65. K2K collaboration, M. H. Alm, et al., *Phys. Rev. D*, 74, 072003 (2006).
66. MINOS collaboration, P. Adamson, et al., *Phys. Rev. Lett.*, 101, 221804 (2008).
67. OPERA Collaboration, N. Agafonova, et al., *Phys. Lett.*, B691, 138 (2010).
68. D. D. F. Shiao, L. J. Fortmiller, A. H. Herz, *J. Phys. Chem.*, 79, 816 (1975).
69. J. Pouradier, A. Pailliotet, C. R. Berry, in James (ed.), *The Theory of the Photographic Process*, 6.
70. A. P. Marchetti, K. P. Johansson, G. L. McLendone, *Phys. Rev.*, 47, 4268 (1993).
71. C. R. Berry, *Phys. Rev.*, 161, 848 (1967).
72. T. Kubota, *Appl. Opt.*, 27, 4358 (1988).
73. S. J. Zacharovas, D. B. Ratcliffe, G. R. Skokov, S. P. Vorobiov, P. I. Kumonko, Y. A. Sazanov, in T. H. Joeng, W. K. Sobotka (eds), *Holography 2000*, SPIE, Vol. 4149 (2000), 73.
74. H. Kanzaki, Y. Tadakuma, *Solid State Commun.*, 80, 33 (1991).
75. K. P. Johannson, G. McLendon, A. P. Marchetti, *Chem. Phys. Lett.*, 179, 321 (1991).
76. H. Mifune, T. Ohzeki, K. Yamane, F. Nariyuki, K. Watanabe, M. Yoshikan, M. Nakanishi, T. Maekawa, T. Funakubo, *J. Imaging Sci. Technol.*, 51, 207 (2007).
77. M. Ueta, H. Kanzaki, K. Kobayashi, E. Hanamura, *Excitonic Processes in Solids* (Berlin: Springer, 1986), Chapter 6.
78. C. R. Berry, *J. Photogr. Sci.*, 21, 202 (1970).
79. M. T. Olm, R. S. Eachus, W. G. McDugle, *Bulgarian Chem. Commun.*, 26, 350 (1993).

8

Preparation and Performance of Silver Nanoparticles

Metal nanoparticles, whose size ranges from several nanometers to several tens of nanometers, have been attracting keen interest from many investigators as one of the most promising areas to study because they exhibit a variety of interesting properties and behavior [1–3] that will allow them to be used in catalysts [4], electronic devices [5–7], optical devices [8,9], sensors [10,11], and color materials [12]. Ag nanoparticles of varying size and shape is one of the group's most important members.

As described in the preceding chapters, silver halide (AgX) grains as well as Ag clusters are formed and used for various purposes in AgX photography. Although AgX grains in photography and Ag nanoparticles in the area of metal clusters are observable by electron microscopy and are comparable, most Ag clusters formed in photographic materials are too small to be observed by electron microscopy. Since AgX grains of photographic interest and Ag nanoparticles in metal clusters have face-centered cubic (fcc) crystal structure in common, there are intimate similarities in their structure and formation. In this chapter, the structure and formation of AgX grains in photography and Ag nanoparticles in metal clusters are comparatively analyzed.

Although Ag clusters formed in photographic materials are not observable, they have been characterized extensively enough to ascertain their structure, formation mechanism, and performance. On the other hand, the structure and formation mechanism of the seeds for Ag nanoparticles in metal clusters have not been determined as they are still too small to be observed by electron microscopy. Therefore, the structure and formation mechanism of Ag clusters in photography are described in Section 8.1 as reference.

8.1 Structure and Formation of Silver Nanoparticles

8.1.1. Structure of Silver Nanoparticles and Silver Halide Grains

Silver halide grains of photographic interest (i.e., AgBr and AgCl) and Ag nanoparticles in metal clusters have fcc crystal structure in common, although the former forms ionic crystals whereas the latter forms metallic crystals. In this section, the structure of AgX grains in photography are compared with that of Ag nanoparticles in metal clusters.

As described in Section 2.1.1, AgX grains are classified into two groups, i.e., regular and irregular. The former grains are single-crystalline, while the latter contain crystal defects that influence their shapes. As shown in Figure 2.1, all the possible shapes of regular AgX grains include cubic, octahedral, rhombic dodecahedral, trisoctahedral, icositetrahedral, tetrahexahedral, and hexactahedral, and can actually be prepared [13]. Grains consisting of mixed face types such as tetradecahedral with both {100} and {111} faces on the surface are not included as members of this classification. Since only {100} and {111} faces are intrinsically stable on the surfaces of AgX grains, cubes and octahedrons among others in

Figure 2.1 are formed in the absence of any particular crystal-habit modifier (i.e., capping agent), as shown in Figure 2.2, whereas regular grains with other shapes featured in Figure 2.1 can be formed only in the presence of appropriate crystal-habit modifiers [13].

As shown in Figure 8.1, single-crystalline Ag nanoparticles are, in principle, cubes, octahedrons, and tetradecahedrons (i.e., cuboctahedral) with {100} and/or {111} faces on their surfaces [16]. Taking into account the similarity in crystal structure between AgX and Ag, it is anticipated that there is the possibility of forming single-crystalline Ag nanoparticles with rhombic dodecahedral, trisoctahedral, icositetrahedral, tetrahexahedral, and hexactahedral shapes in the presence of appropriate capping agents.

For both AgX grains in photography and Ag nanoparticles in metal clusters, crystal defects, which lead to the formation of irregular shapes, are important. However, there is some difference in matters of concern. In photography, tabular AgX grains, as shown in Figures 1.2, 2.5, and 6.2, are extremely important especially for color films, as described in Section 6.2.1, whereas rods and needle-like grains of AgX [17,18] are not important and have scarcely been studied in the past. In metal clusters, Ag nano-rods and nano-wires in addition to tabular Ag nanoparticles (i.e., Ag nanoplates), as illustrated in Figure 8.2, are very important and have been extensively studied by many investigators [15,16,19–26].

As described in Sections 2.1.2 and 2.2.6, there are two groups of tabular AgX grains. The first consists of tabular grains with a hexagonal or triangle profile with {111} faces on the main surfaces, and the second consists of those having a square or rectangular profile with {100} faces on the main surfaces, as shown in Figure 2.9. The former is characterized by the presence of doubly twin planes, which are parallel with the main surfaces, and the latter by the presence of two screw dislocations that extend out in different directions from one of the corners in each grain, as shown in Figure 2.9. In the case of twinned tabular AgX grains, each side surface is divided into three parts by its lines intersecting with two parallel twin planes. In the past all three parts were considered to be composed of {111} faces, making the side surface rough with a re-entrant groove and a ridge, as seen in Figure 2.7A. However, it has been revealed recently that the central part is composed of a {100} face, making the side surface rather smooth, as seen in Figure 2. 7B [27–29]. It is obvious that the side surfaces of square or rectangular tabular AgX grains are composed of {100} faces. Based on the similarity in crystal structure between AgX and Ag there is a possibility of forming tabular Ag nanoparticles with a square or rectangular profile with {100} faces on the main surfaces by introducing screw dislocations into seeds of Ag nanoparticles.

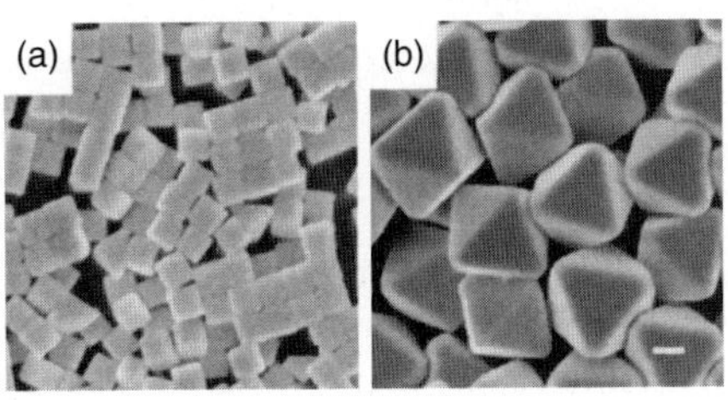

Figure 8.1 Scanning electron micrographs of regular nanoparticles of Ag bounded by the fcc {100} and {111} planes, cubes (a) and octahedral (b) with scale bar = 100 nm. Reproduced with permission from Reference [15], copyright 2008 Wiley-VCH.

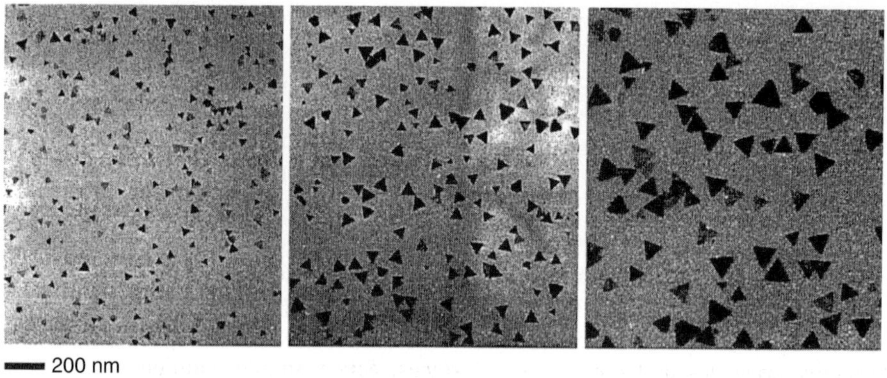

Figure 8.2 TEM images of tabular Ag nanoparticles (i.e., Ag nanoprisms) with average length of 38 ± 7 nm (a), 72 ± 8 nm (b), and 120 ± 14 nm (c). Reproduced with permission from Reference [20]. Copyright 2003 Nature.

There are some similarities in structure between twinned tabular AgX grains and tabular Ag nanoparticles. Their main surfaces are hexagonal or triangle, and composed of {111} faces [22]. Stacking faults are present in parallel with the main surfaces. Although stacking faults are twin planes in the tabular AgX grains, in the tabular Ag nanoparticles they are not [22].

Based on the study of the structure of tabular Ag nanoparticles by means of electron microscopy and X-ray diffraction, it has been proposed that the stacking faults in Ag nanoparticles bring about the formation of a hexagonal-close-packed (hcp) structure [30]. Namely, the structure of tabular Ag nanoparticles is similar to that of tabular AgX grains as compared in Figures 8.3 and 2.7B in that the upper and lower side surfaces are composed of {111} face. However, there is a difference in that the structure of the middle part of the former is hcp, while that of the latter is fcc.

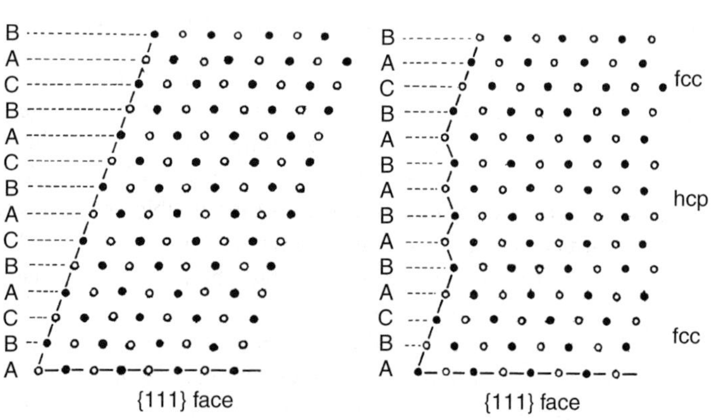

Figure 8.3 Successive stacking of the ABCABC layers on {111} face of a fcc crystal (left) and that with stacking faults giving the ABABAB layer with hcp structure [30]. The closed circles represent atoms in the {110} plane while the open circles represent atoms immediately below.

While the fcc structure is represented by the arrangement of abcabc, as illustrated in Figure 2.6, the hcp structure is represented by the arrangement of ababab. The hcp structure is considered to be unstable for AgX with ionic crystal structure, while it can be stable for Ag with metallic crystal structure. This difference should bring about the difference in the mechanism of anisotropic growth, as described in the next section.

Although tabular Ag nanoparticles containing two parallel twin planes have not been observed, twin planes can be introduced, not only into AgX grains, but also into Ag nanoparticles [16,23,25,31,32]. It is commonly observed for both an AgX grain and an Ag nanoparticle that the introduction of a single twin plane into a grain or a nanoparticle causes a triangular plate to be formed [16]. In the early stage of growth, the presence of a twin plane can cause six side surfaces to be created, where the twin plane ends to form alternating concave- and convex-type surfaces. Since atomic addition to the former surface is more favorable than that to the latter from the viewpoint of the stabilizing energy for attaching atoms to these surfaces, the concave-type surface grows faster than the convex-type one, and vanishes to form a triangular plate. This makes the aspect ratio (i.e., the ratio of the diameter of the main surface to the thickness) of the formed triangular plate limited and therefore useless for photography. Multiple twinned seeds are likely to grow to form rods of AgX and Ag. Although AgX rods [17,18] are not important in photography, Ag rods are important and have been subjected to extensive studies of their structure and formation [15,16,23–25,31–33].

Silver halide grains are randomly arrayed in photographic emulsions, and are not expected to interact with each other for the modification of their photographic performance. Although regular AgX grains can be arrayed to form Langmuir–Blodgett-type mono-layers, as shown in Figure 6.5 [34], they have not been put to practical use. Since the interaction among Ag nanoparticles in metal clusters is very important from the viewpoint of their plasmon resonance, extensive studies have been undertaken to analyze and improve the interaction among nanoparticles by controlling their assemblies [15,16, 31–33] including their Langmuir–Blodgett-type layers [31,33] and regularly arrayed Ag nanoparticles prepared by electron-beam lithography [35].

8.1.2 Formation of Regular Silver Nanoparticles and Silver Halide Grains

Silver halide grains in photography and Ag nanoparticles in metal clusters are formed in aqueous solutions in the presence of protective colloids. In photography, gelatin is solely used as a protective colloid for AgX grains. In metal clusters, poly(vinylpyrrolidone) (PVP) is often used as a protective colloid for Ag nanoparticles [14,15,21,25,26,31]. Silver halide grains in photography are formed as a result of the double decomposition of silver salt and halide salt in aqueous solutions, while Ag nanoparticles are formed as a result of the reduction of silver ions by reducing agents. Despite the difference in the above-stated reactions, both AgX and Ag form fcc crystals.

Metal nanoparticles are shape-controlled from thermodynamic and kinetic viewpoints [16,26]. Thermodynamically favored nanoparticles include regular ones such as cubes, while the surface energy of tabular Ag nanoparticles is higher than that of regular nanoparticles. When metal atoms are generated at a sufficiently high rate, the final product will have no choice but to take the thermodynamically favored shapes. As the reduction becomes substantially slower, the nucleation and growth will be turned into kinetic control and the final product can take a range of shapes that deviate from the thermodynamic ones. In the case of the formation of AgX grains, it is necessary to add reactants at a sufficiently high rate for efficient production of AgX photographic emulsions to meet thermodynamic control.

Both AgX grains in photography and Ag nanoparticles in metal clusters are prepared through nucleation and growth processes according to Lamer's diagram, as illustrated in Figure 2.14 [15,16]. There are two types of nucleation processes: homogeneous and heterogeneous. As described in Section 2.2.3, the nucleation of AgX grains is homogeneous because the addition rate of reactants for the production of AgX grains in photography is too high for heterogeneous nucleation to take place. On the other hand, both homogeneous and heterogeneous nucleation processes are probable for the formation of Ag nanoclusters in metal clusters, depending on the experimental conditions.

The size distribution and shape of AgX grains in photography can be controlled by means of the controlled double-jet method, according to which the concentration of silver ions in the reaction solution is controlled during the precipitation of AgX grains, as described in Sections 2.2.4 and 2.2.5. Usually, the concentration of silver ions in a reaction solution is electrochemically measured by means of an Ag electrode and used to adjust the addition rate of an aqueous solution of halide salt to keep the concentration of silver ions in the solution at the desired value during the precipitation process. The apparatuses for the preparation of AgX grains as described in Section 2.2.2 are particularly important for the control of the size distribution and shape of the grains, making it possible to prepare mono-dispersed AgX grains with desired shape and size. It is anticipated that the methods and apparatuses as stated above will contribute to the progress in the preparation of Ag nanoparticles in metal clusters.

The shapes of regular (i.e., single-crystalline) AgX grains in photography depend on the faces, which appear on their surfaces in the presence of crystal-habit modifiers (i.e., capping agents). Halide ions are regarded as one of the most popular and useful capping agents for AgX grains. They are preferentially adsorbed to {111} faces on the grains and stabilize them, as illustrated in Figure 2.22. Thus, the shape of AgBr grains changes from cubic to octahedral as the concentration of bromide ions in the reaction solutions increases. It is particularly important that halide ions are constituents of AgX grains, and do not therefore disturb photographic performances of the grains. Various kinds of capping agents that make it possible to control the shapes of single-crystalline AgX grains in photography, as illustrated in Figure 2.1, have been developed. Many capping agents for controlling the shape of Ag nanoparticles in metal clusters are being developed [14–16, 19–21,24–26]. One of the most popular capping agents is PVP, which is believed to preferentially bind to {100} and {111} planes on Ag nanoparticles. It is known that citrate is the key component in the transformation from spheres into plates bounded by {111} faces [19–21]. It is anticipated that the development of appropriate capping agents will make it possible to widen the variety of shapes of Ag nanoparticles in the future.

8.1.3 Formation of Tabular Silver Nanoparticles

As for irregular AgX grains in photography, tabular grains are most important especially for color films. One of the two groups of tabular AgX grains is composed of twinned tabular grains with hexagonal and/or triangular profile. The anisotropic growth of these grains is caused by the presence of two twin planes, which are parallel with their main surfaces, and divide each side surface into three parts by their lines intersecting with it. The {100} plane in the central part is the key for anisotropic growth, since it grows much faster than the main surfaces with {111} faces on each grain when the concentration of halide ions in a reaction solution is sufficiently high.

Although twin planes have not been observed in tabular Ag nanoparticles [22], stacking faults have been found to be in parallel with the main surfaces of the tabular nanoparticles, and are believed to cause their anisotropic growth. According to the proposed mechanism, as illustrated in Figure 8.3 [30], the middle part of the surfaces of a tabular nanoparticle is rough in atomic scale, and should support a much faster rate for the addition of Ag atoms during growth than the {111} faces on the main surfaces, bringing about the anisotropic growth of the nanoparticle.

It has been revealed that the formation of a twin plane on a {111} plane in an AgX grain takes place by the coagulation of AgX seeds with {111} faces on the surface [36–39], and can be controlled by adjusting the ability of gelatin to retard the coagulation. The hcp structure in metallic crystals of Ag is considered to be more stable than that in ionic crystals of AgX, although it is less stable than the fcc structure in Ag crystals. Accordingly, the possibility of the hcp structure forming in Ag crystals should be controlled by experimental conditions such as the formation rate of Ag atoms [16,19,20,26,30]. It has also been proposed that four triangular Ag nanoplates coagulate with each other via their side surfaces to form a larger triangular Ag nanoplate [20]. This sort of coagulation has not been observed for the growth of tabular AgX grains in photography.

In photography, it is important to prepare photographic emulsions in which almost all the AgX grains are tabular with a large aspect ratio, as described in Section 2.2.6 and seen in Figure 2.5. However, the formation of twin planes takes place by chance and cannot be controlled by introducing two parallel twin planes into each AgX seed. Therefore, it is not possible to prepare photographic emulsions with tabular AgX grains only without using the Ostwald ripening, which can enable the growth of twinned tabular AgX seeds along with the dissolution of other seeds. Fortunately, this strategy can be realized for the formation of photographic emulsions with tabular AgX grains only. Due to the anisotropic growth of tabular AgX seeds, their growth rate is much larger than those of other AgX seeds, making it possible for tabular seeds to grow and for other seeds to dissolve by carefully controlling the condition for the Ostwald ripening.

Similar procedures need to take place for the formation of an aqueous suspension, in which almost all the Ag nanoparticles are plate-like, through the conversion of initially formed Ag nano-spheres into nanoplates. The realization of this procedure needs the preferential growth of tabular nanoparticles along with the oxidation of initially formed Ag nanospheres. Taking into account the Ostwald ripening, which allows the preferential growth of tabular AgX seeds along with the dissolution of other seeds, it is considered that the origin for the anisotropic growth of tabular Ag nanoparticles should be also the origin for the preferential growth of tabular Ag nanoparticles among other nanoparticles.

Instead of the dissolution of AgX seeds other than tabular ones during the Ostwald ripening, the oxidative etching should take place for the disappearance of initially formed Ag nano-spheres and the supply of silver ions for the growth of tabular Ag nanoparticles. The oxidative etching of Ag nanoparticles proceeds in the presence of oxidizing agents and ligands for silver ions [15,16,19,20]. One of the most popular oxidizing agents is oxygen, while one of the most popular ligands for silver ions is chloride ion. However, the slowness of this process in the Ag system hinders large-scale synthesis [16].

Mirkin et al. reported the photo-induced approach to transfer small Ag nano-spheres into nanoplates with aid of citrate and a ligand [19,20]. The mechanism of photo-induced synthesis of tabular Ag nanoparticles has been studied extensively. The critical step in the synthesis is considered to be the oxidative etching of Ag nano-spheres, and can be enhanced by the irradiation of light on Ag nanoparticles to bring about surface plasmon resonance (SPR) in the nanoparticles. It is known that the SPR of metal nanoclusters exerts a very strong electric field in the area near the nanoclusters [40] as described in Section 8.3.1. The oxidative etching of Ag nanoparticles should be initiated by the electron transfer from a nanoparticle to its exterior, which can be enhanced by an electric field according to the thermionic emission model [41], as described in Section 10.8 and in Figure 10.10. An electron is stabilized in a bulk of water and is called a hydrated electron. It is estimated that the stabilization energy is as large as 3.27 eV [42]. Since the work function of Ag is 4.26 eV, the energy barrier for the thermionic emission from Ag nanoparticle to a hydrated electron is about 1 eV in the absence of an electric field, and should be much decreased by the electric field caused by the SPR of Ag nanoparticles in light of the analysis on the thermionic emission from a metal electrode to an organic layer [43].

Although the mechanism of anisotropic growth of tabular Ag nanoparticles is not necessarily clear [19–22] as compared with that of anisotropic growth of tabular AgX grains, Aherne et al. have proposed that a stacking fault initiates the layer of hcp structure, which is followed by the re-formation of the layer of the fcc structure, as illustrated in Figure 8.3 [30]. It is therefore probable that the formation of Ag nanoparticles with the structure shown in Figure 8.3 is not by chance, and can even be enhanced to bring about the selective formation of tabular nanoparticles of Ag by controlling the conditions for the nanoparticles such as the formation rate of Ag atoms.

Silver nano-rods and wires are important and have been studied extensively in metal clusters. Although AgX rods and needle-like crystals have often been observed among AgX grains in photographic emulsions [17,18], they are not important in photography and have scarcely been studied in the past.

8.1.4 Structure and Formation of Seeds for Silver Nanoparticles

As described in the preceding three sections, knowledge on the structure and formation of Ag nanoparticles in the field of metal clusters is based on experimental results with those that were observable by electron microscope, and is therefore insufficient to ascertain what happens before they become observable. As described in Chapter 4, Ag clusters, which are too small to be observed by electron microscope, play essentially important roles in photographic processes, and have been therefore studied so extensively to reveal their structure and formation mechanism. In this section, knowledge on the structure and formation of Ag clusters in photography is summarized to serve as reference for the structure and growth mechanism of Ag clusters at the stage before they become observable.

In photography, an Ag cluster is formed and grows on the surface and/or in the interior of an AgX grain, being supplied with electrons and silver ions. Silver ions are usually supplied as interstitial silver ions by AgX grains. When AgX grains are in contact with an aqueous solution containing silver ions, an AgX cluster can be supplied with silver ions in the solution [44]. Electrons are usually supplied as photo-electrons by AgX grains to form latent image centers composed of Ag clusters, while they are supplied by reducing agent to form reduction sensitization centers composed of Ag dimers and as fog centers composed of Ag clusters. Silver clusters grow through nucleation and growth stages at special sites on the surface (e.g., kink sites, chemical sensitization centers) and/or in the interior (e.g., jog sites, point defects) of AgX grains. Namely, the nucleation of Ag clusters on the surface and/or in the interior of AgX grains is heterogeneous, although the nucleation of AgX grains is homogeneous, as described in Section 8.1.2. The characterization and design of the sites for the heterogeneous nucleation processes are therefore very important, as described below.

A latent image center is usually formed at a positively charged kink site or at a chemical sensitization center on the surface of an AgX grain, and grows one atom at a time by alternately receiving an electron and a silver ion. It is noted that only one latent image center is formed on an AgX grain with equivalent diameter of less than 0.5 μm. A latent image center can be either an Ag cluster with odd number of atoms (i.e., an odd cluster) or that with even number of atoms (i.e., an even cluster). Although the smallest latent image center is composed of three atoms, it can grow to become large enough to be observed by electron microscopy, as shown in Figure 8.4(a) [45,48]. A latent image center is a deep trap for an electron in the conduction band of an AgX grain and acts as a catalyst for the reduction of the grain by receiving an electron from a developer.

Reduction sensitization centers are composed of dimers of Ag atoms, and formed at kink sites on the surface of AgX grains by reducing agents named reduction sensitizers. Their concentration can be as large as $2800/\mu m^2$ on the surface of an AgBr grain [46,47]. There are two types of reduction sensitization centers. The first is composed of a dimer of Ag atoms formed at an electrically neutral kink site and acts as a trap, not for an electron in

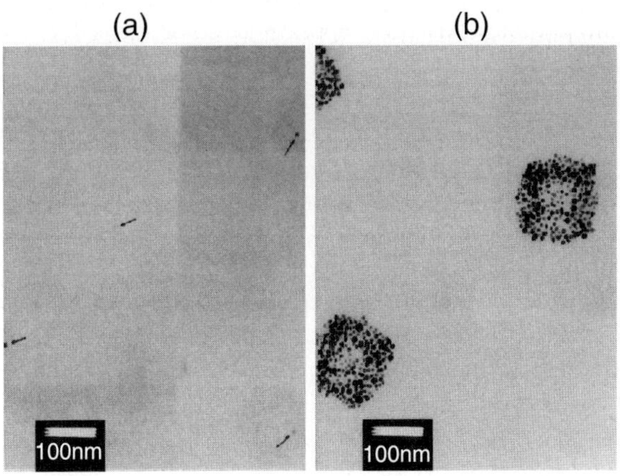

Figure 8.4 (a) An electron micrograph of gelatinate shells of cubic AgBr grains with edge length of 200 nm, where dark areas are gelatinate shells, arrows indicate Ag clusters in the shells, and the length of the bar is 100 nm. These clusters were formed by light. (b) The same except that the clusters were formed by reduction [48].

the conduction band, but for a positive hole in the valence band of an AgX grain. The second is composed of a dimer of Ag atoms formed at a positively charged kink site on the surface of an AgX grain and acts as a trap for an electron in the conduction band of an AgX grain. The former is especially effective for increasing the sensitivity of AgX grains, as described in Section 4.5.4.

When the amount of a reduction sensitizer is larger than that required to fill all the surface kink sites with dimers of Ag atoms, the dimers coagulate with each other to form Ag nanoparticles, which can grow for their observation by electron microscope, as shown in Figure 8.4(b) [48,45], and are called fog centers, since they can initiate photographic development by receiving electrons from a developer. Since the fog centers are formed as a result of the coagulation of dimers of Ag atoms, they are even clusters and therefore diamagnetic, in contrast to latent image centers, which are paramagnetic [48].

It is expected that the structure and formation mechanism of Ag clusters and nanoparticles formed on AgX grain by reducing agents can serve as references when considering the structure and formation mechanism of the seeds for Ag clusters and nanoparticles by reducing agents in aqueous solutions. Since the sites at which Ag clusters are formed play important roles, it is especially important to characterize the sites for heterogeneous nucleation of Ag nanoparticles in the field of metal clusters.

8.2 Silver Nanoparticles in Photography

Various Ag clusters and nanoparticles are used in photographic materials including color films. As described in the previous section and in Chapter 4, Ag clusters, which are too small to be observed by electron microscopy, are formed as latent image centers, reduction sensitization centers, and fog centers.

While metal nanoparticles have recently attracted the keen interest of many investigators from the viewpoint of their SPR, which result from their light absorption [5–12,49], the SPR of Ag nanoparticles has been applied to color films for many years [12,49]. Although pigments are widely used as color materials, their size cannot be smaller than 100 nm. Since metal nanoclusters used for color films are one order of magnitude smaller than pigments, they are transparent due to the absence of light scattering. They are stable for a long time despite their small size. Silver nanoparticles of about 20 nm in diameter have even been used for yellow filters in color films, and are described in the next section.

Many efforts have been made to prepare mono-dispersed Ag nanoparticles. One of the most successful results was reported by Klein and Metz [50]. They mixed an AgX emulsion and an emulsion of silver nuclei, which were capable of initiating development, and coated the mixed emulsion on a film base. Then, the coated emulsion was subjected to solution physical development, according to which electrons were supplied to silver nuclei by a developer, while AgX grains were dissolving to provide the nuclei with silver ions. The size of silver nanoparticles can be controlled by the ratio of the amount of an AgX emulsion to that of an Ag emulsion. Thus, they can prepare mono-dispersed silver clusters varying from 20 to 180 nm in diameter, as shown in Figure 8.5. The following conditions are believed to have realized the formation of mono-dispersed silver nanoclusters. Namely, the growth of each silver nucleus was determined by the amount of AgX grains

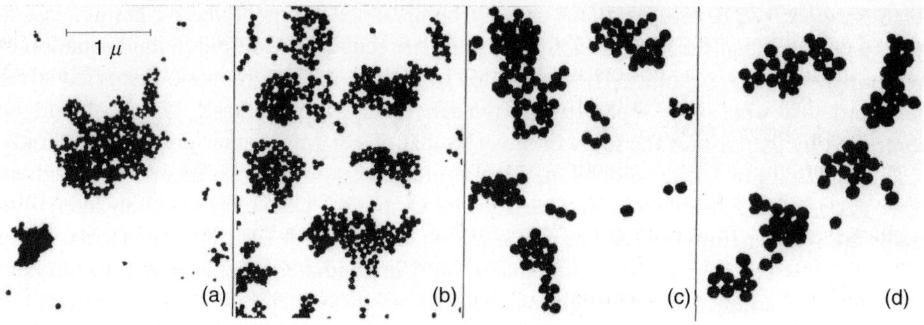

Figure 8.5 Electron micrographs of physically developed silver nanoparticles: (a) 60 nm, (b) 100 nm, (c) 140 nm, and (d) 180 nm. Reproduced with permission from Reference [50], copyright 1961 IS&T, The Society for Imaging Science and Technology.

around the nucleus, while the growth rate was limited by the diffusion of silver ions from the AgX grains to the nucleus.

As seen in Figure 2.2 with description in Chapter 2, mono-dispersed AgX grains can be prepared and separated from each other in a photographic emulsion. Photographic development brings about the reduction of an AgX grain with a latent image center or fog center

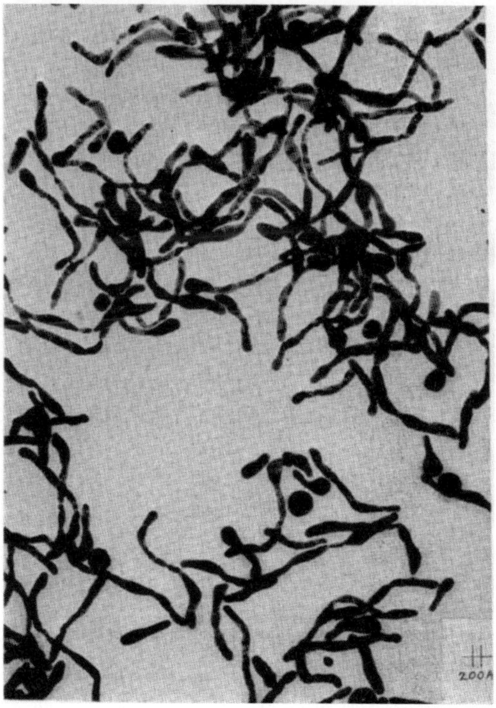

Figure 8.6 Filamentary Ag produced by developing a photographic emulsion. Reproduced with permission from [51], copyright 1969 IS&T, The Society for Imaging and Technology.

to an Ag grain by a developer. In the case of chemical development, the Ag grains cease to grow when all the silver ions in each AgX grain are reduced, resulting in the formation of mono-dispersed silver grains. These facts led to an idea to prepare mono-dispersed silver nanoparticles, which are separated from each other with distance, which should be controlled by the ratio of the mass of silver to that of gelatin or other protective colloid.

During photographic development, a latent image center grows at its interface with an AgX grain, where an electron provided to the center by a developer and an interstitial silver ion coming from the interior of the grain combine with each other to form a silver atom [50,51]. The center grows as a sphere until its diameter reaches about 20 nm, and then continues to grow as a filament of about 20 nm in diameter, as shown in Figure 8.6 [51]. The diameter of a filament is determined by the distance to which a silver atom freshly formed on the above-stated interface can migrate on the surface of a silver center before it finds a place to stay. Filaments of uniform diameter (i.e., 20 nm) are effective for producing efficient black-and-white pictures [52], and are expected to provide unique silver nano-structures useful for electronic and optical devices [53].

8.3 Properties of Silver Nanoparticles

8.3.1 Absorption Spectra of Silver Nanoparticles and Surface Plasmon Resonance

As shown in Figure 8.7, a metal nanocluster absorbs incident light to excite wholely the free electrons in it for their collective oscillation, and exhibits an absorption band peaked at the frequency for meeting the resonance of the oscillation. This condition is named SPR, and is characterized by unique features. First of all, they confine photons in the nano-space (i.e., the space close to the surface of a nanoparticle), which is much narrower than the space limited by the refraction of light. Together with the metal surfaces with nano-structure, metal nanoparticles create a strong electric field in the nano-space and strongly enhance the excitation of materials in the field. Since the condition for SPR of metal nanoparticles is very sensitive to the refractive index (i.e., optical dielectric constant) of their surroundings [49], they can provide us with highly sensitive sensors.

It is predicted from the Mie theory that the peak frequency and half-width of the absorption band due to the SPR of spherical metal clusters depend on the size of

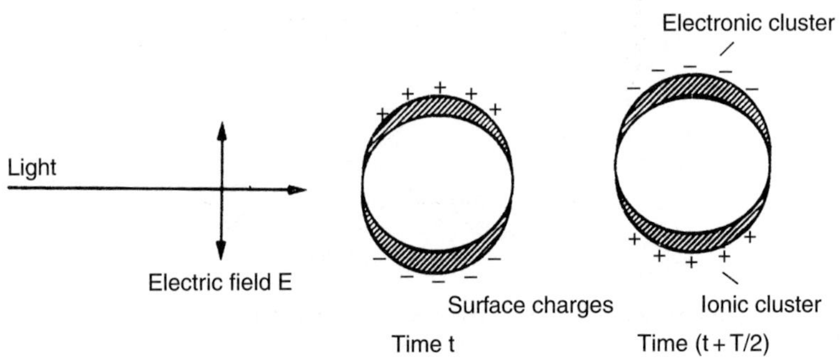

Figure 8.7 The excitation of surface plasmon by an incident light wave.

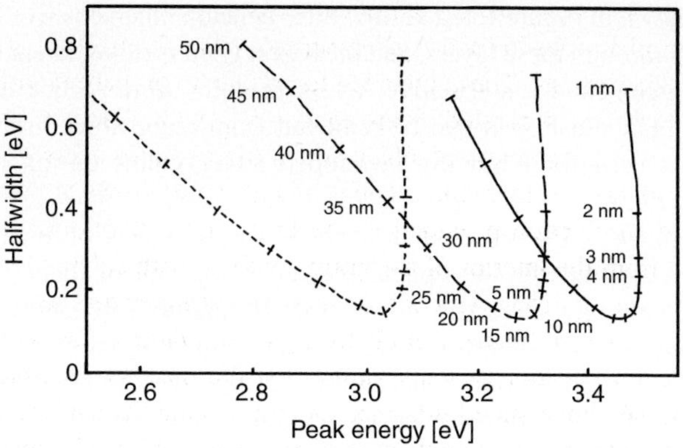

Figure 8.8. Half-width and peak energy of the SPR of Ag nanoparticles suspended in media with variation of optical dielectric constant (1 for vacuum, 2.25 for glass, and 6.25 for TiO_2).

the clusters and the optical dielectric constant of their surroundings [49], as shown in Figure 8.8. As seen in in this figure, it is predicted that the peak frequency corresponding to the SPR of spherical metal nanoparticles is nearly independent of their size when they are dilute and smaller than about 20 nm in diameter, and that these metal nanoclusters should therefore exhibit a sharp absorption band even when their size distribution is wide. As seen in Figure 8.9, Ag nanoclusters of about 20 nm in diameter actually gave a sharp absorption band, exhibiting yellow color in gelatin [50]. This result is in good agreement with the fact that spherical Ag clusters of about 20 nm in diameter have been used for yellow filters in

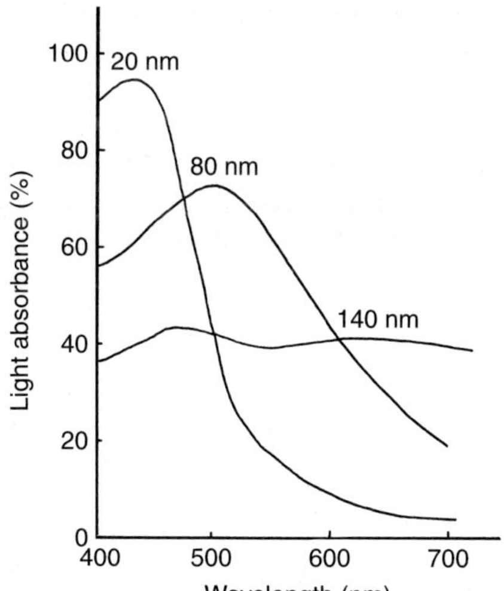

Figure 8.9 Absorption spectra of Ag nanoparticles of varied diameter in gelatin layer (0.08 g Ag/m^2). The average diameter of Ag nanoparticles is indicated on the respective curve. Reproduced is cited with permission from Reference [50], copyright 1961 IS&T, The Society for Imaging Science and Technology.

color films. As shown in Figure 1.1, a yellow filter beneath blue-sensitive layers absorb the blue light passing through these layers, and prevent it from forming latent image centers on AgX grains in green- and red-sensitive layers in color films. Since a yellow filter is composed of Ag nanoclusters, it can be removed from color films during the bleach-fix process, which is used for removing developed silver grains during the development process for color films.

As seen in Figure 8.8, the wavelength of the absorption peak of metal nanoclusters at their SPR sharply depends on the optical dielectric constant of their surroundings, and increases with increasing dielectric constant [49]. Figure 8.10 shows the absorption spectra of Ag clusters and nanoparticles that were formed on the surfaces of octahedral AgBr grains [46]. The absorption band around 550 nm was ascribed to the SPR of fog centers composed of Ag nanoparticles of several nanometers in diameter, while the absorption band at 474 nm was ascribed to reduction sensitization P centers composed of Ag dimers. In agreement with the above-stated prediction together with the fact that the optical dielectric constant of AgBr is larger than that of gelatin, Ag nanoparticles on the surface of AgBr grains absorbed green light [46] in contrast to those in gelatin, which absorbed blue light as seen in Figure 8.9. It is noted that P centers gave rise to the absorption band, which was attributed to their HOMO–LUMO transition.

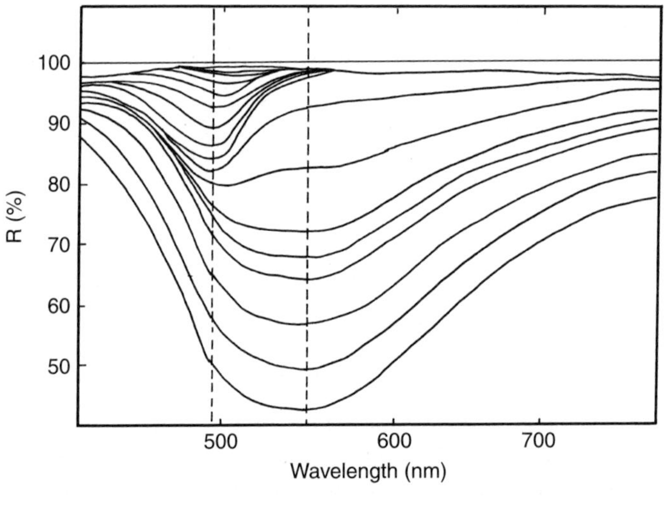

Figure 8.10 Diffuse reflectance spectra of reduction-sensitized octahedral AgBr emulsions (the same as those in Figure 3.32) with the corresponding unsensitized one as a reference to indicate the absorption spectra of Ag clusters and nanoparticles on the grains. The emulsions were reduction-sensitized by use of DMAB with amounts of 0 and 6.3×10^{-7}, 1.0, 1.6, 2.5, 4.0, and 6.3×10^{-6}, 1.0, 1.6, 2.5, 4.0, and 6.3×10^{-5}, 1.0, 1.6, 2.5, 4.0, and 6.3×10^{-4} mol/mol AgBr. The absorption bands appearing in regions at shorter and longer wavelength were assigned to reduction sensitization P centers composed of Ag_2 and fog centers composed of Ag nanoparticles. Reproduced with permission from Reference [46], copyright 1994 IS&T, The Society for Imaging Science and Technology.

In a black-and-white photograph, developed Ag grains are displayed as they are, and are required to exhibit black color (or another color depending on the purpose) with large covering power. The modification of the color of developed Ag grains is named toning, and has been studied by many investigators. The covering power of Ag nanoparticles is defined as the ratio of their optical density to the mass of Ag. Although Ag nanoparticles used for yellow filters in color films exhibit large covering power, which is independent of their size, they are not suitable for black-and-white photographs since they exhibit yellow color. It is known that the wavelength for the SPR of Ag clusters increases with increasing size and aspect ratio (i.e., ratio of long axis to short). Although Ag clusters absorb light of longer wavelength and appear black as their size increases, as seen in Figures 8.8 and 8.9 [49,50], their covering power decreases linearly with increasing size [54]. On the other hand, the increase in aspect ratio is useful for black-and-white photographs by making developed silver appear black without deteriorating its covering power, and is achieved by forming filaments of Ag during development, as described in the previous section.

The idea of changing the color of a black-and-white photograph is called toning, and has a long history [55]. Studies on toning has revealed that the absorption spectra of Ag nanoparticles due to their SPR are influenced by their size and morphology [56–60], interaction among the nanoparticles [58,59], and surface modification [57–61]. The surfaces of Ag nanoparticles were modified by replacing Ag atoms on the grain surface with other metal atoms [62], and by adsorbing organic compounds such as dyes and mercapto-compounds to the nanoparticles [57–60]. The effects of these treatments were significant and explained from the viewpoint of the SPR of metal nanoparticles.

In agreement with the above-stated prediction, yellow color exhibited by dilute silver nanoparticles of about 20 nm in diameter in gelatin became black when the concentration of the nanoparticles was increased [63]. It is known that metal nano-rods exhibit two absorption bands corresponding to the SPR with the directions of its long and short axes [64] and therefore are more suited to exhibiting black color than spherical metal clusters. Since Ag filaments [51–53] are regarded as the extension of Ag rods, they can provide black-and-white photographs with appropriate Ag nanoparticles in terms of black color and high covering power.

8.3.2 Electronic Structures of Silver Nanoparticles

As described in Sections 3.4.2 and 4.4.2, the electronic structures of Ag clusters and nanoparticles depend on various factors including their size (i.e., the number of silver atoms in a cluster), sites at which they are formed, their surface states, and the dielectric constant of their surroundings. Silver clusters acting as latent image centers and reduction sensitization centers are too small to be regarded as metals, and are treated by molecular orbital theory with discrete electronic energy levels, such as the highest occupied molecular orbital (HOMO) and the lowest unoccupied orbital (LUMO) [65–68], and the shell model with discrete electronic energy levels of 1*s*, 1*p*, 1*d*, etc. [1,69,70]. On the other hand, Ag nanoparticles of several nanometers in diameter contain several hundred Ag atoms per nanoparticle, and have electronic energy levels that distribute almost continuously [71–73]. Namely, the heights of the HOMO and LUMO of these nanoparticles are nearly equal. It was observed that the Fermi level of these nanoparticles, as expressed by their redox potential, became higher with decreasing size according to the

Gibbs–Thomson effect [71–73]. Namely, the surface energy of the Ag nanoparticles determined their electronic structure.

Figure 4.5 shows the electronic energy levels of linear Ag clusters as calculated by means of the Hueckel-approximation molecular orbital method [67]. As described in Section 4.4.2, the difference in the electronic energy level between Ag clusters with an odd number of valence electrons (odd clusters) and those with an even number of valence electrons (even clusters) is significant, named the even–odd alternation [65–69,74], and has been supported by experiments [74]. However, it does not seem that the even–odd alternation clearly influences the processes leading to latent image formation in various photographic phenomena. It was proposed that a P center Ag_n combined with a silver ion at a surface kink site to form a silver cluster ion Ag_{n+1}^{+}, and that this combination should diminish the effect of the even–odd alternation [47].

As stated above, a small silver cluster is regarded as a molecule, and is characterized by its frontier orbitals (i.e., HOMO and LUMO). The energy for the transition of an electron from a molecule's HOMO to the vacuum level is the ionization energy of the molecule, while the energy liberated by the transition of an electron from the vacuum level to a molecule's LUMO is the electron affinity of the molecule. It has been pointed out that the ionization energy of an organic semiconductor molecule in its crystal was smaller than that of the molecule in vacuum [75]. The explanation for this difference was extended to the difference in ionization energy between aromatic hydrocarbon molecules in solid and those in vacuum [76]. From this viewpoint, it is considered that the ionization energy of an Ag cluster in solid is smaller due to the polarization energy of its surroundings than that of the cluster in vacuum, and that the electron affinity of an Ag cluster in solid is larger due to the polarization energy of its surroundings than that of the cluster in vacuum. The polarization energy can be evaluated by $(1-1/D)e^2/2r$, where r is the equivalent radius of the ionized portion in a molecule and D is the optical dielectric constant of its surroundings. As a result, this evaluation has been successfully used to explain the difference in ionization energy between aromatic hydrocarbons in vacuum and those in solid [76].

Since the optical dielectric constant of AgX is larger than that of gelatin, it is predicted that the electron affinity is in the order of an Ag cluster in the interior of an AgX grain, the cluster on the surface of the grain, and the cluster in gelatin. Similarly, it is predicted that the ionization energy is in the order of an Ag cluster in gelatin, the cluster on the surface of an AgX grain, and the cluster in the interior of the grain. It is therefore predicted that an electron trap provided by an Ag cluster in the interior of an AgX grain is deeper than that provided by a silver cluster on the grain surface. This prediction is in good agreement with the fact that under illumination Ag clusters in the interior of grains continue to grow, while those on the surface degrade and finally disappear [77].

It is known that the ionization energy of metal is influenced by various factors. Even for such metals as Pt and Au, which are not subjected to any chemical reaction under room atmosphere, their ionization energies (i.e., work functions) have been confirmed to decrease significantly when they are in contact with layers of inert organic materials [78–81] and when they are displaced from vacuum to room atmosphere [82,83]. One of the most probable reasons for this phenomenon is the push-back effect [79]. In vacuum, some fraction of electrons near the surface of metals are spilled out to vacuum, and are pushed back by layers of inert organic materials or contaminating hydrocarbons to form an effective layer of dipoles on the metal surface, which thus decreases the ionization energy of the metal. Metals with small ionization energy, including Ag, are likely to be oxidized under room atmosphere to form layers of oxides on their surfaces, which thus changes their ionization energy.

As stated above, the ionization energy and electron affinity of Ag clusters in photographic films should be influenced by many factors and therefore cannot be evaluated precisely. However, the electronic structure of Ag clusters is essentially important, not only for understanding the mechanisms of latent image formation on AgX grains in photographic films, but also their effect on such devices as solar cells [7] and organic light-emitting diode [5,6], in which they are expected to enhance the excitation and emission of materials concerned without disturbing the behaviors of electronic carriers and excitons. For this purpose, it was proposed to control the electronic structure of metal clusters by modifying their surfaces [84,85].

In organic devices, Ag nanoparticles are surrounded by organic molecules and polymers. It is therefore desirable to measure the electronic structure of Ag nanoparticles covered by them. In photographic films and papers, Ag nanoparticles as well as AgX grains are always covered by gelatin. For the study on photographic films, it was proposed to use ultraviolet photoelectron spectroscopy (UPS) to measure the ionization energy (i.e., the difference in electronic energy between the vacuum level and the top of the valence band) of AgX films covered with thin gelatin membranes, which were thin and did not disturb the UPS measurement, as shown in Figure 3.5 [86,87].

As stated above, small Ag clusters can be also treated via the shell model [1,68,69]. One of the most important and experimentally confirmed characteristics of the shell model for metal clusters is the magic number, which is the number of valence electrons at the closed shell configurations [88,89]. Thus, the clusters with a closed shell, such as Ag_2 with a closed $1s$ shell and Ag_8 with closed $1s$ and $1p$ shells, are more stable and liable to be formed more frequently than clusters with an open shell. This prediction is in good agreement with the observation that reduction sensitization produced as many as 2800 dimers of silver atoms per squared micrometer before it produced fog centers composed of large Ag clusters [46,47].

Metal clusters and small metal nanoparticles are also the subject of the Kubo theory [90]. According to the theory, a metal cluster or a small metal nanoparticle is not likely to lose or receive an electron from its exterior, since it needs large energy to do so. The energy required for the acceptance or release of an electron by a cluster (E) is expressed by $e_2/2\pi\epsilon_0 d$, where d is the diameter of a cluster and ϵ_0 is the dielectric constant of its surroundings and increases with its decreasing size. When d is 2 nm, E is as large as 1 eV. Therefore a Ag cluster and a small nanocluster are considered to keep the number of their valence electrons unchanged after forming on an AgX grain. Then, Ag clusters and small nanoparticles, which were formed photolytically and chemically on the surface of AgX grains, were subjected to electron paramagnetic resonance (EPR) measurements. It was revealed that all the silver nanoclusters formed by reduction of AgX grains were diamagnetic, while many of the nanoclusters formed by photolysis of the grains were paramagnetic. Since Ag clusters grow one atom at a time under illumination to give even and odd clusters alternately, it would be reasonable to observe that nearly half of the clusters were paramagnetic. The observation that all the chemically formed silver nanoclusters were diamagnetic indicates that they were even clusters that formed as a result of the coagulation of dimers of Ag atoms. In general, an organic molecule gives two electrons to an electrode on its oxidation since it has two electrons in its HOMO.

Hoffmann et al. have proposed that the Jahn–Teller effect takes place on the surface of AgX to cause the pairing distortion of silver ions as shown in Figure 2.3 [91]. As a result, two adjacent silver ions on the surface of AgX are expected to approach each other and give a deepened electronic energy level. It has been suggested that the transfer of two electrons to the electronic energy levels provided by the two silver ions causes the direct formation of an Ag dimer.

Metal nanoparticles with SPR are expected to enhance the light absorption of photosensitive materials in solar cells and photo-sensors [7], and the emission of organic molecules in organic light-emitting diodes (OLED) [5,6]. For this purpose, size, structure, and arrangement of metal nanoparticles should be designed so that the molecules can be placed within reach of the confined light or electric field. In addition, they should not disturb the electronic processes in these devices by trapping electronic carriers. It is difficult to determine the electronic structure of metal nanoparticles due to various factors, as stated in this section. However, it is likely that the electronic structure of Ag nanoclusters in organic devices should be modified for this purpose since the work function of Ag is 4.26 eV and is likely to allow Ag nanoparticles to capture electrons and/or positive holes. On the other hand, it seems that Au nanoparticles in a hole transporting layer in an OLED do not disturb the transport of positive holes in the layer [92] possibly because the work function of Au is as large as more than 5 eV, and could prevent Au nanoparticles from disturbing the hole transport in the layer.

8.4 Performance of Silver Nanoparticles

8.4.1 Silver Nanoparticles as Catalysts

Metal clusters act as catalysts for various reactions [93,94]. It has been noted that they include gold clusters [95]. As described in Chapter 4, Ag clusters are photolytically formed on AgX grains as latent image centers, and exhibit catalytic phenomenon by acting as traps for electrons in the grains and in a developer. On the other hand, Ag clusters formed as reduction sensitization centers trap positive holes to disappear on exposure to light [46,47]. Silver clusters acting as electron traps and positive hole traps on AgX grains are named as P centers and R centers, respectively [68,46]. These two kinds of silver clusters with different functions can be experimentally confirmed by means of microwave and radio frequency photoconductivity measurements [96], as shown in Figure 3.32.

It has been proposed that Ag clusters formed at positively charged kink sites on grain surfaces act as electron traps and that silver clusters formed at neutral kink sites act as positive hole traps [97,68,46,47], as shown in Figure 4.3. The former includes latent image centers, which act as catalysis in photographic development. These results indicate that special sites on the substrates play important roles for the function of catalysts. As described in Section 8.3.2, the electronic structure of Ag clusters is essentially important, and has been subjected to examination by molecular orbital calculation and photoelectron spectroscopy.

Catalytic action of a latent image center on an AgX grain is initiated by the capture of an electron in a developer by the center. Then, a captured electron combines with a silver ion to form and add a silver atom to the center. The repetition of these processes brings about the growth of the center. The center is supplied with a silver ion from the interior of the grain in the case of chemical development, and from a developer in the case of physical development. In the case of chemical development, the growth of each center is therefore limited by the number of silver ions in the grain, while it is not limited in the case of physical development.

The size of Ag clusters is also essentially important for their action as latent image centers. In particular, the evaluation of the size of the smallest latent image center is one of the most important subjects in photographic science and technology and has been carried out via various methods by many investigators, as described in Section 4.4.1. Fayet et al. applied quadrupole mass spectroscopy to Ag cluster cations, and deposited size-selected cluster cations on AgBr grains [98]. It was believed that each Ag cluster cation became a neutral Ag cluster by extracting an electron from an AgBr grain. They observed that only Ag cluster cations of Ag_4^+ and larger were able to initiate the development of AgBr grains by a developer, and concluded that the smallest latent image center was composed of four Ag atoms on the assumption that Ag_n^+ became Ag_n when it was deposited on the surface of an AgBr grain. This methodology and result strongly impacted not only photographic science and technology, but also the science and technology of catalysis of metal nanoclusters.

The effectiveness of the catalytic action of a latent image center can be expressed by the degree of amplification brought about by photographic development. As shown in Figure 1.6 with a cubic AgBr grain with edge length of 1 μm, which contain 2×10^{10} silver ions, the smallest latent image composed of four silver atoms on the grain brings about a degree of amplification of 5×10^9. Although the degree of amplification should decrease with increasing size of an actual latent image center, it is still large enough to provide photographic films with high sensitivity.

As described in Chapter 4, the catalytic reaction of a latent image center during photographic development can be used to form dyes for color films [99,100,54]. Namely, a developing agent composed of a p-phenylenediamine derivative gives an electron to a latent image center and becomes its oxidized species named a quinonediimine. Then, a quinonediimine reacts with a coupler to form a dye molecule in one of emulsion layers in color films, as shown in Figure 1.5.

8.4.2 Silver Nanoparticles for Spectral Sensitization (Becquerel Effect)

As early as in 1883, Becquerel observed that the photographic sensitivity of AgCl papers could be extended into green and red regions after the papers were exposed to blue light [101]. This phenomenon was named the Becquerel effect, and was studied by several groups of investigators [102–106]. West and Carroll developed a method for measuring the photoconductivity of AgX films, and applied it to the study on the Becquerel effect of AgBr photographic films [107]. Although an AgBr film was originally inert photoconductively to light longer than 500 nm, it became active to long-wavelength illumination (e.g., 578 nm) after exposure to blue light sufficiently strong to produce a slight visible amount of photolytically formed Ag nanoclusters.

A plausible mechanism for extending photoconductivity toward longer wavelengths via pre-exposure of the film to blue light is afforded by the discussion of Mott and Gurney on the photoelectric behavior of metals in contact with photoconducting salts. The photolytic product produced by the blue pre-exposure almost certainly consists of Ag nanoparticles in contact with AgBr grains. Although photoelectric emission from Ag nanoparticles into the vacuum requires excitation by ultraviolet light, Mott and Gurney pointed out that Ag nanoparticles can emit electrons into the conduction band of AgBr grains by visible light.

The Becquerel effect has been extended to inorganic semiconductors such as TiO_2 grains for their application to photocatalytic and photovoltaic devices [108–111]. Tatsuma et al. observed the Becquerel effect (i.e., spectral sensitization) of the photoelectric effect in TiO_2 grains by Ag and Au nanoparticles. Namely, exposing Ag nanoparticles on TiO_2 grains to light, which excited, not TiO_2 grains, but the Ag nanoparticles, caused the appearance of conduction electrons in the TiO_2 grains. The light absorption of an Ag nanoparticle brings about SPR with collective oscillation of all the free electrons in it, and does not give an excited electron, which can be transferred to the conduction band of TiO_2. The likely mechanisms for the Becquerel effect of TiO_2 by Ag nanoparticles are (1) the energy transfer of the surface plasmon to the HOMO–LUMO transition in an Ag nanoparticle, (2) the energy transfer of the surface plasmon to TiO_2, and (3) the electron transfer from an Ag nanoparticle to the conduction band of TiO_2 according to the thermionic emission under a strong electric field exerted by the SRP.

Tatsuma has considered the possibility of (1). Although this idea has not yet been proved, it is supported by the following results. Ozin et al. observed the absorption peaks due to the HOMO–LUMO transitions of Ag_2, Ag_3, Ag_4, Ag_5, Ag_6, Ag_7, and $Ag_{\geqq 8}$ in Ar matrix at 412, 440, 490, 505, 520, 536, and 540 nm, respectively [112]. The present author observed the absorption band due to the HOMO–LUMO transition of reduction sensitization P centers on AgBr grains at 474 nm, as seen in Figure 8.10 [46]. Together with these results, the theoretical result, as seen in Figure 4.5 [67], indicates that the HOMO–LUMO gap should endlessly decrease with increasing size of Ag clusters and nanoparticles.

As described in Section 5.9.1, it has been theoretically and experimentally confirmed that dye sensitization of AgX photographic films can take place according to two different mechanisms: electron transfer and energy transfer. Although the former demands the appearance of an excited electron in a dye molecule, which can be directly transferred to the conduction band of an AgX grain, the latter needs, not an excited electron, but an excited state, which can be transferred to an AgX grain to excite an electron at some intermediate level to its conduction band. It has therefore been proposed that spectral sensitization of photoelectric effects of semiconductor grains by metal nanoparticles takes place according to the energy transfer from a metal nanocluster at the SPR state to semiconductor grains, which brings about the excitation of an electron at an intermediate level to the conduction band of a semiconductor grain.

The energy transfer mechanism was experimentally confirmed by Kuhn et al. By use of Langmuir–Blodgett (LB) membranes, they were able to adjust the distance between a sensitizing dye layer and AgBr and to carry out the experiments at the distance at which the energy transfer mechanism should work, while the electron transfer one should not. Under this condition, they observed dye sensitization, and concluded that the energy transfer mechanism actually took place [113,114], although its rate was smaller than the rate of electron transfer when the dye layer was in contact with AgBr [115].

The electron transfer from an Ag nanoparticle to the conduction band of TiO_2 is considered to take place according to thermionic emission [41,43] due to the following reasons. Although the actual work function of Ag nanoparticles in devices is not clear, it should not be large since the work function of pure bulk Ag is as small as 4.26 eV. The bottom of the conduction band of TiO_2 crystal is situated at 4.3 eV below the vacuum level [116,117]. The electric field exerted in the area near an Ag nanoparticle by its SPR is high enough to decrease the energy barrier for the electron transfer from an Ag nanoparticle to the conduction band of TiO_2, as seen with the thermionic emission of an electron in the case of a metal/organic semiconductor layer [43].

8.4.3 Stability of Silver Nanoparticles

Since the work function of Ag (i.e., 4.26 eV) is not large enough to resist oxidation and sulfidation, it is probable that Ag nanoparticles in devices suffer from corrosion. An image with a black-and-white photographic paper is composed of Ag nanoparticles and nano-filaments of about 20 nm in diameter and is likely to suffer from degradation in the absence of appropriate treatments [118–120]. In the case of sulfidation, sulfur usually comes, not from the outside of papers after its processing (i.e., development, fixation, washing, and drying), but from sulfur-containing compounds such as thiosulfate, which remained in the papers after processing due to insufficient washing. In the case of oxidation, oxidizing agents such as ozone gas and other oxidizing gases come from the outside of the papers after processing. As a result, various treatments for protecting photographic papers from degradation have been proposed [118–120].

Silver nanoparticles deposited on a substrate suffer from corrosion due to sulfidation and oxidation [121]. In contrast to black-and-white photographic papers, it has been proposed that sulfidation is more important than oxidation as to the origin of the corrosion of silver nanoparticles, and that corrodent gases come from the outside of the Ag layers. Gelatin layers in photographic papers are considered capable of preventing sulfidations's corrodent gases from penetrating into the layers.

It is obvious that latent image centers are much smaller than Ag nanoparticles, and that oxidation or sulfidation of one Ag atom has a larger effect on the performance of a latent image center than on that of an Ag nanoparticle. However, it seems that latent image centers on AgX grains in photographic films are more resistive to corrosion than Ag nanoparticles, as described in Chapters 4 and 6. This observation is based on the following reasons. The electronic structure of a latent image center is composed of discrete electronic energy levels with a large energy gap between HOMO and LUMO, while that of a nanoparticle is composed of continuously distributed electronic energy levels. Moreover, a latent image center is formed at a positively charged kink site. It is therefore more difficult to remove an electron from a latent image center than from a nanoparticle. According to the Kubo effect [90], a small metal cluster is highly resistive to degradation due to the large energy required to take an electron from or give an electron to the cluster, as described in Section 8.3.2.

References

1. S. Sugano, *Microcluster Physics* (Berlin, Heidelbelg: Springer-Verlag, 1991).
2. Y. Waseda, A. Muramatsu (eds), *Morphology Control of Materials and Nanoparticles* (Berlin: Springer, 2004).

3. G. Schmid (ed.), *Nanoparticles* (Weinheim, Germany Wiley-VCH, 2004).
4. H. Hirai, N. Toshima, in Y. Iwasawa (ed.), *Tailored Metal Clusters* (Dortrecht: D. Reidel, 1986), 87.
5. D. K. Gifford, D. G. Hall, *Appl. Phys. Lett.*, 81, 4315 (2002).
6. J. R. Lokowitz, *Anal. Biochem.*, 337, 171 (2005).
7. K. R. Catchpole, A. Polman, *Optics Express*, 16, 21793 (2008).
8. W. L. Barnes, A. Dereux, T. W. Ebbesen, *Nature*, 424, 824 (2003).
9. S. Ogawa, Y. Hayashi, N. Kobayashi, T. Tokizaki, A. Nakamura, *Jpn. J. Appl. Phys.*, 33, L331 (1994).
10. J. Wieser, A. Wokaun, H. Hoffmann, *Progr. Colloid Polym. Sci.*, 76, 271 (1988).
11. F. Meriaudeau, T. R. Downey, A. Passian, A. Wig, T. L. Ferrell, *Appl. Opt.*, 37, 8030 (1998).
12. Y. Nakao, K. Kaeriyama, *J. Appl. Polym. Sci.*, 36, 269 (1988).
13. J. E. Maskasky, *J. Imaging Sci.*, 30, 247 (1986).
14. A. Tao, P. Sinsermsuksakul, P. Yang, *Angew. Chem. Int. Ed.*, 45, 4597 (2006).
15. A. R. Tao, S. Habas, P. Yang, *Small*, 4, 310 (2008).
16. Y. Xia, Y. Xiong, B. Lim, S. E. Skrabalak, *Angew. Chem.*, 48, 60 (2009).
17. K. Klein, H. J. Metz, E. Moisar, *Photogr. Korresp.*, 99, 99 (1963).
18. K. Klein, H. J. Metz, E. Moisar, *Photogr. Korresp.*, 100, 55 (1963).
19. R. Jin, Y. Cao, C. A. Mirkin, K. L. Kelly, G. C. Schatz, *J. G. Zheng, Science*, 294, 1901 (2001).
20. R. Jin, Y. C. Cao, E. Hao, G. S. Metraux, G. C. Schatz, C. A. Mirkin, *Nature*, 425, 487 (2003).
21. Y. Sun, Y. Xia, *Adv. Mater.*, 15, 695 (2003).
22. V. Germain, J. Li, D. Ingert, Z. L. Wang, M. P. Pileni, *J. Phys. Chem. B*, 107, 8717 (2003).
23. J. Reyes-Gasga, J. L. Elechiguerra, C. Liu, A. Camacho-Bragado, J. M. Montejano-Carrizales, M. J. Yakaman, *J. Cryst. Growth*, 286, 162 (2006).
24. C. J. Murphy, T. K. Sau, A. M. Gole, C. J. Orendorff, J. Gao, L. Gou, S. E. Hunyadi, T. Li, *J. Phys. Chem. B*, 109, 13857 (2005).
25. B. Wiley, Y. Sun, B. Mayers, Y. Xia, *Chem. Eur. J.*, 11, 454 (2005).
26. I. Washio, Y. Xiong, Y. Yin, T. Xia, *Adv. Mater.*, 18, 1745 (2006).
27. R. Jagannathan, R. V. Mehta, J. A. Timmson, D. L. Black, *Phys. Rev. B*, 48, 13261 (1993).
28. R. V. Mehta, R. Jagannathan, J. A. Timmons, *J. Imaging Sci. Technol.*, 37, 107 (1993).
29. R. V. Mehta, R. Jagannathan, J. A. Timmons, *J. Imaging Sci. Technol.*, 40, 77 (1996).
30. D. Aherne, D. M. Ledwith, M. Gara, J. M. Kelly, *Adv. Funct. Mater.*, 18, 2005 (2008).
31. Z. L. Wang, *J. Phys. Chem. B*, 104, 1153 (2000).
32. J. L. Elechiguerra, L. Larios-Lopez, C. Liu, D. Garcia-Gutierrez, A. Camacho-Bragado, M. J. Yacaman, *Chem. Mater.*, 17, 6042 (2005).
33. M. Brust, C. J. Kiely, *Colloid. Surf. A*, 202, 175 (2002).
34. T. Heki, N. Inoue, Forma, 4, 55–8 (1989).
35. Y. Yokota, K. Ueno, V. Mizeikis, S. Juodkazis, K. Sasaki, H. Misawa, *J. Nanophotonics*, 1, 011594 (2007).
36. C. T. Mumau, E. F. Haugh, *J. Imaging Sci.*, 30, 198 (1986).
37. M. G. Antoniades, J. S. Wey, *J. Imaging Sci.*, 36, 517 (1992).
38. M. G. Antoniades, J. S. Wey, *J. Imaging Sci. Technol.*, 37, 272 (1993).
39. M. G. Antoniades, J. S. Wey, *J. Imaging Sci. Technol.*, 39, 323 (1995).
40. C. Kittel, *Introduction to Solid State Physics*, 7th edn (New York: Wiley, 1996), Chapter 10.
41. S. M. Sze, *Physics of Semiconductor Devices* (New York: Wiley, 1981), Chapter 5.
42. Y. Tang, H. Shen, K. Sekiguchi, N. Kurahashi, T. Mizuno, Y. Suzuki, T. Suzuki, *Phys. Chem. Chem. Phys.*, 12, 3653 (2010).
43. T. Toda, J. Hanna, T. Tani, *J. Appl. Phys.*, 101, 023505 (2007).
44. T. Tani, *J. Soc. Photogr. Sci. Tech. Jpn.*, 72, 88 (2009).
45. T. Tasaka, M. Murofushi, T. Tani, *J. Imaging Sci. Tech.*, 47, 463 (2003).

46. T. Tani, M. Murofushi, *J. Imaging Sci. Tech.*, 38, 1 (1994).
47. T. Tani, *J. Imaging Sci. Tech.*, 41, 577 (1997).
48. T. Tani, *J. Appl. Phys.*, 91, 4595 (2002).
49. U. Kreibig, M. Vollmer, *Optical Properties of Metal Clusters* (Berlin/Heidelberg: Springer-Verlag, 1995).
50. E. Klein, H. J. Metz, *Photogr. Sci. Eng.*, 5, 5 (1961).
51. C. R. Berry, *Photogr. Sci. Eng.*, 13, 65 (1969).
52. C. R. Berry, D. C. Skillman, *Photogr. Sci. Eng.*, 15, 236 (1971).
53. T. Tani, *J. Dispersion Sci. Tech.*, 25, 375 (2004).
54. J. R. Thirtle, L. K. J. Tong, L. J. Fleckenstein, in T. H. James (ed.), *The Theory of the Photographic Process*, 4th edn (New York: Macmillan, 1977), Chapter 12.
55. W. E. Lee, in E. Ostroff (ed.), *Pioneers of Photography, Their Achievements in Science and Technology* (Springfield, VA: Soc. Imaging Sci. Tech., 1987), Chapter 8.
56. D. C. Skillman, C. R. Berry, *J. Chem. Phys.*, 48, 3297 (1968).
57. C. R. Berry, D. C. Skillman, *J. Photogr. Sci.*, 17, 145 (1969).
58. G. R. Bird, M. Morse, H. Rodriguez, P. E. Bastian, J. Johnson, W. E. Gray, *Photogr. Sci. Eng.*, 15, 356 (1971).
59. R. C. Jones, G. R. Bird, *Photogr. Sci. Eng.*, 16, 16 (1972).
60. D. C. Skillman, C. R. Berry, *J. Opt. Soc. Am.*, 63, 707 (1973).
61. C. R. Berry, D. C. Skillman, *J. Appl. Phys.*, 42, 2818 (1971).
62. G. T. Eaton, in C. E. K. Mees, T. H. James (eds), *The Theory of the Photographic Process*, 3rd edn (New York: Macmillan, 1966), Chapter 18.
63. E. Weyde, E. Klein, H. J. Metz, *J. Photogr. Sci.*, 10, 110 (1962).
64. J. F. Hamilton, R. C. Baetzold, *Science*, 205, 4412 (1979).
65. R. C. Baetzold, *J. Solid State Chem.*, 6, 352 (1973).
66. R. C. Baetzold, *Photogr. Sci. Eng.*, 17, 78 (1973).
67. T. Tani, *Phys. Today* (September 1989), 36.
68. J. F. Hamilton, R. C. Baetzold, *Photogr. Sci. Eng.*, 25, 189 (1981).
69. W. D. Knight, K. Clemenger, W. A. de Heer, W. A. Saunders, M. Y. Chou, M. L. Cohen, *Phys. Rev. Lett.*, 52, 2124 (1984).
70. T. Tani, *Photographic Sensitivity: Theory and Mechanisms* (New York/Oxford: Oxford University Press, 1995), Chapter 4.
71. I. Konstantinov, A. Panov, J. Malinowski, *J. Photogr. Sci.*, 21, 250 (1973).
72. I. Konstantinov, J. Malinowski, *J. Photogr. Sci.*, 23, 1 (1975).
73. I. Konstantinov, J. Malinowski, *J. Photogr. Sci.*, 23, 145 (1975).
74. M. Kawasaki, Y. Tsujimura, H. Hada, *Phys. Rev. Lett.*, 57, 2796 (1986).
75. L. E. Lyons, *J. Chem. Soc.*, 5001 (1951).
76. T. Tani, S. Kikuchi, *Photogr. Sci. Eng.*, 11, 129 (1967).
77. R. Matejec, E. Moisar, *Ber. Bunsenges. Physik. Chem.*, 69, 566 (1965).
78. K. Seki, T. Tani, H. Ishii, *Thin Solid Films*, 273, 20 (1996).
79. H. Ishii, K. Sugiyama, E. Ito, K. Seki, *Adv. Mater.*, 11, 605 (1999).
80. H. Ishii, N. Hayashi, E. Ito, Y. Washizu, K. Sugiyama, Y. Kimura, M. Niwano, O. Ouchi, K. Seki, in W. Bruetting (ed.), *Physics of Organic Semiconductors* (Weinheim, Germany: Wiley-VCH, 2005), 69.
81. H. Ishii, K. Seki, in W. R. Salanek, K. Seki, A. Kahn, J.-J. Pireaux (eds), *Conjugated Polymer and Molecular Interfaces Science and Technology for Photonic and Optoelectronic Applications* (New York/Basel, Marcel Dekker, 2001), 293.
82. A. Wan, J. Hwang, F. Amy, A. Kahn, *Organic Electronics*, 6, 47 (1995).
83. D. Kinukawa, T. Toda, J. Hanna, T. Mitsui, T. Tani, Autumn Meeting of Jpn. Appl. Phys., September 2007, Sapporo, Japan.

84. C. Ganzorig, K.-J. Kwak, K. Yagi, M. Fujihira, *Appl. Phys. Lett.*, 79, 272 (2001).
85. T. Toda, J. Hanna, T. Tani, *J. Soc. Photogr. Sci. Tech. Jpn.*, 70, 38 (2007).
86. T. Tani, Y. Inami, *J. Appl. Phys.*, 88, 3601 (2000).
87. T. Tani, *J. Soc. Photogr. Sci. Tech. Jpn.*, 72, 88 (2009).
88. I. Katakuse, T. Ichikawa, Y. Fujita, T. Matsuo, T. Sakurai, H. Matsuda, *Intern. J. Mass. Spectrom. Ion Proc.*, 67, 229 (1985).
89. I. Katakuse, in S. Sugano, Y. Nishina, S. Ohnishi (eds), *Microclusters* (Berlin: Springer-Verlag, 1987), 10.
90. R. Kubo, *J. Phys. Soc. Jpn.*, 17, 740 (1962).
91. A. S. Malik, J. T. Blair, W. A. Bernett, F. J. DiSalvo, R. Hoffmann, *J. Solid State Chem.*, 146, 516 (1999).
92. Y. Tang, H. Shen, K. Sekiguchi, N. Kurahashi, T. Mizuno, Y. Suzuki, T. Suzuki, *Phys. Chem. Chem. Phys.*, 12, 3653 (2010).
93. G. C. Bond, *Catalysis by Metals* (London/New York: Academic Press, 1962).
94. B. C. Gates, L. Guczi, H. Knoezinger (eds), *Metal Clusters in Catalysis* (Amsterdam/New York: Elsevier, 1986).
95. M. Haruta, T. Kobayashi, H. Sano, N. Yamada, *Chem. Lett.*, 405 (1987).
96. T. Tani, T. Tasaka, M. Murofushi, K. Hosoi, A. Hirano, *Imaging Sci. J.*, 47, 1 (1999).
97. T. Tani, *Photogr. Sci. Eng.*, 15, 181 (1971).
98. P. Fayet, F. Granzer, G. Hegenbart, E. Moisar, B. Pischell, L. Woeste, *Phys. Rev. Lett.*, 55, 3002 (1985).
99. P. Krause, in J. M. Sturge, V. Walworth, A. Shepp (eds), *Imaging Processes and Materials* (New York: Van Nostrand Reinhold, 1989), Chapter 4.
100. S. Fujita, *Organic Chemistry of Photography* (Berlin: Springer-Verlag, 2004).
101. E. Becquerel, *Ann. Chim. Phys.*, 9, 257 (1883).
102. S. E. Sheppard, *J. Franklin Inst.*, 210, 578 (1930).
103. J. Eggert, W. Noddack, *Z. Phys.*, 31, 922 (1925).
104. W. Meidinger, *Z. Wiss. Photogr.*, 44, 137 (1949).
105. H. Arens, J. Eggert, F. G. Kleinschrodt, *Z. Wiss. Photogr.*, 42, 33 (1943).
106. W. West, P. B. Gilman, in James (ed.), *The Theory of the Photographic Process*, 269.
107. W. West, B. H. Caroll, *J. Chem. Phys.*, 15, 529 (1947).
108. Y. Ohko, T. Tatsuma, T. Fujii, K. Naoi, C. Niwa, Y. Kubota, A. Fujishima, *Nature Mater.*, 2, 29 (2003).
109. Y. Tian, T. Tatsuma, *Chem. Commun.*, 1810 (2004).
110. Y. Tian, T. Tatsuma, *J. Am. Chem. Soc.*, 127, 7632 (2005).
111. N. Sakai, T. Tatsuma, *Adv. Mater.*, 22, 3185 (2010).
112. G. A. Ozin, H. Huber, *Inorg. Chem.*, 17, 155 (1978).
113. H. Buecher, H. Kuhn, B. Mann, D. Moebius, L. von Szentpaly, P. Tillman, *Photogr. Sci. Eng.*, 11, 233 (1967).
114. L. von Szentpaly, D. Moebius, H. Kuhn, *J. Chem. Phys.*, 52, 4618 (1070).
115. R. Steiger, H. Hediger, P. Junod, H. Kuhn, D. Moebius, *Photogr. Sci. Eng.*, 24, 185 (1980).
116. H. Nozaki, T. Iida, *Hyomen*, 9, 49 (1971).
117. T. Tani, *Photogr. Sci. Eng.*, 17, 11 (1973).
118. K. B. Hendriks, *J. Soc. Photogr. Sci. Technol. Jpn.*, 52, 7 (1989).
119. M. Torigoe, K. Ohmura, S. Yagami, H. Iwano, *Sci. Publ. Fuji Photo Film*, 29, 31 (1988).
120. SPSTJ (ed.), *Principles of Photographic Science and Engineering* (Tokyo: Corona, 1998), Chapter 5.1.
121. J. L. Elechiguerra, L. Larios-Lopez, C. Liu, D. Garcia-Gutierrez, A. Camacho-Bragado, M. J. Yacaman, *Chem. Mater.*, 17, 6042 (2005).

9

Dye Sensitization for Solar Cells and Color Films

Dye sensitization was invented for spectral sensitization of silver halide (AgX) photographic materials in 1883 by Vogel [1], and has become widely used in AgX photographic films and papers [2–4]. Dye sensitization was also applied to semiconductor electrodes [5], and now is indispensable for dye-sensitized solar cells [6–8]. In this chapter, dye sensitizations in color films and dye-sensitized solar cells are analyzed and compared from the viewpoints of systems, sensitizing dyes, adsorption of dyes to substrates, electronic structures of interfaces between sensitizing dyes and substrates, quantum yield of electron transfer from excited dyes to substrates, rates of electron transfer, back electron transfer processes, transport of electrons in substrates, production of devices, and stability.

9.1 Comparison in System between Dye-Sensitized Solar Cells and Color Films

In dye-sensitized systems, sensitizing dyes on the surface of substrates absorb incident light, and dominate the capability of the systems. It is therefore important to develop substrates with large specific surface areas for enhancing the capability of the systems by increasing the amount of the dyes in the systems. For this purpose, very thin tabular AgX grains, as seen in Figure 1.2, were developed for color films [3], while thick sintered layers of nanoparticles of TiO_2 were developed for dye-sensitized solar cells [6–8].

As shown in Figure 3.1, AgX grains, which are used as the substrates for color films, are sensitive only to blue region in visible light, while color films should be sensitive to not only blue light, but also green and red light. Dye sensitization is used to render AgX grains sensitive to green and red light and also to enhance the sensitivity of AgX grains to blue light. Figure 9.1 shows the spectral intensity of the solar light together with the absorption

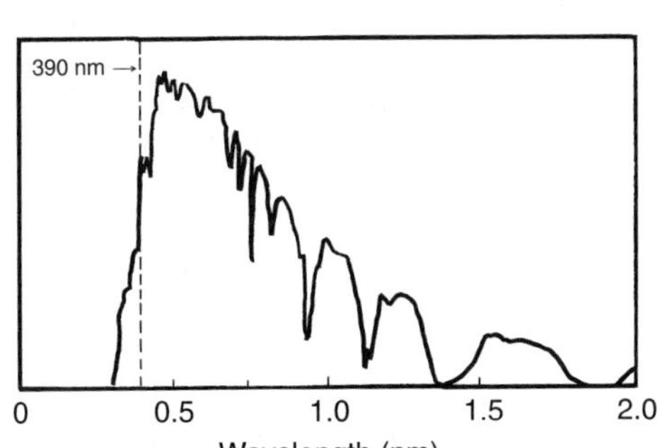

Figure 9.1 Spectral intensity of the solar light and absorption edge of TiO_2 (i.e., 390 nm).

edge of TiO_2. In order to use TiO_2 for solar cells, dye sensitization needs to render TiO_2 sensitive to light of wavelength longer than 390 nm.

As shown in Figure 1.1, a color film is composed of blue-, green-, and red-sensitive layers. While incident light is absorbed by sensitizing dyes on AgX grains, blue light is also absorbed by AgX grains in the blue-sensitive layer. As shown Figure 1.2, tabular AgX grains with a large aspect ratio (i.e., the ratio of the diameter of the main surface to thickness) and therefore large specific surface area are used in color films in order for a grain to adsorb a large amount of sensitizing dyes. As a result, incident light is mostly absorbed by J-aggregated sensitizing dyes on tabular AgX grains even in the blue-sensitive layer. Therefore color films are regarded as a type of organic device with J-aggregated sensitizing dyes as organic semiconductors and tabular AgX grains as substrates [4]. Actually, an exciton formed as a result of the light absorption of a dye molecule in a J-aggregate migrates in it, and dissociates to inject an electron into the conduction band of the grain and a positive hole in the aggregate. Then, transferred electrons collaborate with interstitial silver ions to form silver clusters named latent image centers to record a light image. Positive holes left in dye layers after the electron injection are inclined to hinder the latent image formation, and should be removed. In the process of development, only AgX grains with latent image centers are reduced to display a recorded image.

While various inorganic semiconductors have been extensively studied for use in photosensitive devices, great efforts have been made to apply dye sensitization to them to expand the wavelength region of their activities [9]. Among them, dye-sensitized solar cells attract the keen interest of many investigators due to their possibility of meeting the demands of the times [6–8].

Substrates of dye-sensitized solar cells are in most cases sintered TiO_2 nanocrystals. As shown in Figure 9.1, nanoparticles of TiO_2 absorb light of wavelength shorter than 390 nm, and can therefore absorb only 4% of incident solar light. Graetzel et al. developed a solar cell composed of TiO_2 by extending its sensitivity to a longer wavelength with the aid of dye sensitization [6–8]. Instead of single crystals and thin layers, thick layers of sintered nanoparticles of TiO_2 are used to adsorb large amounts of sensitizing dyes.

The difference in system between photographic film and a dye-sensitized solar cell is the role of positive holes. What is required for photographic films is to prevent positive holes from recombining with electrons and photolytically formed silver clusters. In dye-sensitized solar cells, positive holes should be taken out efficiently as an electric current by an electrode through an appropriate route such as an electrolyte solution with a I^-/I_3^- redox system.

9.2 Sensitizing Dyes

Most sensitizing dyes used in photographic films are cyanine dyes [10,11]. The molecular structures and absorption spectra of a typical cyanine dye are shown in Figure 9.2. The absorption band of a cyanine dye at the longest wavelength is due to the transition of an electron from the highest occupied molecular orbital (HOMO) to the lowest unoccupied molecular orbital (LUMO), and are narrow and intense. Cyanine dyes in photographic films form J-aggregates, whose absorption bands are much narrower than those of the

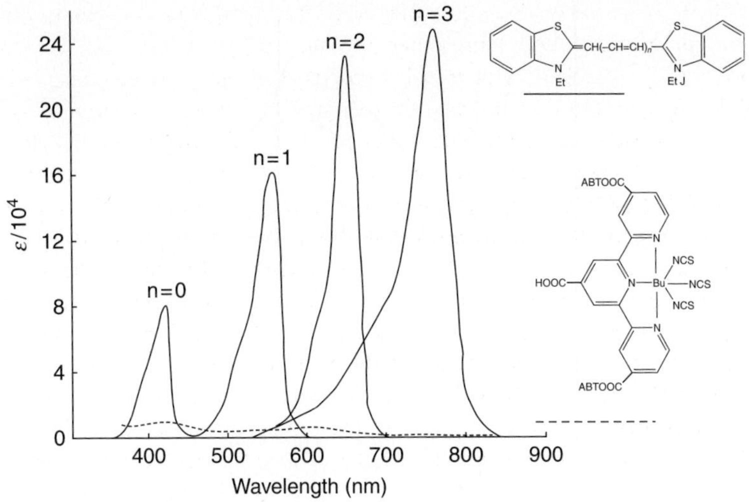

Figure 9.2 Absorption spectra of a series of thiacyanine dyes acting as sensitizing dyes for photographic films and of a Ru dye (i.e., Black Dye) acting as a sensitizing dye for dye-sensitized solar cells.

monomers of the corresponding dyes, as shown in Figure 5.4, being suitable for achieving appropriate spectral sensitivity.

Typical sensitizing dyes proposed for a dye-sensitized solar cell are Ru dyes composed of Ru(II) complex compounds [6–8]. The molecular structure and absorption spectrum of a Ru dye named Black Dye are shown in Figure 9.2. It is noted that a Ru dye is quite different in molecular structure from those of cyanine dyes used for photographic films. A Ru dye is composed of a Ru(II) ion, which is attached by a π-conjugated system with carboxylic groups and SCN, and exhibits an extraordinarily wide absorption band, which is quite suitable for efficiently capturing solar light. The light absorption of this dye takes place as a result of the transition of an electron from a d-orbital of a Ru(II) ion to a π^*-orbital of the conjugated system, and is ascribed to the metal-to-ligand charge transfer (MLCT) transition [12,13].

Figure 9.3 shows the schematic representation of the relation between band width and equilibrium nuclear distance of the excited states, indicating that the band width is proportional to the slope of the potential curve of the excited state as a function of the intermolecular distances, taken at the equilibrium point of the ground state. Namely, the band width tends to increase with increasing distance between the ground state and excited state equilibrium configurations. Since MTCT transition is usually associated with the displacement of an electron for a long distance, it should enhance the difference in the equilibrium nuclear configuration between ground and excited states, and widen the absorption band. In contrast, the HOMO–LUMO transition of a cyanine dye is associated with very small electronic displacement, as seen in Figure 5.3, and gives a sharp and intense absorption band, which is not suitable for solar cells with respect to very wide spectral intensity of solar light as seen in Figure 9.1.

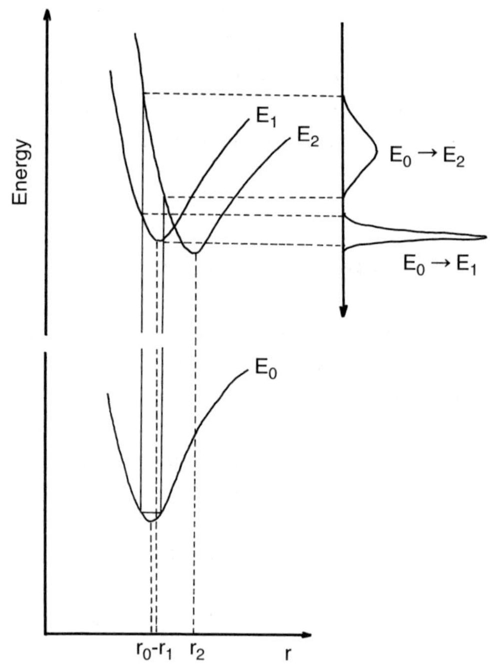

Figure 9.3 Schematic representation of the relation between band width and equilibrium nuclear distance of the excited states.

Although the absorption band of a Ru dye extends over a wide wavelength range, it is much weaker in intensity than that of a cyanine dye, even when it is integrated over the absorption band. The probability of the HOMO–LUMO transition in a dye molecule is proportional to the overlap between the wave functions of the ground and excited states, which are dominated by those of HOMO and LUMO of the molecule, respectively. Since the absorption band of a Ru dye in the visible region results from the MLCT transition, it should be inevitably small in intensity owing to the fact that the overlap of the wave functions between the HOMO and LUMO of the dye is very small [13]. In contrast, the overlap of the wave functions between the HOMO and LUMO of a cyanine dye is large as seen in Figure 5.3, giving a very intense absorption band.

Although the molecular extinction coefficient of a Ru dye is small, its amount in the cell is large enough to sufficiently capture incident light in its sensitive wavelength region since the dye is adsorbed on the surfaces of nanoparticles of a TiO_2 layer, which is as thick as 10 μm. However, it is hard for electrons to migrate a distance of 10 μm through sintered nanoparticles of TiO_2, as described in Section 9.7. It is therefore desirable to develop sensitizing dyes with a wide and more intense absorption band than that of Ru dyes.

A sensitizing dye molecule is adsorbed to the surface of TiO_2 by forming a carboxylate between a carboxylic group attached to a π-conjugated system and the surface of TiO_2, and displacing an electron from a *d*-orbital of a Ru ion to the π-conjugated system according to the MLCT transition [6–8]. Then, an optically excited state of a dye molecule transfers an excited electron from its π-conjugated system to the conduction band of TiO_2 rapidly, leaving a dye positive hole in a d-orbital of a Ru ion in the dye molecule. The positive hole left in the dye molecule is therefore far from the surface of TiO_2, and is not apt to recombine with an electron transferred into the conduction band of TiO_2 [6–8].

9.3 Adsorption of Sensitizing Dyes to Substrates

In both color films and dye-sensitized solar cells, sensitizing dyes in contact with their substrates take part in dye sensitization. Therefore, the adsorption of sensitizing dyes to substrates is essentially important, and is judged on the basis of the nature of the force for the binding of dye molecules to substrates and the adsorption isotherm.

In the case of color films, sensitizing dyes were adsorbed to the grains in photographic emulsions (i.e., aqueous gelatin solutions suspending AgX grains), which were then coated and dried on a film base, as described in Section 5.4. The influence of gelatin on the dye adsorption was not large enough to be considered [14]. As described in the section, J-aggregated cyanine dye molecules are bound to AgX grains due to not only the attractive interaction between the dye molecules and AgX, but also to the attractive interaction among the dye molecules. It is considered that van der Waals force contributes to the attractive force between the dye molecules and AgX in such an atmosphere as in photographic emulsions and films, where the electrostatic interaction between them is shielded by the presence of water molecules and various ions in high concentration.

Although the shapes of the adsorption isotherms depend on the kind of dyes and experimental conditions, they indicate that most sensitizing dyes suitable for dye sensitization of photographic films were adsorbed to AgX grains to cover them with mono-molecular layers according to the Langmuir-type isotherm, as shown in Figure 9.4 [2,14]. By increasing the amount of dye molecules added to an AgX emulsion, almost all the dye molecules can be adsorbed to the grains until the amount of added dye molecules almost equals the saturated amount of adsorbed dye molecules. This condition should be suitable for photographic films, since free dye molecules are not only inactive, but also obstructive to dye sensitization by absorbing incident light. The saturated amount of adsorbed dye molecules with respect to the surface area of the grains indicates that the area occupied by a molecule of a typical sensitizing dye was as small as 0.68 nm^2 due to the edge-on orientation of dye molecules to form their J-aggregates [14]. In agreement with these results, the heat of the adsorption of J-aggregated cyanine dye molecules to AgX grains is as large as about 10 kcal/mol despite the fact that the interaction between cyanine dyes and AgX grains among cyanine dye molecules is physical.

In dye-sensitized solar cells, each Ru dye molecule is chemically adsorbed to a TiO_2 nanoparticle by forming mono-molecular layers of the dye molecules [15]. Although it was initially suggested that the reaction of carboxylic acids with surface hydroxyl groups to form ester linkages contributed to the binding of Ru dyes to TiO_2 [16,17], many experimental data have been consistent with carboxylate binding, rather than ester [18–22], and therefore do not seem to support strong adsorption of Ru dyes to TiO_2

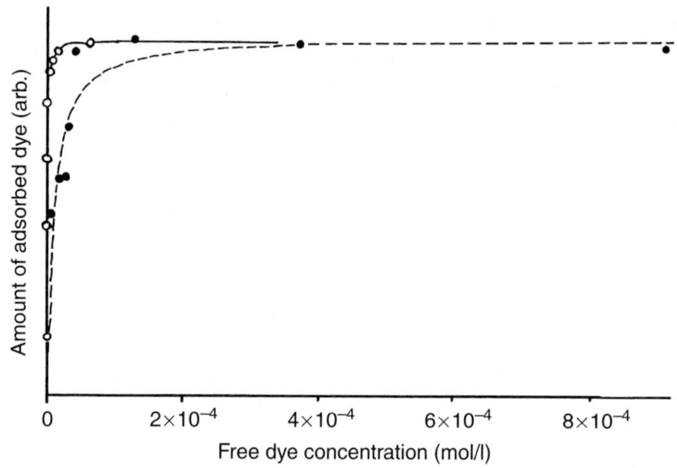

Figure 9.4 Adsorption isotherms of aggregated dye 2 to cubic AgBr grains in a photographic emulsion and of a Ru dye to sintered TiO_2 nanoparticles in a dye-sensitized solar cell.

nanoparticles. As usually observed with this kind of behavior, the adsorption of Ru dye molecules to TiO_2 nanoparticles exhibits a Langmuir-type isotherm, as shown in Figure 9.4 [6–8]. However, the concentration of dye molecules that cannot be adsorbed to the particles was not small even when the amount of adsorbed dye molecules was not close to the amount of saturated ones.

The dye adsorption should therefore be enhanced, since dye molecules that were not adsorbed to the particles are likely to decompose in an electrolyte solution during the operation of a dye-sensitized solar cell and to enhance the desorption of adsorbed dye molecules. In addition, the occupied area by a Ru dye molecule on the surface of TiO_2 has been as large as 1.6 nm^2 [6–8], much larger than that by a cyanine dye molecule on the surface of AgX grains. Namely, a considerable part of the surface of TiO_2 nanoparticles is free from the dye adsorption and in contact with an electrolyte solution containing a redox system, which captures and transfers positive holes to a cathode. It is therefore probable that some fraction of electrons transferred from excited dye molecules to TiO_2 nanoparticles to recombine with positive holes that were transferred from the dye molecules to the redox system. It has been proposed that the electrostatic repulsive interaction among sensitizing dye molecules on the surface of TiO_2 nanoparticles is responsible for the large area occupied by a dye molecule [6–8], and should be improved. It is noted that the attractive interaction among J-aggregated cyanine dye molecules on the surface of AgX grains is responsible for such a small surface area occupied by a dye molecule in color films.

9.4 Electronic Structures of Interfaces between Sensitizing Dyes and Substrates

From the viewpoint of the electronic structure of the interface between sensitizing dyes and substrates, the necessary condition for dye sensitization of usual photographic films is the electron transfer from a sensitizing dye in the excited state to the conduction band of AgX grains [2]. On the other hand, there are in principle two necessary conditions for the operation of a dye-sensitized solar cell [6–8]. One of them is the electron transfer from a sensitizing dye in the excited state to the conduction band of a TiO_2 nanoparticle. The other is the transfer of a positive hole left in a dye molecule to a redox system in an electrolyte solution, which guides it to a cathode.

Figure 9.5 illustrates the electronic structures of the interfaces of sensitizing dyes with AgBr and with TiO_2. Since the bottoms of the conduction bands of AgBr and TiO_2 are located at –3.5 and –4.3 eV, respectively [23], the necessary condition for the light-induced electron transfer in dye sensitization should be severer for photographic films than that for dye-sensitized solar cells, and become severer with increasing wavelength, since the height of an excited electron in a dye should generally decrease with increasing wavelength of its light absorption, as seen in Figure 5.12 [2].

The photosensitive wavelength region of photographic films can be, however, extended beyond 1300 nm to provide infrared-sensitive photographic films [10,11]. This was realized by the following conditions. In general, the stability of a cyanine dye decreases with increasing wavelength of its light absorption, and can be stabilized by the design of its molecular structure [10]. It has been judged that the height of the bottom of the conduction band of AgX decreases when the concentration of free silver ions in photographic emulsions is increased [24]. This can be confirmed directly by measuring the electronic structure of evaporated AgX layers covered with a thin gelatin membrane of varying concentration of silver ions, as seen in Figures 3.4 and 3.5 [25,26]. In addition, the efficiency of the electron transfer in infrared-sensitive photographic films and plates is not necessarily high.

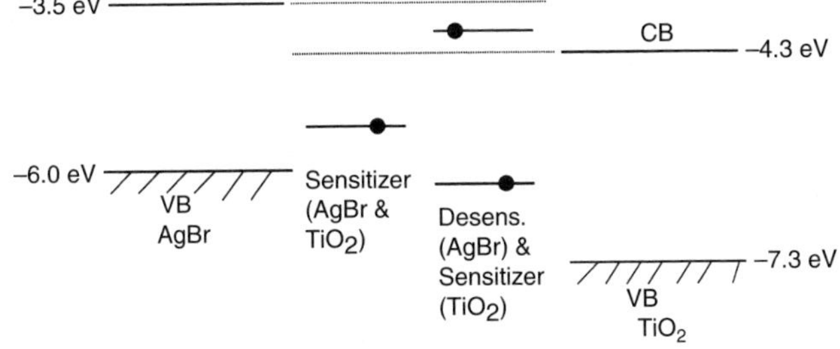

Figure 9.5 Electronic energy levels of two kinds of dyes in excited state with respect to the conduction and valence bands of AgBr and TiO_2.

In principle, the increase in surface area of AgX grains enhances the sensitivity and deteriorates the image quality of color films. It is therefore important to increase the efficiency of latent image formation in tabular AgX grains in color films [2,3]. To meet this requirement, a positive hole left in a dye molecule after the electron injection (i.e., a dye positive hole) is used to create a free electron available for latent image formation by means of two-electron sensitization, as described in Section 4.4.2. Namely, a dye positive hole oxidizes a two-electron sensitizer to form an activated species capable of transferring an electron to an AgX grain. The two-electron sensitizers include reduction sensitization centers composed of dimers of Ag atoms [27,28] and newly developed organic compounds, which were developed based on the progress in photochemistry [29]. The necessary conditions for dye sensitization of color films should be the electron transfer from an excited dye molecule to the conduction band of an AgX grain with high efficiency and the transfer of a dye positive hole to a two-electron sensitizer. It has been judged that the photosensitive wavelength region of photographic emulsions for meeting these conditions is in fact the visible wavelength region, and fitted with that of color films.

Figure 9.6 illustrates the electronic energy level diagram of a dye-sensitized solar cell in terms of electrochemical potential [7], according to which the bottom of the conduction band of TiO_2 corresponds to -0.7 V vs SCE (saturated calomel electrode), while the redox potential of I^-/I_3^- system is 0.2 V. The potential energy barrier for the transfer of an excited electron in a dye to the conduction band of TiO_2 and that for the transfer of a dye positive hole to the redox system of I^-/I_3^- in the absence of the difference in the free

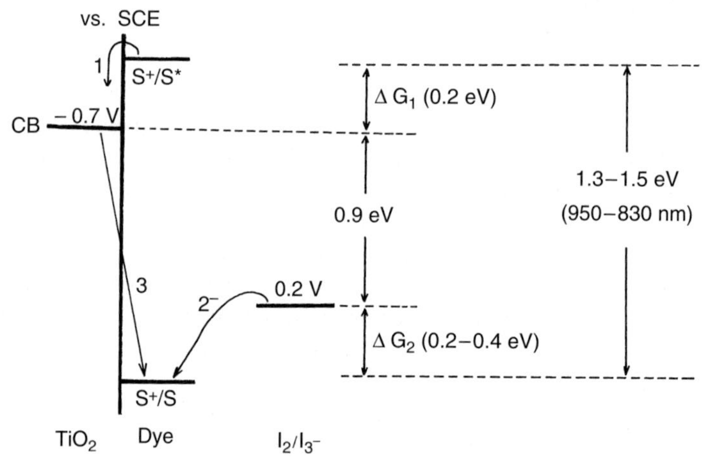

Figure 9.6 Electronic energy level diagram of a dye-sensitized solar cell in electrochemical expression, where CB is the bottom of the conduction band of TiO_2, S, S*, and S^+ are a sensitizing dye in the ground state, optically excited state, and oxidized state, respectively, and ΔG_1 and ΔG_2 are the energy barriers for the transfer processes of an excited electron to the conduction band of TiO_2 and of a positive hole in a dye molecule to the redox system in an electrolyte solution, respectively, under the conditions that the free energies of the states before and after the transfers of electronic carriers are the same.

energy of the states before and after the carrier transfer was 0.2 eV and 0.2–0.4 eV, respectively. Accordingly, the HOMO/LUMO gap of a sensitizing dye available for a dye-sensitized solar cell is evaluated to be 1.3–1.5 eV, which corresponds to a light of wavelength 950–830 nm and the longest wavelength edge of the action spectrum of a black dye. This prediction seems to be in agreement with the present state-of-the art that it is difficult to develop a sensitizing dye for a dye-sensitized solar cell that exhibits a longer photosensitive region of higher efficiency than that of Black Dye.

Although the potential energy barriers for the carrier transfer processes are not significant in color films, they are significant in dye-sensitized solar cells, and should be reduced for the extension of the photosensitive region of dye-sensitized solar cells to a longer wavelength. One point to be considered is that the energy difference between the Frank-Condon state and equilibrium state in an excited Ru dye should be large due to the large distance between the dye's ground state and excited state equilibrium configurations, as seen in Figure 9.3. When a "hot" electron at the Frank-Condon state can be efficiently transferred to the conduction band of TiO_2, the above-stated energy difference is not the cause for the potential energy barriers. Otherwise, the energy difference would be a candidate for the potential energy barrier. The participation of "hot" electrons in dye sensitization is described in Section 9.6.

The rearrangement energy in the Marcus theory for the electron transfer should be another candidate for the potential energy barriers, and is described in Section 9.5.

9.5 Efficiency of Electron Transfer from Excited Dye Molecules to Substrates

Dye sensitization takes place as a result of electron transfer from excited dye molecules to AgX grains and latent image formation by the transferred electrons and interstitial silver ions [2]. As described in Section 5.6, the efficiency of electron transfer is evaluated by the relative quantum yield (RQE), which is the ratio of the quantum yield of latent image formation by the absorption of light of sensitizing dyes in a photographic film to that by the absorption of light of AgX grains in the same film [30,2]. It is noted that RQE was introduced to determine the efficiency of the electron transfer, which should not depend on the degree of recombination of the transferred electrons with positive holes, as described in Section 5.6. Careful examinations were made to confirm the conditions required for RQE to determine the efficiency of the electron transfer irrespective of the degree of recombination [2,30,31], The degree of recombination is given by the degree of desensitization, which is the ratio of the efficiency of latent image formation of dyed emulsions to that of undyed ones [2].

In the case of dye-sensitized solar cells, the efficiency of dye sensitization is evaluated by the incident photon to current conversion efficiency (IPCE), which depends on the absorption of light by sensitizing dyes, the efficiency of the electron transfer from excited dye molecules to TiO_2 grains, and the fraction of transferred electrons that escape from recombining with positive holes. Namely, the value of IPCE is influenced by the degree of recombination between transferred electrons and positive holes, and should make it difficult to analyze the efficiency of dye sensitization based on IPCE. It is thought to be desirable to analyze the efficiency of dye sensitization in dye-sensitized solar cells based on RQE and the degree of recombination separately.

The efficiency and rate of electron transfer from a dye in the excited state to the conduction band of an AgX grain are related to the energy difference (ΔE) in electronic energy levels between the LUMO of a dye and the bottom of the conduction band of AgX grains in the framework of the Marcus theory with rearrangement energy (λ) as a parameter, as described in Section 5.6. Figure 9.7 shows the efficiency of dye sensitization of AgBr grains by cyanine dyes as a function of ΔE in terms of the electrochemical reduction potential E_R of the dyes [2,32]. It was found that the efficiency of dye sensitization was solely dependent on ΔE and reached unity with increasing ΔE. Although the efficiency decreased steeply with decreasing ΔE, it can be improved by supersensitization, as illustrated in Figure 5.18. As described in Section 5.6, the electron transfer for dye sensitization of photographic films was thus characterized by a small value of λ, which was as small as 0.05 eV, and the absence of the inverted region in the Marcus theory [2,32]. The small value of λ was confirmed by the temperature dependence of the electron transfer rate and was valid not for monomeric dye molecules, but for J-aggregated ones, as described in Section 5.6 [33,34]. The absence of the inverted region in the Marcus theory is common for the electron transfer from excited dye molecules to inorganic

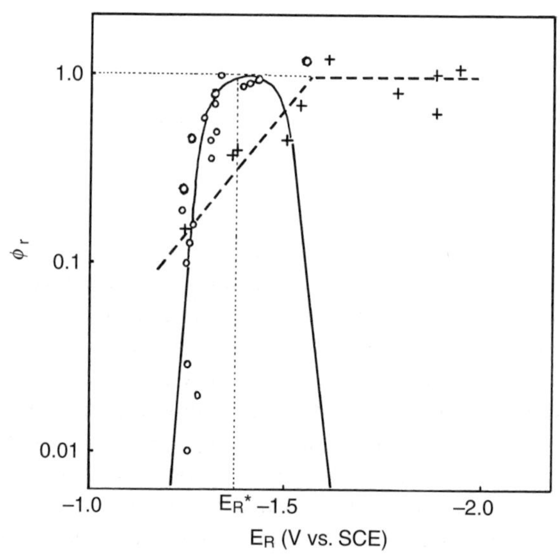

Figure 9.7 The ΔE dependence of the efficiency of the light-induced electron transfer from sensitizing dyes to AgBr grains in photographic films (○), and ZnO particles for dye-sensitized solar cells (+), where ΔE is defined as the energy level of an excited electron in sensitizing dyes with respect to the bottom of the conduction band of the substrates (i.e., AgBr grains for photographic films and ZnO particles for dye-sensitized solar cells) and represented by the reduction potential (E_R) of dyes for photographic films on the abscissa. While E_R^* indicates the reduction potential corresponding to the condition with $\Delta E = 0$ for the AgBr system, the experimental points for the ZnO system were horizontally shifted to share E_R^* as the reduction potential of the dye corresponding to the condition with $\Delta E = 0$.

semiconductors, as also described in Section 5.6 [35,36]. These characteristics should be helpful for achieving the electron transfer from excited dye molecules to AgX grains with high efficiency.

Although it has been confirmed that the efficiency of electron transfer from Ru dyes in the excited state to TiO_2 grains can be unity [37], data on the energy gap dependence of the rate and/or efficiency of the electron transfer from excited dye molecules to TiO_2 in dye-sensitized solar cells does not seem to be available at the present. Kato et al. reported the result with ZnO shown in Figure 9.7, which should reflect the characteristics of electron transfer in dye-sensitized solar cells since the height of the bottom of the conduction band is nearly the same between ZnO and TiO_2 [38,39]. The efficiency can be unity at large values of ΔE.

Applying the Marcus theory to the obtained result, they evaluated that the value of λ for electron transfer in their system was 0.3 eV. Reflecting on this evaluation, the efficiency at $\Delta E = 0$ is still low in this system, in contrast to the case of photographic films, where the efficiency at $\Delta E = 0$ was close to unity. As illustrated in Figure 9.7, the efficiency of unity can be achieved when ΔE is nearly equal to λ. It is considered that λ is a candidate for the origin of ΔG_1 in Figure 9.6, and should be reduced to achieve highly efficient electron transfer to TiO_2 grains from various dyes in the excited state. The above-stated results on photographic films indicate that the J-aggregate formation of sensitizing dyes and the elimination of solvent molecules may contribute to the reduction of λ for electron transfer in dye-sensitized solar cells.

9.6 Rates of Electron Transfer and Back Electron Transfer

While the lifetime of the first excited singlet state of J-aggregated cyanine dye molecules on AgX grains was around several tens of picoseconds, as described in Section 5.7, the frequency of the electron transfer from the excited dye molecules to the grains was evaluated to be relatively small ($\sim 10^{10}$/s) [40], and was in agreement with the following facts. Namely, sensitizing dye molecules form J-aggregates on the surface of AgX grains especially in color films. An exciton (i.e., an excited dye molecule in a J-aggregate) migrates within a J-aggregate and also among J-aggregates in sub-picoseconds [41], reaches a site where an exciton preferably dissociates, and transfers an electron to the conduction band of an AgX grain in several tens of picoseconds [2,40]. The migration of excitons in and among J-aggregates before their dissociation is believed to be one of the reasons why the time required for electron transfer in a color film is longer than that in a dye-sensitized solar cell. In photographic films, the migration of an exciton among sensitizing dye molecules is believed to contribute to its dissociation by allowing it to reach one of the sites or molecules where it can dissociate more efficiently than on other sites or molecules, as described in Section 5.6 [2].

The relatively small frequency of the electron transfer from excited cyanine dye molecules to AgX grains is also in agreement with the fact that the efficiency of electron transfer is independent of the exciting light wavelength within the absorption band of the dye molecules in the same aggregate state on the grains [42–44], indicating the absence of electron transfer in "hot" vibrational level in the excited state of the dye to the grains. It is also noted that cyanine dyes on AgX grains in color films form

J-aggregates with extraordinarily small Stokes shift [45]. Therefore, it is not probable that "hot" electrons significantly contribute to electron transfer in the dye sensitization for color films.

In contrast, Ru dyes on TiO_2 particles in dye-sensitized solar cells are monomeric, and transfer electrons directly to TiO_2 particles without allowing excitons to migrate. In addition, time-resolved near-infrared (IR) spectroscopy revealed that electron transfer from an excited Ru dye to TiO_2 is very fast and takes place in less than 25 ft, prior to the redistribution of the excited-state vibrational energy [46]. According to the measurement of femtosecond mid-IR spectroscopy [47], the excitation of a Ru dye on TiO2 led to the appearance of a broad absorption band assigned to conduction electrons in TiO_2. The rise fitted to a bi-exponential model with a fast 50-fs rate (84%) and a slow 1.5-ps rate (16%). Moser and Graetzel [48] observed the wavelength dependence of the yield of electron transfer in dye sensitization, which can provide evidence of electron transfer from "hot" vibrational states in an excited Ru dye to TiO_2 and other substrates. It is also noted that a Ru dye has a large Stokes shift, as illustrated in Figure 9.3, and has a large possibility of electron transfer from a "hot" vibrational state in an excited Ru dye to TiO_2.

In a dye-sensitized solar cell, the frequency of transfer of a dye positive hole to the redox system of I^-/I_3^- is 10^8/s, much larger than that of the back electron transfer process (i.e., the recombination of a transferred electron with a positive hole), which is ~10^6/s [7]. Such a slow back electron transfer process can be attributed to the molecular design of Ru dyes, on which a dye positive hole is situated far from the surface of TiO_2 in a dye-sensitized solar cell, as described in Section 9.2 [6–8]. However, it does not seem that sensitizing dyes can cover all the surface of TiO_2 grains and prevent the recombination of transferred electrons and positive holes transferred to an electrolyte solution. It should also be pointed out that the high exothermic condition of the back electron transfer can make the system lie deep in the inverted Marcus region, where the rate of the charge transfer process is expected to decrease with increasing driving force [8].

In color films, dye positive holes have been detected and characterized by electron spin resonance spectroscopy [2,49–51] and microwave photoconductivity measurement with two synchronized light pulses as light sources [2,52]. Since dye positive holes move to the valence band of AgX grains and then recombine with photolytically formed silver clusters, the lifetime of dye positive holes is as short as 10^{-6} s when a dye's HOMO height is close to the top of the valence band of AgX grains, and is too long to be measured under the experimental conditions employed when the dye's HOMO height is far from the top of the valence band. It is important to develop technologies for hindering the activity of dye positive holes, and furthermore to utilize dye positive holes to create free electrons for latent image formation.

9.7 Transport of Charge Carriers in Substrates

In color films, each electron is required to reach a growing latent image center at any site in the grain. Although the Hall mobility of photoelectrons in AgBr is 50 cm^2/Vs [53], the drift mobility of photoelectrons in a AgBr grain is ~1 cm^2/Vs [54]. The large difference between them indicates that each electron stays at a trap for a long time. Since the trapping of an electron on the grain surface is the initial step for latent image formation, it is not easy to increase the drift mobility by hindering the activity of the electron traps. The lifetime of photoelectrons in a AgBr grain is ~100 ns, and ascribed to the time for a trapped electron to combine with an interstitial silver ion to form an Ag atom [55]. It is therefore difficult to increase the electron lifetime. The diffusion length of photoelectrons, as judged from the drift mobility and lifetime, is ~0.7 μm and comparable to the size of AgX grains available for color films. Under this condition, an electron can barely reach a growing image center on a AgX grain.

In dye-sensitized solar cells, a donor (sensitizing dyes) and an accepter (a layer of sintered TiO_2 nanoparticles) are in direct contact with each other, and enable the charge separation of an exciton to create an electron with a positive hole with high efficiency. A positive hole is transferred to a redox system of I^-/I_3^-, moves through it, and reaches an anode. On the other hand, an electron migrates through a layer of sintered TiO_2 nanoparticles, which is as thick as 10 μm and is capable of absorbing nearly all the incident light in the sensitizing dye's absorption bands [6–8].

While the electron mobility in a large single crystal of TiO_2 at room temperature is more than 1 cm^2/Vs [56], that in a layer of sintered TiO_2 nanoparticles is as small as ~10^{-5} cm^2/Vs [57]. The observed large difference between them indicates that photoelectrons stay at traps for most of the time during their migration. Photoelectrons with such a small drift mobility in a layer of sintered TiO_2 nanoparticles migrate to a distance of 10 μm due to their lifetime [6–8], which is as long as several seconds and much longer than that in color films. The lifetime of photoelectrons in AgBr grains is less than 1 μs because photoelectrons at traps combine with interstitial silver ions to form silver clusters, recombine with positive holes, and are captured by oxygen molecules [2,3]. In dye-sensitized solar cells, photoelectrons are alive for a long time, being attacked by neither positive holes nor oxygen molecules despite the fact that they stay at traps in sintered TiO_2 nanoparticles for a long time. A positive hole is considered to be situated at a *d*-orbital of a Ru ion, which is apart from the surface of a TiO_2 nanoparticle and close to an electrolyte solution, to which a positive hole should be transferred. However, it is still anticipated that photoelectrons transferred to sintered TiO_2 nanoparticles recombine with positive holes transferred to an electrolyte solution on the parts of their surfaces not covered by sensitizing dyes.

A central question still remains: how can an initially very poorly conducting network of TiO_2 nanoparticles attain the excellent photocurrent-voltage characteristics in a dye-sensitized liquid-junction solar cell [8]?

9.8 Light Stability of Sensitizing Dyes

There are several aspects to be considered for the stability of organic devices. First, the stability of sensitizing dyes under illumination is one of the most important aspects for the stability of dye-sensitized solar cells. It is noted that sensitizing dyes on sintered TiO_2 nanoparticles in contact with an electrolyte solution under strong illumination can survive for a long time such as required for a solar cell to work [6–8]. In color films, sensitizing dyes are not exposed to light before use and therefore do not need to be stable under illumination. There is no problem even if sensitizing dye molecules decompose when they absorb incident light and inject electrons into AgX grains. However, sensitizing dyes in color films should be stable in the dark within the term specified for a guarantee of quality, which is usually two years.

It is known that sensitizing dyes for AgX photographic materials decompose rapidly when they are excited in an electrolytic solution [3]. In addition, it is known that a conventional polarography does not give the accurate value of a sensitizing dye's reduction or oxidation potential, since the reduction or oxidation is followed by its succeeding reaction, which is rapid enough to influence the ratio of concentration of reduced or oxidized molecules to that of the original amount [58], and takes place at a frequency of around 10^3/s [59].

It is known that the frequency of decomposition of an oxidized Ru dye molecule on the surface of TiO_2 is several times per second and much smaller than that of an oxidized cyanine dye molecule in a solution. Since the frequency of electron transfer from the redox system in an electrolytic solution to an oxidized dye molecule on a TiO_2 nanoparticle is 10^8/s, it is estimated that a Ru dye molecule on the surface of TiO_2 can work more than 10^7 times per second [7]. This condition might be one of the reasons for the stability of dye-sensitized solar cells. It is, however, anticipated that Ru dye molecules, which are not adsorbed to the nanoparticles of TiO_2, are present in an electrolyte solution, as described in Section 9.3, and apt to decompose under strong illumination for a long time, and to cause the desorption of Ru dye molecules from the surface of TiO_2 according to the equilibrium between adsorbed dye molecules on TiO_2 nanoparticles and free ones in a solution.

Although the stability of sensitizing dye molecules is one of the most important problems to be solved, limited information on the stability of dye-sensitized solar cells has been published, probably because most information on the stability of dye-sensitized solar cells is being created within private companies.

9.9 Mechanism of Dye Sensitization

In the history of dye sensitization, thorough discussions about the electron transfer mechanism and the energy transfer mechanism have been made for many years, as described in details in Section 5.9. Although it has been proved that both mechanisms can actually work, the rate of electron transfer is one order of magnitude larger than that of energy transfer when sensitizing dye molecules are in contact with the surface of AgX grains. Actually, it has been confirmed that dye sensitization in AgX photographic materials takes place rapidly and efficiently, showing the energy gap dependence of the quantum yield, which strongly supports the adequacy of the electron transfer mechanism.

The design of dye sensitization for dye-sensitized solar cells has been made according to the electron transfer mechanism without examining its validity in comparison to the energy transfer mechanism. In light of the result of extensive examination on the mechanism of dye sensitization in AgX photographic materials, dye sensitization in dye-sensitized solar cells is considered to take place according to the electron transfer mechanism due to the following reasons.

It is obvious that sensitizing dye molecules in contact with substrates bring about dye sensitization in dye-sensitized solar cells. In addition, it has been confirmed that dye sensitization in dye-sensitized solar cells takes place rapidly [46] and efficiently, showing the energy gap dependence of the quantum yield, which strongly supports the adequacy of the electron transfer mechanism [38,39]. Since the bottom of the conduction band of the substrates for dye-sensitized solar cell (i.e., TiO_2 and ZnO) is much lower than that of AgX [23], the electron transfer from excited dye molecules to the substrates for solar cells is more likely to take place for a category of dyes wider than that from excited dye molecules to AgX. Although AgBr and AgCl have long-wavelength absorption tails according to Urbach's rule [60], indicating the presence of forbidden transitions available for energy transfer from excited dye molecules to AgX, it is not certain whether the substrates for dye-sensitized solar cells (i.e., TiO_2 and ZnO) have such forbidden transitions in the visible region.

For dye-sensitized solar cells and photocatalysts, it is essentially important to develop such technologies as to render their substrates sensitive to light wavelength longer than that of their absorption edge by spectral sensitization. As described above, the spectral sensitization by dyes (i.e., dye sensitization) is one of the most promising technologies. Other new technologies include spectral sensitization by metal nanoparticles with surface plasmon resonance [61,62] and quantum dots of inorganic semiconductors [63], and should be examined in light of the result of discussions concerning the electron transfer and energy transfer mechanisms for dye sensitization of AgX photographic materials, since a metal nanoparticle under surface plasmon resonance condition does not have any excited electron available for electron transfer to the substrates. More considerations are described in Section 8.4.2.

References

1. H. Vogel, *Ber.*, 6, 1302 (1873).
2. T. Tani, *Photographic Sensitivity: Theory and Mechanisms* (New York/Oxford: Oxford University Press, 1995).
3. T. Tani, *J. Imaging Sci. Tech.*, 51, 110 (2007).
4. T. Tani, IEICE Tech. Rep. 109, ED2009–12, 49 (2009).
5. R. Memming, *Semiconductor Electrochemistry* (Weiheim, Germany: Wiley-VCH, 2001).
6. B. O'Regan, M. Graetzel, *Nature*, 353, 737 (1991).
7. A. Hagfeldt, M. Graetzel, *Chem. Rev.*, 95, 49 (1995).
8. M. Graetzel, J.-E. Moser, inV. Balzani (ed.), *Electron Transfer in Chemistry*, vol. 5 (Weinheim, Germany: Wiley-VCH, 2001), 589.
9. H. Meier, *Spectral Sensitization* (London/New York: Focal Press, 1968).
10. L. G. S. Brooker, in C. E. K. Mees, T. H. James (eds), *The Theory of the Photographic Process*, 3rd edn (New York: Macmillan, 1966), Chapter 11.
11. D. M. Sturmer, D. W. Heseltine, in T. H. James (ed.), *The Theory of the Photographic Process*, 4th edn (New York: Macmillan, 1977), Chapter 8.
12. M. K. Nazeeruddin, P. Pechy, M. Graetzel, *Chem. Commun.*, 1705 (1997).
13. M. K. Nazeeruddin, F. D. Angelis, S. Fantacci, A. Selloni, G. Viscard, P. Liska, S. Ito, T. Bessho, M. Graetzel, *J. Am. Chem. Soc.*, 127, 16835 (2005).
14. T. Tani, T. Suzumoto, *J. Appl. Phys.*, 70, 3626 (1991).

15. P. Qu, G. Meyer, in V. Balzani (ed.), *Electron Transfer in Chemistry*, vol. 4 (Weinheim, Germany: Wiley-VCH, 2001), Chapter 2.
16. J. B. Goodenough, A. Hammett, M. P. Dare-Edwards, G. Campet, R. D. Wright, *Surf. Sci.*, 101, 531 (1980).
17. T. Osa, M. Fujihira, *Nature*, 264, 349 (1976).
18. S. Umpathy, A. M. Cartner, A. W. Parker, R. E. Hester, *J. Phys. Chem.*, 94, 1357 (1990).
19. K. S. Finnie, J. R. Bartlett, J. L. Woolfrey, *Langmuir*, 14, 2744 (1998).
20. R. Argazzi, C. A. Bignozzi, T. A. Heimer, F. N. Castellano, G. J. Meyer, *Inorg. Chem.*, 33, 5741 (1994).
21. P. Qu, G. J. Meyer, *Langmuir*, 16, 4662 (2000).
22. G. B. Deacon, R. J. Phillips, *Coord. Chem. Rev.*, 33, 227 (1980).
23. T. Tani, *Photogr. Sci. Eng.*, 17, 11 (1973).
24. R. W. Berriman, P. B. Gilman, Jr., *Photogr. Sci. Eng.*, 17, 235 (1973).
25. T. Tani, Y. Inami, *J. Appl. Phys.*, 88, 3601 (2000).
26. T. Tani, *J. Soc. Photogr. Sci. Tech. Jpn.*, 72, 88 (2009).
27. M. Harbison, H. E. Spencer, in James (ed.), *The Theory of the Photographic Process*, 152.
28. T. Tani, *J. Imaging Sci.*, 30, 41 (1986).
29. I. R. Gould, J. R. Lenhard, A. A. Muenter, S. A. Godleski, S. Y. Farid, *J. Am. Chem. Soc.*, 122, 11934 (2000).
30. J. Spence, B. H. Carroll, *J. Phys. Colloid Chem.*, 52, 1090 (1948).
31. T. Suzumoto, T. Tani, *J. Imaging Sci. Tech.*, 40, 56 (1948).
32. T. Tani, T. Suzumoto, K. Ohzeki, *J. Phys. Chem.*, 94, 1298 (1990).
33. M. T Spitler, A. Ehert, R. Kietzmann, F. Willig, *J. Phys. Chem. B*, 101, 2552 (1997).
34. F. Willg, M. T. Spitler, *J. Imaging Sci. Tech.*, 41, 272 (1997).
35. K. Hashimoto, M. Hiramoto, A. B. P. Lever, T. Sakata, *J. Phys. Chem.*, 92, 1016 (1988).
36. K. Hashimoto, M. Hiramoto, T. Sakata, *Chem. Phys. Lett.*, 148, 215 (1988).
37. R. Katoh, A. Furube, T. Yoshihara, K. Hara, G. Fujihashi, S. Takano, S. Murata, H. Arakawa, M. Tatsuya, *J. Phys. Chem. B*, 108, 4818 (2004).
38. R. Katoh, A. Furube, K. Hara, S. Murata, H. Sugihara, H. Arakawa, M. Tachiya, *J. Phys. Chem. B*, 106, 12975 (2002).
39. R. Katoh, A. Furube, A. V. Barzykin, H. Arakawa, M. Tachiya, *Coord. Chem. Rev.*, 248, 1195 (2004).
40. T. Tani, T. Suzumoto, Klaus Kemnitz, K. Yoshihara, *J. Appl. Phys.*, 96, 2778 (1992).
41. J.-W. Oh, S. Kumazaki, I. V. Rubtsov, T. Suzumoto, T. Tani, K. Yoshihara, *Chem. Phys. Lett.*, 352, 357 (2002).
42. J. Spence, B. H. Carroll, *J. Phys. Colloid Chem.*, 52, 1090 (1948).
43. J. A. Leermakers, *J. Chem. Phys.*, 5, 889 (1937).
44. W. West, B. H. Carroll, in Mees, James (eds), *The Theory of the Photographic Process*, Chapter 12.
45. I. V. Rubtsov, K. Ebina, F. Satou, J. Oh, S. Kumazaki, T. Suzumoto, T. Tani, K. Yoshihara, *J. Phys. Chem. A*, 106, 2795 (2002).
46. T. Hannapel, B. Burfeidt, W. Storek, F. Willig, *J. Chem. Phys. B*, 101, 6799 (1997).
47. J. B. Asbury, R. J. Ellington, N. H. Chosh, S. Ferrere, A. Nozik, T. Lian, *J. Phys. Chem. B*, 103, 3110 (1999).
48. J. E. Moser, M. Graetzel, *Chimia*, 52, 160 (1998).
49. T. Tani, *Photogr. Sci. Eng.*, 19, 356 (1975).
50. T. Tani, *J. Appl. Phys.*, 62, 2456 (1987).
51. T. Tani, Y. Sano, *J. Appl. Phys.*, 69, 4391 (1991).
52. K. Hosoi, A. Hirano, T. Tani, *J. Appl. Phys.*, 90, 6197 (2001).
53. R. K. Ahrenkiel, *Phys. Rev.*, 180, 859 (1969).

54. R. J. Deri, J. P. Spoonhower, *J. Appl. Phys.*, 57, 2806 (1985).
55. T. Oikawa, N. Saeki, T. Kaneda, A. Hirano, T. Tani, *J. Imaging Sci. Technol.*, 39, 233 (1995).
56. L. Forro, O. Chauvet, D. Emin, L. Zuppiroll, H. Berger, F. Levy, *J. Appl. Phys.*, 75(1), 633 (1994).
57. A. Solbrand, H. Lindstrom, H. Rensmo, A. Hagfeldt, S.-E. Lindquist, *J. Phys. Chem. B*, 101, 2514 (1997).
58. J. Lenhard, *J. Imaging Sci.*, 30, 27 (1986).
59. T. Tani, K. Ohzeki, K. Seki, *J. Electrochem. Soc.*, 138, 1411 (1991).
60. F. Uebach, *Phys. Rev.*, 92, 1324 (1953).
61. Y. Tian, T. Tatsuma, *Chem. Commun.*, 1810 (2004).
62. Y. Tian, T. Tatsuma, *J. Am. Chem. Soc.*, 127, 7632 (2005).
63. Y. Tachibana, H. Y. Akiyama, Y. Ohtsuka, T. Torimoto, S. Kuwabata, *Chem. Lett.*, 36, 88 (2007).

10

Organic Semiconductors and J-Aggregated Dyes

10.1 Introduction

One of the most important aspects of silver halide (AgX) photographic materials is the function of J-aggregated sensitizing dye layers on their substrates (i.e., tabular AgX grains), as described in Chapters 1, 4, 5, and 6 [1,2]. This aspect has been extended to dye-sensitized solar cells, as described in Chapter 9 [3–5], and is expected to serve as reference to organic semiconductors, since these sensitizing dyes can be regarded as organic semiconductors [6].

In contrast to inorganic semiconductors, organic semiconductors are composed of organic molecules with π-conjugated systems, which form thin layers due to the physical interaction among them. Therefore, the frontier orbitals, i.e., the highest occupied molecular orbital (HOMO) and the lowest unoccupied molecular orbital (LUMO), determine not only the property of the molecules, which constitute an organic semiconductor layer, but also the property of the layer [7]. The light absorption of an organic semiconductor layer takes place as a result of the molecules' electronic transition from HOMO to LUMO. The transport of electronic charge carriers in an organic semiconductor layer takes place through the frontier orbital of the molecules in the layer. In many cases, organic semiconductor layers are in contact with substrates and take part in the exchange of electronic carriers with them in organic devices. The above-stated aspects are analyzed in relation to color films and described in the following sections.

10.2 Systems of Organic Devices

In the first place, the systems of typical organic devices are outlined in this section. As seen in Figures 1.1, 1.2, and 1.4, a color film is composed of J-aggregated cyanine dye layers as light-absorbing organic semiconductors and tabular AgX grains as their substrates for treating the result of the absorbed light [1,2]. As described in Chapter 9, this system was transferred to dye-sensitized solar cells, which are composed of Ru dye layers as light-absorbing organic semiconductors and sintered layers of TiO_2 nanoparticles as their substrates for transporting electrons to an anode [3–5]. Figure 10.1 illustrates the electronic structure of the interface between light-absorbing dye layers and substrates of color films and dye-sensitized solar cells.

Figure 10.1 also illustrates the electronic structure of the interface between a donor and an acceptor in a organic thin film solar cell [8,9]. A donor is a light-absorbing organic semiconductor phase, while an acceptor is a substrate phase available for transporting electrons and usually is composed of fullerenes. The electronic energy gap between a donor and an acceptor is necessary for the dissociation of an exciton light-created in

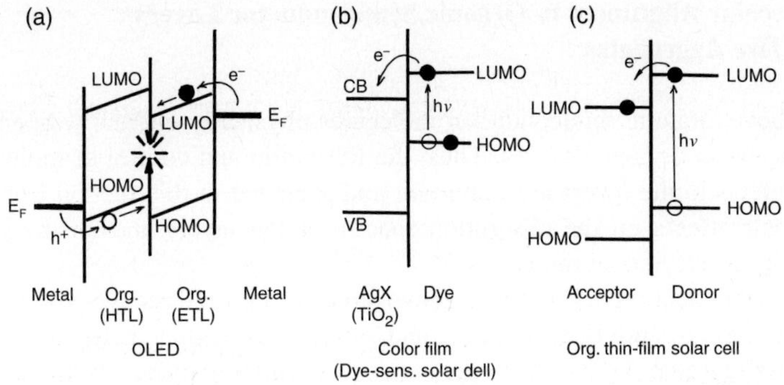

Figure 10.1 The electronic structures of an organic light-emitting diode with an electron-transporting layer (ETL) and a hole-transporting layer (HTL) (a), a color film (a dye-sensitized solar cell) with sensitizing dyes and AgX grains (a sintered layer of nanoparticles of TiO_2) as their substrates (b), and a organic thin film solar cell with a donor and an acceptor (c).

a donor layer, since the binding energy of an exciton in an organic layer is too large for it to dissociate and create an conductive electron and a positive hole at room temperature [10,11]. Thus, a created electron migrates through an acceptor layer and reaches an anode, while a positive hole migrates through a donor layer and reaches a cathode. Photosensitive layers in copiers are composed of organic photoconductors with charge transfer complexes between donors and acceptors for charge generation on exposure [12,13].

Figure 10.1 also illustrates the electronic structure of an organic light-emitting diode (OLED) with an organic electron-transporting layer (ETL) and an organic hole-transporting layer (HTL) [14]. By applying an electric field to these layers, electrons are injected from an electrode with small work function to the LUMO level of ETL, while positive holes are injected from a transparent electrode with large work function (i.e., ITO in many cases) to the HOMO level of HTL. Injected electrons and positive holes combine with each other to emit photons.

One of other important organic devices is an organic field-effect transistor [15,16] with an organic semiconductor layer, which is attached to source and drain electrodes, and is separated from a gate electrode by an insulator layer. While the applied electric field between a gate electrode and a source electrode enhances the injection of electronic charge carriers from a source electrode into an organic semiconductor layer, the applied electric field between source and drain electrodes causes the transport of the injected carriers from a source electrode to a drain electrode.

As seen in Figure 10.1, all these organic devices including color films are characterized in common by the presence of the interfaces between organic semiconductor layers and substrates, which play important roles in the functions of these devices. They enhance the charge separation of excitons in organic semiconductors, and the charge injection from substrates into organic semiconductors.

10.3 Molecular Alignment in Organic Semiconductor Layers and Dye Aggregates

As stated above, organic semiconductor molecules physically interact with each other to form thin layers in organic devices. Thus, the formation and control of molecular alignment and defects in the layers are important and analyzed in this section from the viewpoints of their effects on the absorption spectra of the layers and on the transport of electronic charge carriers in the layers.

As described in Section 5.3, sensitizing dye molecules form three types of aggregates on AgX grains, i.e., J- and H-aggregates, and aggregates with a herring-bone structure (herring-bone aggregate) [17–19]. J-aggregates of sensitizing dyes, which are particularly useful in color films as described in Chapter 5, were discovered in 1936 by Jelly and Scheibe [20,21].

Dye molecules are adsorbed to AgX grains due to attractive forces between the AgX grains and dye molecules in the layer on the grains and among dye molecules. Since these attractive forces are remarkably larger than that between dye molecules in the first and second layers, only mono-molecular layers of sensitizing dyes are formed on the AgX grain surface. Mono-molecular layers of J-aggregated sensitizing dyes on AgX grains have been judged to be single-crystalline according to the microscopic analysis with polarized light by Maskasky, as shown in Figure 10.2 [22]. It was found that each J-aggregate can grow to be several micrometers in edge length, and that all the molecules in each aggregate were aligned in the same direction. This occurs because each molecule was repeatedly adsorbed to and desorbed from the AgX grains until the most stable place and alignment were found, and because the adsorbed dye molecules can smooth the grain surfaces by repelling silver ions at kink sites into interstitial positions [23,1], as described in Section 5.4.

It appears that the three kinds of aggregates correspond to head-to-tail, parallel, and oblique alignments in Kasha's classification, as illustrated in Figure 5.7 [24]. Corresponding to an H-aggregate, a parallel alignment has an absorption band with a single peak, which appears at a wavelength shorter than that of the corresponding monomer band. Corresponding to a J-aggregate, a head-to-tail alignment has an absorption band with a single peak, which appears at a wavelength longer than that of the corresponding monomer band. These aggregates and alignments are considered to have the structure of only one molecule in a unit cell.

An oblique alignment is similar to a herring-bone aggregate in that they have absorption bands with two peaks, as illustrated in Figure 5.5. The interaction among organic semiconductor molecules in the layers often brings about the splitting of their absorption band (i.e., the Dividov splitting), exhibiting a band with two peaks [25]. One appears at a wavelength longer than that of the corresponding monomer band, while the other appears at the shorter wavelength. They are considered to have in common a structure of two molecules per unit cell. However, an oblique alignment does not necessarily correspond to a herring-bone aggregate. In a herring-bone aggregate, molecular planes are in a face-to-face arrangement, as modeled [17–19] and directly observed [26]. It does not seem that molecular planes in an oblique alignment are always in face-to-face arrangements. In fact, aromatic hydrocarbon molecules, which are typical organic semiconductors, have crystal

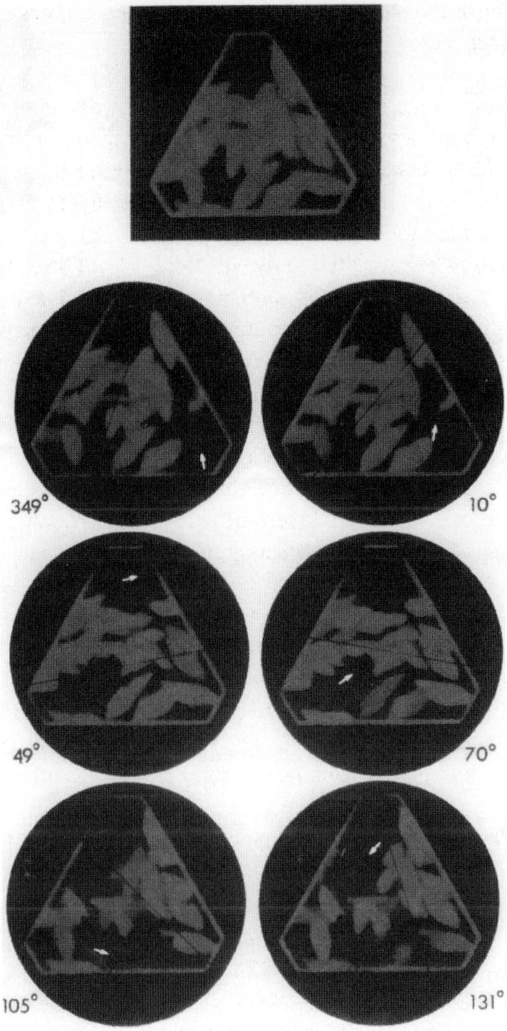

Figure 10.2. A series of photomicrographs of one AgBr tabular grain showing the luminescence of aggregates of 3, 3'-disulfopropyl-5, 5'-dichloro-9-ethyl-oxacarbocyanine dye at 77K. The first photo is without a polarizer in the emitted light path. The remaining series of photos shows aggregate disappearance (location indicated by white arrows) at six different polarization angles. The number next to each photo is the angle between the lower left edge of the tabular grain (<110> direction) and the reticle that indicates the polarizer axis and was set at the estimated mid-range of sample rotation over which an aggregate was no longer visible. These aggregates are an example of the cigar-shaped type and measure ~7.5 × 2μm. The grain is 20.0μm across as measured between parallel edges and is estimated to be 0.1–0.2μm in thickness. Reproduced with permission from Reference [22], copyright 1991 American Chemical Society.

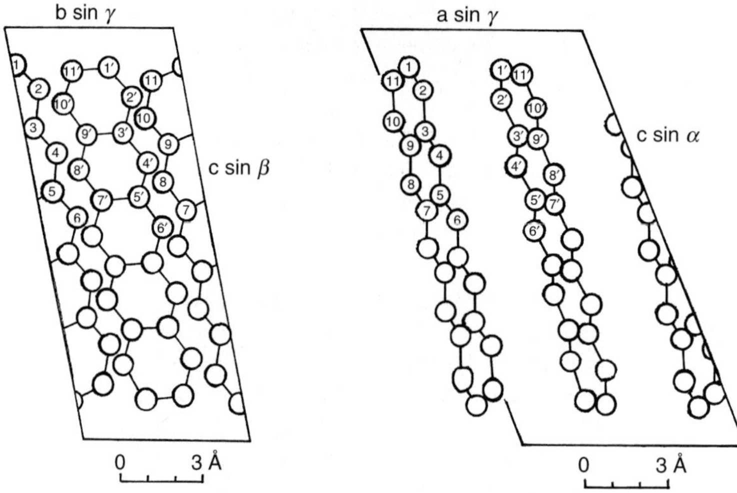

Figure 10.3 Projection of the crystal structure of pentacene along [100] (left) and that along [010] (right). Reproduced with permission from Reference [28], copyright 1961 International Union of Crystallography.

structures in which adjacent molecules are in edge-to-face arrangements, as exemplified by the crystal structure of pentacene in Figure 10.3 [28].

The closest distance of approach of two molecules is determined by the point and alignment at which the attractive forces are balanced by the repulsive ones, which arise when the electron clouds of two molecules begin to penetrate each other. One of the repulsive forces arises because two electrons cannot occupy the same volume element of space, according to the Pauli principle, and therefore tends to prevent two adjacent molecules from aligning in face-to-face arrangements [27,24], as seen in aromatic hydrocarbon molecules. On the other hand, the planar π-conjugated systems in adjacent molecules in J-, H-, and herring-bone aggregates of cyanine dyes are in face-to-face alignment due to the presence of attractive electrostatic interaction, which tends to make them face each other.

From a general and quantitative point of view, the slip angle between adjacent molecules (α), as defined in Figure 5.8, is important and substantial. It has been proposed that the aggregates with α, which can be smaller than 54.7° and larger than 54.7°, correspond to J- and H-aggregates, respectively, while head-to-tail and parallel aggregates are included with the former when $\alpha = 0°$ and with the latter when $\alpha = 90°$, respectively. These properties are used for the design of organic semiconductors for the capture of incident light, as described in the next section.

The face-to-face arrangement of organic molecules is considered to be suitable for the transport of electronic charge carriers, since the degree of overlap of π-electrons in adjacent molecules should be large especially as they form an H-aggregate. Attention has been paid to the application of J- and H-aggregated organic semiconductors to organic devices [29–31]. Thus, a comparison of the transport of electronic charge carriers between the layers of organic semiconductor molecules with and without face-to-face arrangements needs to be carried out, although it is difficult to compare molecules with the same π-conjugated system. Namely, molecules with π-conjugated systems, which form aggregates with face-to-face arrangements, should have dipoles as stated above, whereas aromatic hydrocarbon molecules, which are in edge-to-face arrangements, do not.

From the viewpoint that the overlapping of the frontier orbital should be larger in H-aggregates than in J-aggregates, it is probable that the transport of electronic charge carriers in H-aggregates is superior to that in J-aggregates. For AgX photographic films, J-aggregates are more preferable than H-aggregates because the energy for the excitation of the former is smaller than that for the latter and that the energy of relaxation taking place before an excited dye molecule injects an electron into an AgX grain is much smaller for the former than that for the latter. Thus, it is probable that H-aggregates as well as J-aggregates are useful for organic semiconductor layers. Hiramoto and Ukei studied the generation and transport of electronic charge carriers in organic solar cells with phthalocyanines as donors, which formed J- and H-aggregates, depending on the kind of central metal ions. In order to explain their observations, they proposed that J-aggregates were suitable for charge generation, while H-aggregates were suitable for charge transport [29,30].

In the layers composed of polycrystalline organic semiconductors, the control of grain boundaries in addition to molecular alignment in the layers is important for the transport of electronic charge carriers. A method for decreasing the closest distance of approach between π-conjugated systems in two adjacent molecules by their substituents was proposed and named "molecular fastener" effect [32,33]. By putting substituents with long alkyl chains into π-conjugated systems, hydrophobic bond among the substituents forces the π-conjugated systems to align and the distance between two π-conjugated systems to decrease with increasing length of alkyl groups. This idea was extended and made it possible to develop layers of polycrystalline organic semiconductors with electronic carrier mobility, which is even larger than those of single-crystalline aromatic hydrocarbons [34–36].

It was demonstrated that liquid-crystalline organic semiconductors, which are composed of π-conjugated systems with substituents having long alkyl chains, exhibit a much larger mobility of electronic carriers than amorphous ones [37–41], indicating that the molecular alignment caused by the hydrophobic bond among the substituents enhanced the molecular alignment and the electronic transport. Iino, Hannna, and others prepared their polycrystalline layers by cooling them through liquid-crystalline conditions [42–44]. It was postulated that grain boundaries were introduced into the layers in such a direction that they hardly disturbed the anisotropic transport of electronic carriers in the layers.

Although a tremendous number of merocyanine dyes were developed for photographic films [45], they could hardly form J-aggregates on the surface of AgX grains. However, many merocyanine dyes form J-aggregates in their Langmuir–Blodgett (LB) membranes [46,47]. Since substituents with long alkyl chains are needed for the formation of the LB membranes of merocyanine dyes, the hydrophobic force among the substituents is considered to enhance the molecular alignment and the formation of J-aggregate of merocyanine dyes.

10.4 Light Absorption of Organic Semiconductors and Dye Aggregates

Figure 9.2 shows two extremely different absorption spectra of sensitizing dyes. Cyanine dyes exhibit strong and sharp absorption bands to meet the demand of color films with delicately designed spectral sensitivity. On the other hand, Ru dyes exhibit very wide absorption bands to meet the demand of solar cells, which should efficiently absorb solar light with spectral intensity extending over a wide wavelength region, as seen in Figure 9.1. As illustrated in Figure 9.3, the width of an absorption band increases with increasing distance between the ground state and excited state equilibrium configurations, and therefore with increasing Stokes shift. The distance between the ground state and excited state equilibrium configurations is brought about by the displacement of an electron on its transition from a molecule's HOMO to LUMO, as is the case of a Ru dye with metal to ligand charge transfer (MLCT) transition, and is therefore likely to decrease the overlap between the HOMO and LUMO. Since the absorption intensity increases with increasing overlap between HOMO and LUMO, it is likely to decrease with increasing band width. These considerations could explain the difference in absorption band width and intensity between cyanine dyes and Ru dyes, as described in Section 9.2.

Another organic device in which the light absorption of organic semiconductors plays an important role is a organic thin film solar cell. One of the typical light-absorbing organic semiconductors in organic solar cells is a group of phthalocyanines. As shown in Figure 10.4, monomeric phthalocyanine dye molecules in an solution exhibit a sharp absorption band in the visible region [48,49], which is not necessarily suitable on its own to collect incident solar light with spectral intensity extending over a wide wavelength region, as seen in Figure 9.1.

Since the above-stated light-absorbing organic semiconductor molecules are used as condensed states in corresponding devices, it is possible to adjust their absorption spectra by making them form appropriate types of aggregates and alignments. As described in

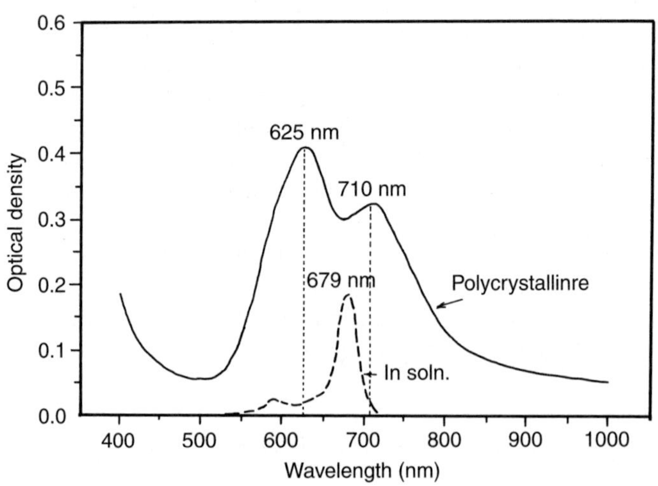

Figure 10.4 Absorption spectra of zinc phthalocyanine in a solution and in thin polycrystalline layer with thickness of 50μm at room temperature [49].

Table 10.1 Analysis of Light Absorption in Some Organic Devices

		Color film	Dye-sensitized solar cell	Organic thin film solar cell
Demand for absorption band of organic semiconductor		Sharp	Very wide	Very wide
Organic semiconductor for light-absaroption		Cyanine dye	Ru dye	Phthalocyanine
Transition of organic semiconductor	Transition	$\pi \rightarrow \pi^*$	MLCT	$\pi \rightarrow \pi^*$
	Stokes' shift [a]	Small	Large	Small
	HOMO/LUMO overlap	Large	Small	Large
	Absorption band	Sharp and strong	Wide and weak	Sharp and strong
Aggregation	Type of aggregate	J	Monomer	Herring-bone
	Effect on absorption band	Sharpening	No effect	Widening

[a] In relation to the distance between the ground state and the excited state equilibrium configurations and to the slope of the potential energy curve of the excited state at the equilibrium ground state, as illustrated in Figure 9.3.

Sections 5.3 and the previous section, there are three types of aggregates: J-aggregates, with a sharp absorption band peaked at longer wavelength region; H-aggregates, with an absorption band peaked at shorter wavelength region; and herring-bone aggregates, with a double band or a band with Davidov splitting, which is much wider than a corresponding monomer band, since it is peaked at both longer and shorter wavelength sides of the monomer band. Cyanine dyes forming J-aggregates are used for color films, which need sensitizing dyes with a sharp absorption band. As seen in Figure 10.4, phthalocyanine dye molecules in their layer form an aggregate with a double band or a band with Davidov splitting, which can enhance the capture of incident solar light with spectral intensity extending over a wide wavelength region. The mixture of J- and H-aggregates can also be used to collect incident light with a wide wavelength range, and was proposed by Hiramoto and Ukei to be applied to phthalocyanines in organic thin film solar cells [29,30], as described in the previous section.

Table 10.1 summarizes the results of analyzing the characteristics of light-absorbing organic semiconductors used in color films, dye-sensitized solar cells, and organic thin film solar cells. This analysis is expected to provide ideas on how to design light-absorbing organic semiconductor layers for organic devices.

10.5 Electronic Structure of the Interface between Substrate and Organic Layer

One of the most important subjects in the electronic structure of the interface between substrates and organic layers is the alignment of their vacuum levels. It was believed that the electronic structure of the interface between substrates and organic layers always obeyed the Schottky–Mott model, according to which the vacuum levels of a substrate and an organic layer coincided with each other when they were in contact [27].

The first observation of the displacement of the vacuum levels between a substrate and an organic layer was made via the collaboration between the laboratories of Seki and of the present author to carefully study the electronic structure between AgX and sensitizing dye layers (merocyanine dyes) by means of ultraviolet photoelectron spectroscopy (UPS) [50,51]. It was revealed that the vacuum level of a dye layer was situated below that of an AgBr layer for all the dyes studied (Figure 5.13), and that the difference in sensitizing ability among dyes cannot be explained without taking into account the above-stated displacement of the vacuum levels.

Following the above-stated result, Seki et al. extensively studied the electronic structure between metal substrates and organic layers with a wide variety of electronic energy levels. They observed that the vacuum level of an organic layer was situated below the vacuum level of a metal substrate for most of the combinations between metal substrates and organic layers, indicating the formation of dipole moments at the interface between them [52,7]. One of the most probable and common causes for the vacuum level shift is the push-back effect. Usually some fraction of electrons are spilled out of the surface of a metal electrode to vacuum, and are pushed back by an organic layer, substantially forming a layer of arrayed dipoles on the interface between the electrode and the organic layer, as illustrated in Figure 10.5. The shift of the vacuum level of an organic layer with respect to that of a metal substrate can be more than 1 eV, and become indispensable to understanding the performance of organic devices.

It is noted that the above-stated results were observed for pure materials under ultra-high vacuum conditions. Following the proposal by Seki and others, many groups of investigators have studied the electronic structure of the interfaces between substrates and organic layers under various conditions. In general, the degree of the vacuum level shift is likely to decrease under practical conditions for the preparation and operation of organic

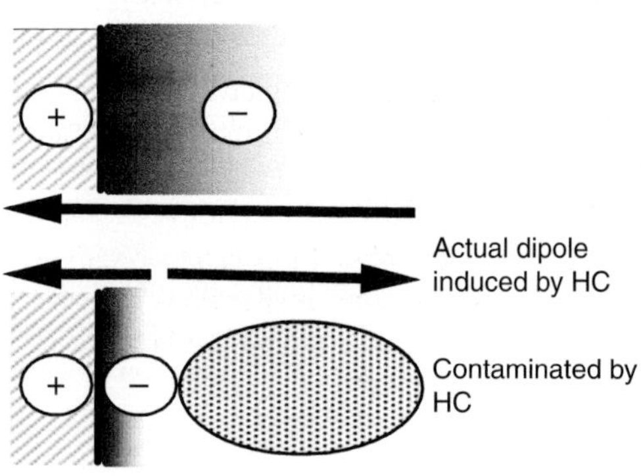

Figure 10.5 The push-back effect. The upper illustration indicates that some fraction of electrons in a metal electrode are spilled out of its surface to vacuum. The lower illustration indicates that an organic layer such as a contaminating hydrocabon (HC) layer pushes the spilled electrons back to the electrode, creating a layer of arrayed dipoles on the surface of the electrode, which lowers the vacuum level of the organic layer with respect to that of the metal.

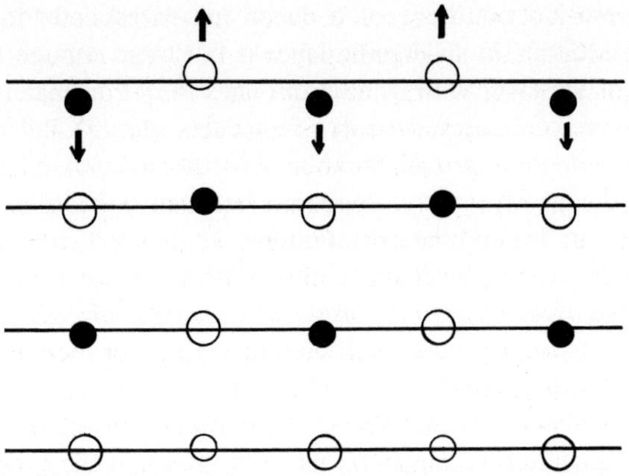

Figure 10.6 A side view of the relaxed AgBr {100} surface with the atomic displacements exaggerated. The Br ions (○) and Ag ions (●) on the top are displayed above and below the normal surface plane, respectively.

devices. At the extreme, Salaneck et al. did not observe any substantial shift of the vacuum level at the interface between substrates and polymeric semiconductor layers under room atmosphere, indicating the absence of dipoles at the interface [53]. Malliaras et al. considered that the electric fields due to dipoles at the interfaces are likely to be neutralized by the presence of polar species such as water, oxygen, and solvent molecules coming from their surroundings [54].

The shift of the vacuum level of merocyanine dye layers with respect to that of AgBr, as seen in Figure 5.13 [50,51], indicates a decrease in ionization energy (i.e., the energy difference between the top of the valence band and the vacuum level) of an AgBr layer by the presence of dye layers on it. Similarly, the ionization energy of an AgBr layer was decreased by the presence of a thin gelatin membrane on it, as seen in Figures 3.4 and 3.5 [55]. It is considered that an array of dipoles is present on the surface of an AgBr layer, and was relaxed by organic layers having dipoles. One of the candidates of the arrayed dipoles on the surface of an AgBr layer might be a "rumpled effect" with a displacement of Br ions slightly above the normal surface plane and Ag ions below the surface plane, as illustrated in Figure 10.6 [56–58].

The heights of the HOMO and LUMO of a series of cyanine dyes available for dye sensitization of AgX grains in photographic films can be relatively given by their electrochemical oxidation and reduction potentials, respectively [59], and evaluated with respect to the heights of the top of the valence band and the bottom of the conduction band of silver halides in photographic films [59,60].

10.6 Diffusion of Excitons in Organic Semiconductors and J-Aggregated Dyes

Most organic devices are faced with the difficulty of achieving efficient dissociation of excitons in organic layers because the binding energy of an exciton is too large for it to spontaneously dissociate at room temperature without the aid of the energy gap at the interface of the hetero-junction between a donor and an acceptor [61,62], and that the diffusion length of excitons in an organic layer is not large enough for each exciton to reach the interface of the layer with its adjacent ones [63]. Fortunately, a color film with mono-molecular layers of J-aggregated dye molecules (donor) and tabular AgX grains (accepter) can overcome these problems. Thus, the diffusion and dissociation of excitons in organic semiconductor layers in comparison between color films and other organic devices are described in this and the next sections.

Although an exciton in a dye layer is in contact with an accepter in color films, there are preferable sites for the dissociation of excitons on the surface of AgX grains. It is therefore desirable for excitons to migrate and reach preferable sites for their efficient dissociation. A model experiment was carried out to indicate that the migration of excitons among sensitizing dye molecules on the surface of AgX grains can increase the yield of their dissociation for the injection of electrons into the grains in the presence of preferable sites [64]. An example of such preferable sites is the molecule at the end of a J-aggregate [65–68], as described in Section 5.7 and in Figure 5.20. It was proposed that the quantum yield of unity can be realized by controlling the size of J-aggregates so that every exciton in a J-aggregate can reach one of the molecules at its ends [67,68]. Another example is a hole-trapping supersensitizer in contact with a J-aggregate. The migration of an exciton in a J-aggregate makes it possible to realize the hole-trapping supersensitization [69,70] with a small amount of supersensitizers, as illustrated in Figure 5.18.

A model experiment also indicated that the migration of excitons should decrease their dissociation in the presence of sites that capture excitons and prevent them from dissociating for the injection of electrons into AgX grains [64]. Excitons are captured by sites whose excitation energy is smaller than that of the excitons. For excitons in a layer of organic molecules, there are various states including J-aggregates, H-aggregates, herring-bone aggregates, and monomers, as illustrated in Figures 5.7 and 5.8. Among them, the energy of an exciton in a J-aggregate is the lowest. Therefore the probability of excitons being captured by traps before their dissociation is considered to be the least in J-aggregates among various states.

Although the diffusion length of excitons in J-aggregates cannot be determined quantitatively, there have been some observations to indicate that excitons in J-aggregated dye molecules on AgX grains diffuse and contribute to their dissociation. Namely, they diffused among J-aggregated cyanine dye molecules with different orientations to change the direction of their transition moments in sub-picoseconds [71] until they decompose to inject electrons into AgX grains in several tens of picoseconds [72–74]. It was observed that the luminescence of J-aggregated cyanine dyes quenched in the presence of one quencher molecule among 10,000 dye molecules [69]. This is consistent with the observation that excitons can migrate to the distance of several micrometers in single crystals of organic semiconductors [75] by taking into account the fact that J-aggregates of cyanine dyes on AgX grains were judged as single crystals, as seen in Figure 10.2 [22].

It is suggested from the above that the diffusion length of excitons in an organic semiconductor layer should be improved by removing defects, impurities, and such states as to capture excitons. It seems that efforts to increase the diffusion length of excitons in organic semiconductor layers are not enough to improve the performance of organic thin film solar cells despite the fact that they should be important.

10.7 Dissociation of Excitons in Organic Semiconductors and J-Aggregated Dyes

Figure 10.7 illustrates the electronic structure of the interface of the hetero-junction between a donor and an acceptor. Although the energy levels of both an excited electron and a conduction electron in a molecule correspond to its LUMO, the former is lower by the binding energy of an exciton than the latter. For efficient dissociation of an exciton in a donor at the interface of the hetero-junction, the LUMO level of a donor should be higher than that of an acceptor by the binding energy of an exciton in an accepter, which is comparable to or more than 0.4eV in the case of organic molecules [10,11].

However, the binding energies of excitons are as small as 0.022eV in AgBr and 0.04eV in AgCl [76,77]. It is therefore expected that the dissociation of excitons should be unity when the energy gap between the LUMO level of a dye and the bottom of the conduction band of an AgX grain (ΔE) is larger than 0.022eV for AgBr grains and larger than 0.04eV for AgCl grains.

The dissociation of excitons in sensitizing dye layers should also be enhanced by small rearrangement energy (λ) in the Marcus theory [78,79]. The electron transfer processes among monomeric species in solutions are generally associated with the inverted regions and λ, which is more than 0.3eV [80,81].

As described in Section 5.6 with Figures 5.14 and 5.15, the electron transfer from sensitizing dyes in the excited state to AgX grains in photographic films takes place in the absence of the inverted region, and is convenient for dye sensitization with high efficiency [82]. Similar results have been observed with dye sensitization of inorganic semiconductors [83,84]. As illustrated in Figure 5.15, the absence of the inverted region of the Marcus theory in these results was explained by the idea that the accepter in these systems was composed of widely and densely distributed electronic energy levels, and can efficiently accept an electron from excited sensitizing dye molecules with high ΔE at one of the above-stated energy levels without any barrier. In the case of hetero-junctions, in which both donor and accepter are composed of organic materials, as illustrated in Figure 10.7,

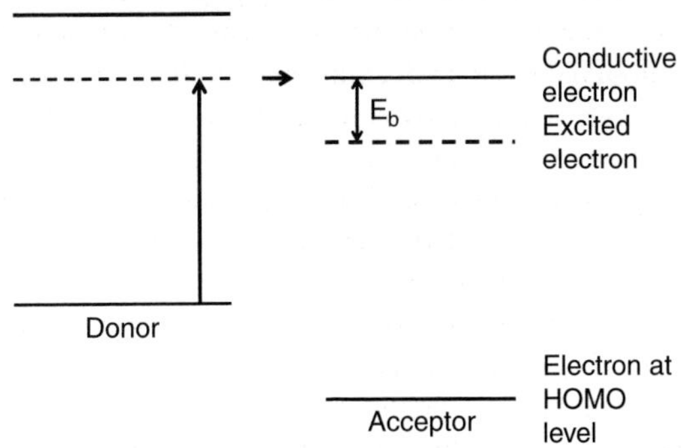

Figure 10.7 The necessary condition for charge separation at the hetero-junction between a donor and an acceptor. Electronic energy level diagram indicates those of an electron at HOMO (lower solid line), an excited electron (broken line), and a conductive electron (upper solid line) in a donor and an acceptor. When the binding energy of an exciton (E_b) is the same between a donor and an acceptor, the LUMO of a donor should be higher than that of an acceptor by E_b.

the accepter is the LUMO level of an organic molecule, which should be too narrow to escape from the inverted region.

Applying the Marcus theory to the relation between the quantum yield of the electron transfer and ΔE (Figures 5.11 and 5.14) gave ~0.05eV as λ for dye sensitization of AgBr grains by cyanine dyes with similar molecular structure [82]. The temperature-dependent rate of light-induced electron transfer from a cyanine dye to AgBr grains revealed that the value of λ was as small as 0.02eV for J-aggregated dye molecules, while it was about 0.3eV for monomeric dye molecules [85,86]. Katoh et al. measured the energy gap-dependent efficiency of the light-induced electron transfer to ZnO powder from monomeric dye molecules in contact with an electrolyte solution, and found that the value of λ was about 0.3eV [87]. It seems that the reduced value of λ can be realized by eliminating solvent molecules from the layer and/or forming a rigid arrangement of molecules in it, and is desirable for the charge separation at the hetero-junction with high efficiency.

When ΔE is not enough to dissociate an exciton at the interface of the hetero-junction between a layer of J-aggregated sensitizing dye molecules and an AgX grain in photographic film, another hetero-junction is formed between a J-aggregate of a dye and a supersensitizer in order to enhance the dissociation of an exciton (Figure 5.18) [69,70]. Only one supersensitizer molecule is needed for each J-aggregate, since an exciton in a J-aggregate can migrate and reach a supersensitizer molecule anywhere on a moderate-sized aggregate.

The quantum yield of the dissociation of excitons in J-aggregates of dyes on AgX grains should increase with decreasing the binding energy of exitons. In addition to theoretical methods [11], an experimental method for evaluating the binding energies of excitons in organic semiconductors by means of inverse photoelectron spectroscopy (IPES) and UPS was proposed [10], although IPES is neither easily accessible nor commonly available for many materials. It was found that the binding energy of excitons is usually more than 0.4eV.

In the research on color films, the binding energy of excitons in mono-molecular layers of many cyanine dyes, which have similar molecular structures and different electronic structures (Table 5.2 in reference 1), was evaluated under the assumption that the binding energies of excitons in these dyes are similar to each other [82]. As illustrated in Figure 5.11, the relative height of a dye's LUMO was given by its reduction potential. Since dye sensitization takes place as a result of electron transfer from an excited dye molecule's LUMO to the conduction band (CB) of AgX, the energy level of an electron in an excited dye molecule's LUMO was evaluated with respect to the bottom of the CB of AgBr by analyzing its sensitizing ability, as illustrated in Figure 5.16. Since the electron transfer from the CB of AgX to the LUMO of a dye in ground state brings about a decrease in intrinsic sensitivity (i.e., desensitization), the energy level of an electron in the LUMO in a dye in ground state was evaluated with respect to the bottom of the CB of AgBr by analyzing the degree of the desensitization, as shown in Figure 5.16.

For many cyanine dyes with similar structure, the difference in energy level of an electron in the LUMO between a dye in ground state and an excited dye was evaluated to be 0.17eV, and ascribed to the binding energy of excitons in these dyes on AgBr, as shown in Figure 5.17. The binding energy of excitons in mono-molecular layers of dyes on AgX grains was much smaller than those evaluated for excitons in layers of various organic semiconductors (>0.4eV) [10,11], and can be responsible for achieving efficient

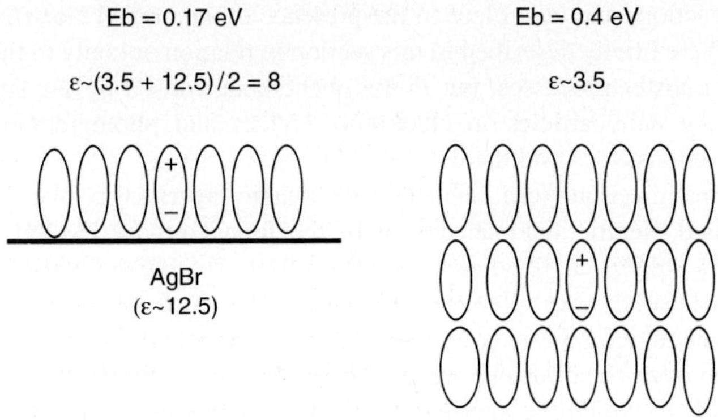

Figure 10.8 The binding energy of excitons in a mono-molecular layer of sensitizing dye molecules on AgBr grains (left) and in a layer of organic molecules (right) in relation to the dielectric constant (ϵ) of their surroundings.

dissociation of excitons in mono-molecular layers of sensitizing dyes on AgX grains. The binding energy of an exciton decreases with increasing dielectric constant of its surroundings [11,29]. The presence of AgBr with a large dielectric constant (12.5) beneath mono-molecular dye layers, as illustrated in Figure 10.8, is considered to be responsible for why the binding energy of excitons in this system is as small as 0.17eV.

The above-stated result indicates that the binding energy of an exciton in an organic layer varies, depending on the dielectric constant of its surroundings, and should be smaller when the layer is in contact with a metal electrode and inorganic semiconductor layer with a large dielectric constant. It is therefore important to evaluate the binding energy of excitons in organic layers in actual organic devices. The above-stated method by the present author, which is based on the processes to put an electron in and take it out of a molecule in a layer from the bottom of the CB of AgBr, is therefore more useful for photographic phenomena than the method with UPS and IPES by Kahn and others [10], and would be applicable to excitons in organic devices having interfaces with inorganic semiconductors.

10.8 Charge Carrier Transfer across Interfaces

Organic devices involve various charge carrier transfer processes across interfaces, as seen in Figure 10.1. As described in the previous section, dissociating an exciton at the interface between a donor layer and an accepter layer results in the electron transfer from the former to the latter. As described in Chapters in 5 and 9, dye sensitization results in electron transfer from sensitizing dye molecules in the excited state to substrates such as tabular AgX grains in photography and a sintered layer of TiO_2 nanoparticles in dye-sensitized solar cells. These charge carrier transfer processes can be exothermic reactions and take place efficiently in the absence of an applied electric field.

Some organic devices such as OLED and organic field effect transistors (FET) involve the charge carrier injection processes from electrodes to organic layers, which are endothermic reactions and take place in the presence of an applied electric field. Data on these processes are briefly described in this section in relation not only to the above-stated charge carrier transfer processes, but to the phenomena caused by the surface plasmon resonance of Ag nanoparticles in electronic devices and photographic materials, as described in Sections 8.1.3 and 8.4.2.

Charge carrier injection from electrodes to organic layers takes place between them under an applied electric field according to the tunnel model [88,89] or thermionic emission model [88,90], as illustrated in Figurc 10.9. The charge carrier injection with the tunnel model usually takes place in the presence of an high electric field, which makes the barrier thin enough for a carrier to pass through by tunnel effect. In many cases, charge carrier injection from electrodes to organic layers takes place by the thermionic emission model, according to which the applied electric field is in concert with the mirror image potential of a charge carrier to decrease the barrier height and enhance the carrier injection. The tunnel model and thermionic emission model are judged in the light of the Fowler–Nordheim and Richardson–Schottky equations, respectively.

Evaluating the barrier height by analyzing current–voltage characteristics should be carried out by getting the blanch, where the current is injection-limited, and by analyzing it at varying electric field and temperature. For this purpose, it is therefore necessary to have such conditions where the mobility of charge carriers is high enough to realize injection-limited current, and independent of electric field and temperature. Fortunately, liquid-crystalline organic semiconductors can meet this condition [37–41]. Thus, the Richardson–Schottky equation was applied to the current due to the injection of positive holes from electrodes to liquid-crystalline organic semiconductors [90]. Figure 10.10 shows the barrier height for the positive hole injection from a Pt electrode to a

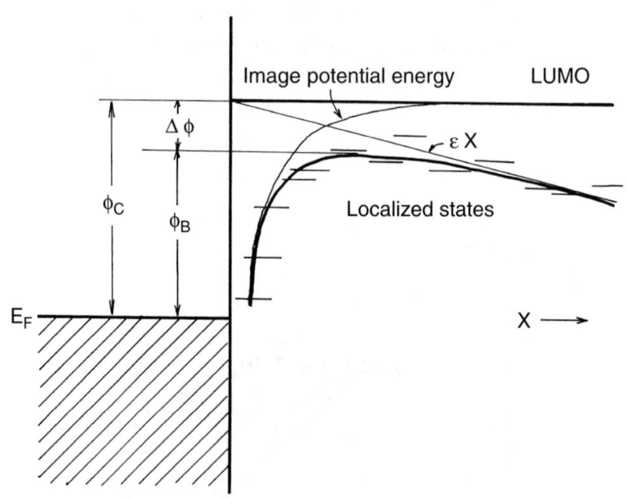

Figure 10.9 The principle of thermionic emission of an electron from metal to organic electron-transporting layer (ETL) over a barrier composed of the potential energy difference between the Fermi level of the metal and the LUMO level of the ETL (Φ_C), mirror image potential, and applied electric field (ϵ). The barrier height (Φ_B) decreases by $\Delta\Phi$, which is in linear relationship with the square root of the applied electric field.

liquid-crystalline organic semiconductor, 2-(4'-octylphenyl)-6-dodecyloxynaphthalene, in smectic A and B phases as a function of the square root of the electric field. It is noted that the barrier height decreased linearly with the square root of applied electric field according to the Richardson–Schottky equation, and can become fairly small in the presence of a high electric field, as seen in this figure.

10.9 Transport of Electronic Carriers in Organic Devices

The transport of electronic carriers in organic devices is important, and includes those in organic layers and in substrates. In the case of color films, electrons do not migrate in organic layers with sensitizing dyes, since they are transferred to substrates with tabular AgX grains immediately after the optical excitation of the layers, as described in Chapter 5. However, positive holes remain in the layers after the electrons transfer to the substrates, migrate in the layer, and attack latent image centers. As described in Section 5.8, the migration of positive holes is enhanced by temperature. The result in Figure 10.10 can be revealed by use of the unique property of liquid-crystalline organic semiconductors, in which the mobility of electronic charge carriers is independent of electronic field and temperature [37–41,91], being suitable for analyzing the current–voltage characteristics of organic devices. Namely, the transport of charge carriers in organic layers with various structures is characterized by their dependence on temperature and applied electric field.

From the viewpoint of structure, organic layers are classified into four groups; single-crystalline, polycrystalline, liquid-crystalline, and amorphous. The mobility of electronic carriers in organic layers with amorphous and polycrystalline structures increases with the square root of electric field (i.e., Pool–Frenkel model) [92,93] due to the decrease in barrier height for the release of the carriers by electronic field, and increases with

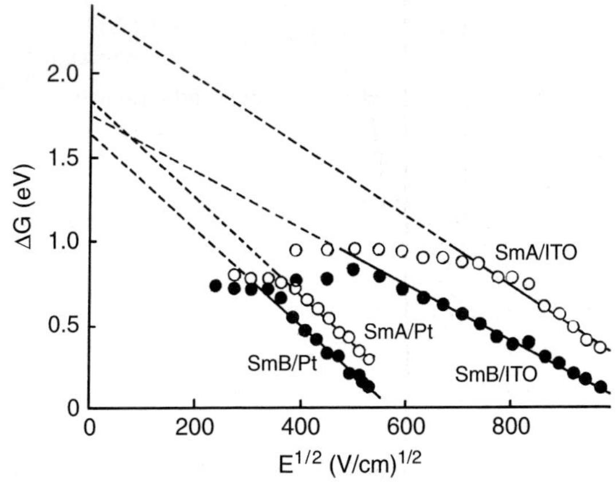

Figure 10.10 The barrier height for the injection of a positive hole from a Pt and ITO electrodes to a liquid-crystalline organic semiconductor (a phenyl-naphthalene derivative) in the phases of smectic A and B as a function of the square root of the applied electric field (E).

temperature due to the enhancement of the release of the carriers by thermal stimulation. In the case of polycrystalline organic layers, charge carriers migrate with multiple trapping model [94] in the presence of traps provided by defects and boundaries of polycrystals. Thus, the mobility of charge carriers (μ) is enhanced by temperature (T) with Arrhenius-type dependence (i.e., linear relationship between $\log\mu$ and $1/T$). In the case of amorphous organic layers, the molecules' HOMO and LUMO electronic energy levels are distributed due to their disordered structure[95], and correlate with each other to form potential valleys, which bound charge carriers for their hopping [96–98]. In this case, μ is enhanced by T with a linear relationship between $\log\mu$ and $1/T^2$.

In single-crystalline organic layers, the mobility of electronic carriers is independent of electric field due to the absence of disordered structure, and decreases with temperature due to the collision of the carriers with phonons (i.e., lattice vibration) [99]. In liquid-crystalline organic layers, the mobility of electronic carriers is independent of the electric field due to the absence of disordered structure, and independent of temperature because thermal energy is converted, not to lattice vibration, but to molecular rotation, which is so slow [100] in comparison with the transport of an electronic carrier from a molecule to its adjacent one that the increase in the rotation rate with temperature hardly influence the carrier transport.

In color films, the dissociation of an exciton in a mono-molecular layer of a sensitizing dye on a tabular AgX grain gives an electron in the conduction band of the grain and a positive hole in the dye layer (Dye^+; a dye positive hole).

Knowledge of Dye^+ in dye layers is important for the design of color films and has been given by an electron spin resonance (ESR) method as early as in 1975 [101–103]. As shown in Figures 5.22 and 5.23, the decay rates of both Dye^+ and latent image centers were observed with similar time constants, and similarly increased with second-order kinetics as the dyes' HOMO height decreases, indicating that dye positive holes were responsible for their recombination with latent image centers. The sharpness and intensity of ESR spectra of Dye^+ in the mono-molecular layers of J-aggregated cyanine dyes on AgBr grains in photographic films increased with increasing aggregates size, and were therefore subjected to motional narrowing due to the movement of dye positive holes in the layers, although the diffusion length of dye positive holes in the layers cannot be determined.

The study of dye positive holes by means of ESR spectra was applied to the study of electronic carriers in the layers of organic semiconductors. Marumoto, Kuroda, and others analyzed the ESR spectra of positive holes in the single crystals of pentacene, and supported the band structure of its HOMO by observing that a positive hole was delocalized over about 10 molecules.in the crystal [104–106].

In the hetero-junction between sensitizing dye layers and AgX grains, an exciton in the dye layer is subjected to the dissociation to give an electron in the conduction band in the grain and positive hole in the dye layer, as described in Section 5.8. In contrast to the old theory of dye sensitization [107,108], an electron appears and migrates in an AgX grain, neither appearing nor migrating in the dye layer. The study of the electronic transport in single crystals of silver halides was initiated by Haynes and Shockley [109,110] and followed by many investigators, as described in Section 3.3. The lifetime and mobility of electrons in AgX grains were measured by means of the microwave photoconductivity method [111–113] and radio frequency photoconductivity method [114,115], which were developed in 1972 and improved in 1988.

In the layers of organic semiconductors, the creation of electronic carriers on exposure to light with their band-gap energy is difficult because the binding energy of excitons in the layers is too large for excitons to be dissociated at room temperature [10,11], and needs aids such as applied electronic field [11] and energy gap at the interface of the hetero-junction between a donor and an acceptor [116–120]. Therefore, it is usually difficult to apply the microwave and radio frequency photoconductivity methods to layers of organic semiconductors since any electronic field cannot be applied to the sample layers in these methods. Hiramoto et al. successfully applied the microwave photoconductivity method to organic photovoltaic layers with hetero-junctions, which enabled the dissociation of excitons to create electronic carriers in the absence of applied electric field [118–120]. Warman and van de Croats applied the microwave photoconductivity method to measure the mobility and lifetime of electronic carriers in organic layers by use of pulses of radiation, whose energy was high enough to create electronic carriers instantaneously [121].

The mobility of electronic carriers in organic layers is usually measured by means of the time-of-flight (TOF) method [122,123] and FET method [124], both of which require electronic carriers to move a long distance and are not available for samples with deep carrier traps in high concentration. On the other hand, electronic carriers are required to move short distances during the measurements of the microwave and radio frequency methods, respectively. These methods are therefore available for samples with deep carrier traps in high concentration. Warman and van de Croats can measure the mobility of positive holes in polycrystalline triphenylene derivative by the microwave photoconductivity method, while they cannot by the TOF method [121].

Organic devices, in which the dissociation of excitons for electronic charge generation is successful, include color films, dye-sensitized solar cells, and organic thin film solar cells. As a result of the dissociation of excitons, positive holes appear in donors (i.e., J-aggregated cyanine dye layers in color films, Ru dyes in dye-sensitized solar cells, and phthalocyanines in organic thin film solar cells), while electrons appear in substrates (i.e., AgX grains in color films, sintered layer of TiO_2 nanoparticles in dye-sensitized solar cells, and fullerene in organic thin film solar cells). The electronic transports in these devices are summarized in Table 10.2.

As seen in Table 10.2, the drift mobility of electrons in the grains of AgX is considerably smaller than that in its single crystal, and ascribed to the fact that on the surface of AgX grains there are many shallow electron traps at which electrons take rests during their migration. Although the enhancement of the mobility and diffusion length of electrons are desirable for the growth of latent image centers, they are faced with difficulty because many sensitization centers acting as shallow electron traps are necessary for efficient latent image formation, as described in Section 4.3.2.

The values of electron mobility in dye-sensitized solar cells and organic thin film solar cells were markedly smaller than that in color films, indicating that electrons took rests for long time at traps during their migration. This condition is not desirable for electrons to migrate and surely reach anodes in these devices, and should be improved. Particularly in the case of dye-sensitized solar cells, electrons with such a small mobility as seen in Table 10.2 are demanded to migrate to a distance of 10μm in a sintered layer of TiO_2 nanoparticles, and can be barely coupled with an extremely long electron lifetime of several seconds [3–5] to meet the demand.

Table 10.2 Transport of electrons (e^-) in color films, dye-sensitized solar cells (DSSC), and organic photovoltaic cells (OPV) with hetero-junction between donor (D; light-absorbing) and acceptor (A; substrate for electron transport)

		Color film	DSSC	OPV
Hetero junction	D; light-abs.	Cyanine dyes	Ru dyes	Phthalocyanine
	A; substrate	Tabular AgX	TiO_2 particle	Fullerene
Mobility of e^- in device <that in single crystal>		~1cm^2/Vs [112]<50 [125]>	~10^{-3} [126]<10 [127]>	~10^{-4} [128] <0.5 [129] ~5 [130]>
Diffusion constant of e^-		2×10^{-2}cm^2/s	~10^{-5}	~10^{-6}
Lifetime of e^-		~100ns[113]	Several seconds	Increase by purification
Diffusion length of e^-		~0.7 μm	~10[3–5]	0.13→≥1.0 [131]

The conductivity of organic devices in the dark is important for their design. The conductivity of an organic layer in the dark is usually given as its current-voltage characteristics. The conductivity of AgX grains in the dark has been observed with interstitial silver ions as carriers by means of the dielectric loss method since 1970, as described in Section 3.2.1 [132], which has not yet been applied to organic devices.

10.10 Production of Organic Devices and Color Films

Color films are composed of single-crystalline mono-molecular layers of sensitizing dyes on AgX grains as described in Chapter 5 and Section 10.3. It is noted that the dye layers are formed through the adsorption of dye molecules to the grains, according to which each molecule is always situated at the most stable site, as described in Section 5.4. On the other hand, organic layers in most devices are formed on substrates by means of evaporation, coating, and printing, through which each molecule cannot always find its most stable site and is likely to be involved in the formation of layers with disordered structure.

In typical color negative films, more than 10 aqueous gelatin solutions and suspensions with different functions are coated in pile at once on a very wide film base with high precision and speed by means of a simultaneous multilayer coating method [133,134]. On drying, these layers are about 20nm thick in total. It is noted that this coating method has been elaborated for more than a half-century for color films and other photographic films and papers.

Figure 10.11 illustrate a coating machine, which is called a slide coater, for small-scale experiments to indicate the principle of multilayer coating of color films. A slide coater basically consists of a multilayer die with an inclined plane, placed at a small distance from a moving web, supported by a backing roll. Various aqueous gelatin solutions with functional materials including AgX grains are fed through distribution cavities and slots, forming a multilayer stack on the incline without convective mixing. The solutions are bounded sideways by edge guides. At the end of the inclined plane, the solution bridges the narrow gap between coater and web, wets the web and gets entrained by it.

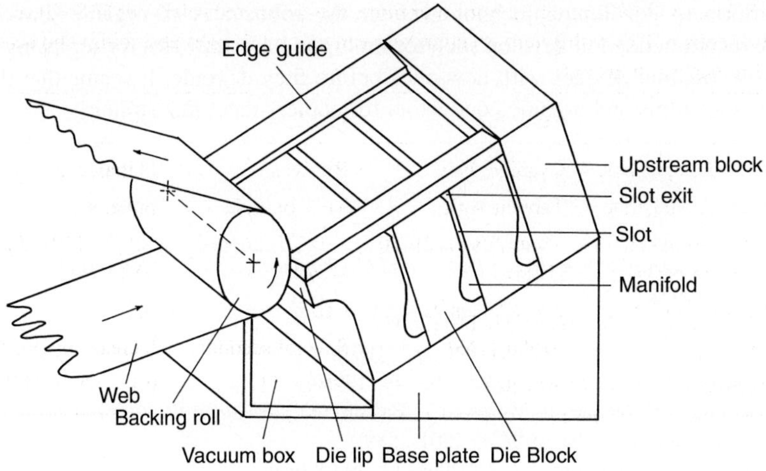

Figure 10.11 A slide coater, which basically consists of a multilayer die with an inclined plane, placed at a small distance from a moving web, supported by a backing roll. Solutions are fed through distribution cavities and slots, forming a multilayer stack on the incline without convective mixing. The solutions are bounded sideways by edge guides. At the end of the inclined plane, the solution bridges the narrow gap between coater and web, wets the web, and gets entrained by it.

As seen in Figure 1.1, tabular AgX grains and other materials with anisotropic shape can be aligned in parallel with the film base by making the entrained rate much larger than the flowing rate on the incline. This coating technique is applicable to multilayers, each of which can be as thin as several tens of nanometers. It is expected that this kind of multilayer coating methods is also useful for preparing organic devices with multilayers and alignment of materials with anisotropic shape.

10.11 Durability of Organic Devices and Color Films

Since organic molecules are usually fragile, organic devices are always faced with the difficulty of securing their durability. Consumers are recommended to use color films within two years of their production. Similarly, consumers must replace organic photo conductors in copiers with new ones when they degrade usually every several months. Extensive efforts are being made to develop new technologies for securing the durability of various organic devices. They include the development of robust organic materials used in the devices and technologies for maintaining that robustness.

From the viewpoint of durability, the best model should be a living thing. The durability of a living thing is realized not by the robustness of organic materials acting as its constituents, but by the dynamic equilibrium always taking place in it. Schoenheimer has found that all the elements in a living thing are completely replaced by new ones within several months [135,136]. Namely, all the constituents are replaced by new ones before they degrade.

In addition to developing technologies for the robustness of organic devices, it is therefore recommended to develop such technologies and strategies for replacing organic materials in organic devices with new ones before they degrade. It seems that the treatments of color films and organic conductors for copiers meet this strategy.

References

1. T. Tani, *Photographic Sensitivity: Theory and Mechanisms* (New York/Oxford: Oxford University Press, 1995).
2. T. Tani, *J. Imaging Sci. Tech.*, 51, 110 (2007).
3. B. O'Regan, M. Graetzel, *Nature*, 353, 737 (1991).
4. A. Hagfeldt, M. Graetzel, *Chem. Rev.*, 95, 49 (1995).
5. Michael Graetzel, Jacques-E. Moser, in V. Balzani (ed.), *Electron Transfer in Chemistry*, vol. 5 (Weinheim, Germany: Wiley-VCH, 2001), 589.
6. T. Tani, IEICE Techn. Rep. 109, ED2009–12, 49 (2009).
7. H. Ishii, K. Seki, in W. R. Salaneck, K. Seki, A. Kahn, J.-J. Pireaux (eds.), *Conjugated Polymer and Molecular Interfaces Science and Technology for Photonic and Optoelectronic Applications* (New York/Basel: Marcel Dekker, 2001), 293.
8. C. W. Tang, *Appl. Phys. Lett.*, 48, 183 (1986).
9. P. Peumans, A. Yakimov, S. R. Forrest, *J. Appl. Phys.*, 93, 3693 (2003).
10. G. Hill, A. Kahn, Z. G. Soos, R. A. Parcel, Jr., *Chem. Phys. Lett.*, 327, 181 (2000).
11. V. I. Arkhipov, H. Baessler, in W. Bruetting (ed.), *Physics of Organic Semiconductors* (Weinheim, Germany: Wiley-VCH, 2005), 183.
12. D. S. Weiss, J. R. Cowdery, R. H. Young, in Balzani (ed.), *Electron Transfer in Chemistry*, vol. 5, Chapter 2.
13. P. M. Borsenberger, S. O. Weiss, *Organic Photoconductors for Imaging Systems* (New York: Marcel Dekker, 1993).
14. C. W. Tang, S. A. VanSlyke, *Appl. Phys. Lett.*, 51, 913 (1987).
15. K. Kudo, M. Yamashina, T. Moriizumi, *Jpn. J. Appl. Phys.*, 23, 130 (1984).
16. C. R. Newman, C. D. Frisbie, D. A. da Silva Filho, J.-L. Bredas, P. C. Ewbank, K. R. Mann, *Chem. Mater.*, 16, 4436 (2004).
17. J. M. Simson, *Photogr. Sci. Eng.*, 18, 302 (1974).
18. C. Reich, *Photogr. Sci. Eng.*, 18, 335 (1974).
19. S. L. Scrutton, *J. Photogr. Sci.*, 22, 69 (1974).
20. E. E. Jelly, *Nature*, 138, 1009 (1936).
21. G. Scheibe, *Angew. Chem.*, 49, 563 (1936).
22. J. E. Maskasky, *Langmuir*, 7, 407 (1991).
23. S. Takada, T. Tani, *J. Appl. Phys.*, 45, 4767 (1974).
24. E. G. McRae, M. Kasha, *J. Chem. Phys.*, 28, 721 (1958).
25. A. S. Davidov, *Theory of Molecular Excitons*, trans. M. Kasha, M. Oppenheimer, Jr. (New York: McGraw-Hill, 1962).
26. G. Jannssens, F. Touhari, J. W. Gerritsen, H. van Kempen, P. Callant, G. Deroover, D. Vanderbroucke, *Chem. Phys. Lett.*, 344, 1 (2001).
27. J. D. Wright, *Molecular Crystals* (Cambridge: Cambridge University Press, 1987).
28. R. B. Campbell, *Acta Cryst.*, 14, 705 (1961).
29. M. Hiramoto, *J. Soc. Photogr. Sci. Tech. Jpn.*, 72, 337 (2009).
30. K. Ukei, *J. Phys. Soc. Jpn.*, 40, 140 (1976).
31. S. Okada, H. Segawa, *J. Am. Chem. Soc.*, 125, 2792 (2003).

32. H. Inokuchi, G. Saito, P. Wu, K. Seki, T. B. Tang, T. Mori, K. Imaeda, T. Enoki, Y. Higuchi, K. Inaka, N. Yasuoka, *Chem. Lett.*, 1263 (1986).
33. S. Ukai, S. Igarashi, M. Nakajima, K. Marumoto, H. Ito, S. Kuroda, K. Nishimura, Y. Enomoto, G. Saito, *Colloids Surf.*, A284–5, 589 (2006).
34. Y. Kunugi, K. Takimiya, Y. Yamane, K. Yamashita, Y. Aso, T. Otsubo, *Chem. Mater.*, 15, 6 (2003).
35. T. Yamamoto, K. Takimiya, *J. Am. Chem. Soc.*, 129, 2224 (2007).
36. T. Izawa, E. Miyazaki, K. Takimiya, *Adv. Mater.*, 20, 3388 (2008).
37. D. Adam, F. Coss, T. Frey, D. Funhoff, D. Haarer, H. Ringsdorf, P. Schuhmacher, K. Siemensmeyer, *Phys. Rev. Lett.*, 70, 460 (1993).
38. D. Adam, P. Schuhmacher, J. Simmerer, L. Haeussling, K. Slemensmeyer, K. H. Etzbach, H. Ringsdorf, D. Haarer, *Nature*, 371, 141 (1994).
39. M. Funahashi, J. Hanna, *Jpn. J. Appl. Phys.*, 35, L703 (1996).
40. M. Funahashi, J. Hanna, *Phys. Rev. Lett.*, 78, 2184 (1997).
41. M. Funahashi, J. Hanna, *Appl. Phys. Lett.*, 71, 602 (1997).
42. H. Iino, Doctoral Thesis, Tokyo Institute of Technology, 2006.
43. H. Iino, J. Hanna, J. Bushby, B. Movaghar, B. J. Whitaker, M. J. Cook, *Appl. Phys. Lett.*, 87, 132102 (2005).
44. H. Iino, Y. Takayashiki, J. Hanna, R. Bushby, *Jpn. J. Appl. Phys.*, 44, L1310 (2005).
45. D. M. Sturmer, D. M. Heseltine, in T. H. James (ed.), *The Theory of the Photographic Process*, Fourth edn (New York: Macmillan, 1977), Chapter 8.
46. H. Nakahara, K. Fukuda, D. Moebius, H. Kuhn, *J. Phys. Chem.*, 90, 6144 (1986).
47. H. Nakahara, D. Moebius, *J. Colloid Interface Sci.*, 114, 363 (1986).
48. J. H. Sharp, M. Lardon, *J. Phys. Chem.*, 78, 3230 (1968).
49. B. Maenning, M. Pfeiffer, A. Nollau, X. Zhou, K. Leo, *Phys. Rev. B*, 64, 195208 (2001).
50. K. Seki, T. Tani, H. Ishii, *Thin Solid Films*, 273, 20 (1996).
51. K. Seki, H. Yanagi, Y. Kobayashi, T. Ohta, T. Tani, *Phys. Rev. B*, 49, 2760 (1994).
52. H. Ishii, K. Sugiyama, E. Ito, K. Seki, *Adv. Mater.*, 11, 605 (1999).
53. G. Greczynski, M. Fahlman, W. R. Salaneck, *Chem. Phys. Lett.*, 321, 379 (2000).
54. Y. Shen, A. R. Hosseini, M. H. Wong, G. G. Malliaras, *Chem. Phys. Chem.*, 5, 16 (2004).
55. T. Tani, *J. Soc. Photogr. Sci. Tech. Jpn.*,72, 88 (2009).
56. R. C. Baetzold, Y. T. Tan, P. W. Tasker, *Surface Sci.*, 195, 579 (1988).
57. P. Tangyunyong, T. N. Rhodin, Y. T. Tan, K. J. Lushington, *Surface Sci.*, 255, 259 (1991).
58. Y. T. Tan, K. J. Lushington, P. Tangyunyong, T. N. Rhodin, *J. Imaging Sci. Tech.*, 36, 118 (1992).
59. T. Tani, S. Kikuchi, *Photogr. Sci. Eng.*, 11, 129 (1967).
60. R. W. Berriman, P. B. Gilman, Jr., *Photogr. Sci. Eng.*, 17, 235 (1973).
61. I. G. Hill, A. Kahn, Z. G. Soos, R. A. Pascal, Jr., *Chem. Phys. Lett.*, 327, 181 (2000).
62. V. I. Arkhipov, H. Baessler, in Bruetting (ed.), *Physics of Organic Semiconductors*, 183.
63. P. Peumans, A. Yakimov, S. R. Forrest, *J. Appl. Phys.*, 93, 3693 (2003).
64. A. A. Muenter, J. S. Kittel, in *Papers from Internation Congress of Photographic Science 1982, University of Cambridge, September 6–10, 1982*, p. 196.
65. K. Kemnitz, K. Yoshihara, T. Tani, *J. Phys. Chem.*, 94, 3099 (1990).
66. T. Tani, T. Suzumoto, K. Kemnitz, K. Yoshihara, *J. Phys. Chem.*, 96, 2778 (1992).
67. W. West, B. H. Carroll, *J. Chem. Phys.*, 15, 529 (1947).
68. W. West, B. H. Carroll, *J. Chem. Phys.*, 19, 417 (1951).
69. W. West, P. B. Gilman, in James (ed.), *The Theory of Photographic Sensitivity*, Chapter 10.
70. P. B. Gilman, Jr., *Photogr. Sci. Eng.*, 18, 418 (1974).
71. J. W. Oh, S. Kumazaki, I. V. Rubtsov, T. Suzumoto, T. Tani, K. Yoshihara, *Chem. Phys. Lett.*, 352, 357 (2002).

72. K. Kemnitz, K. Yoshihara, T. Tani, *J. Phys. Chem.*, 94, 3099 (1990).
73. T. Tani, T. Suzumoto, K. Kemnitz, K. Yoshihara, *J. Phys. Chem.*, 96, 2778 (1992).
74. A. A. Muenter, D. V. Brumbaugh, J. Apolito, L. A. Horn, F. C. Spano, S. Mukamel, *J. Phys. Chem.*, 96, 2783 (1992).
75. J. Tsutsumi, J. Yamada, H. Matsui, S. Haas, T. Hasegawa, in the preprint of *The 57th Meeting of Applied Physics (March 2010, Tokai University)*, p. 172.
76. H. Kanzaki, S. Sakuragi, *J. Phys. Soc. Jpn.*, 24, 652 (1968).
77. H. Kanzaki, S. Sakuragi, *J. Phys. Soc. Jpn.*, 29, 924 (1970).
78. R. A. Marcus, *Ann. Rev. Phys. Chem.*, 15, 155 (1964).
79. V. G. Levich, in H. Eyring, D. Henderson, W. Jost (eds), *Physical Chemistry*, vol. IXB (New York: Academic Press, 1970), 986.
80. J. R. Miller, L. T. Calcaterra, G. L. Closs, *J. Am. Chem. Soc.*, 106, 3047 (1984).
81. L. Eberson, *Adv. Phys. Org. Chem.*, 18, 79 (1982).
82. T. Tani, T. Suzumoto, K. Ohzeki, *J. Phys. Chem.*, 94, 1298 (1990).
83. K. Hashimoto, M. Hiramoto, A. B. P. Lever, T. Sakata, *J. Phys. Chem.*, 92, 1016 (1988).
84. K. Hashimoto, M. Hiramoto, T. Sakata, *Chem. Phys. Lett.*, 148, 215 (1988).
85. M. T Spitler, A. Ehert, R. Kietzmann, F. Willig, *J. Phys. Chem. B*, 101, 2552 (1997).
86. F. Willg, M. T. Spitler, *J. Imaging Sci. Tech.*, 41, 272 (1997).
87. R. Katoh, A. Furube, A. V. Barzykin, H. Arakawa, M. Tachiya, *Coordination Chem. Rev.*, 248, 1195 (2004).
88. S. M. Sze, *Physics of Semiconductor Devices*, 2nd edn (New York: Wiley, 1981).
89. I. D. Parker, *J. Appl. Phys.*, 75, 1656 (1994).
90. T. Toda, J. Hanna, T. Tani, *J. Appl. Phys.*, 101, 024505 (2007).
91. A. Ohno, J. Hanna, *Appl. Phys. Lett.*, 82, 751 (2003).
92. F. Frenkel, *Phys. Rev.*, 51, 647 (1938).
93. W. D. Gill, *J. Appl. Phys.*, 43, 5033 (1972).
94. D. M. Pai, *J. Chem. Phys.*, 52, 2285 (1970).
95. H. Baessler, *Phys. Stat. Sol. b*, 175, 15 (1993).
96. D. H. Dunlap, P. E. Parris, V. M. Kenkre, *Phys. Rev. Lett.*, 77, 542 (1996).
97. S. V. Novikov, D. H. Dunlap, V. M. Kenkre, P. E. Parris, A. V. Vannikov, *Phys. Rev. Lett.*, 81, 4472 (1998).
98. P. E. Parris, V. M. Kentre, D. H. Dunlap, *Phys. Rev. Lett.*, 87, 126601 (2001).
99. W. Warta, N. Karl, *Phys. Rev. B*, 32, 1172 (1985).
100. S. Mitra, V. S. S. Sastry, K. Venu, R. Mukhopadhyay, *Chem. Rev. Lett.*, 406, 263 (2005).
101. T. Tani, *Photogr. Sci. Eng.*, 19, 356 (1975)
102. T. Tani, *J. Appl. Phys.*, 62, 2456 (1987).
103. T. Tani, Y. Sano, *J. Appl. Phys.*, 69, 4391 (1991).
104. K. Marumoto, Y. Muramatsu, S. Ukai, H. Ito, S. Kuroda, *J. Phys. Soc. Jpn.*, 73, 1673 (2004).
105. K. Marumoto, Y. Muramatsu, Y. Nagano, T. Iwata, S. Ukai, H. Ito, S. Kuroda, Y. Shimori, S. Abe, *J. Phys. Soc. Jpn.*, 74, 3066 (2006).
106. K. Marumoto, S. Kuroda, T. Takenobu, Y. Iwasa, *Phys. Rev. Lett.*, 97, 256603 (2006).
107. H. Meier, *Spectral Sensitization* (London/New York: Focal Press, 1968).
108. A. Terenin, I. Akimov, *J. Phys. Chem.*, 69, 730 (1965).
109. J. R. Hayes, W. Scockley, in *Report on a Conference on the Strength of Solids, University of Bristol, 1947* (London: The Physical Society, 1948), 151.
110. J. R. Hayes, W. Shockley, *Phys. Rev.*, 82, 935 (1951).
111. L. M. Kellogg, N. B. Liebert, T. H. James, *Photogr. Sci. Eng.*, 16, 115 (1972).
112. R. J. Deri, J. P. Spoonhower, *J. Appl. Phys.*, 57, 2806 (1985).
113. T. Oikawa, N. Saeki, T. Kaneda, A. Hirano, T. Tani, *J. Imaging Sci. Technol.*, 39, 233 (1995).

114. L. M. Kellogg, J. Hodes, J. Muenter, in *IS&T and SPSTJ's Proc. of the Intern. East-West Symposium* (Springfield, VA: IS&T, 1988), C11.
115. K. Hosoi, A. Hirano, T. Tani, *J. Imaging Sci. Technol.*, 46, 76 (2002).
116. C. W. Tang, *Appl. Phys. Lett.*, 48, 183 (1986).
117. N. S. Sariciftci, L. Smilowitz, A. J. Heeger, F. Wudl, *Science*, 258, 1474 (1992).
118. M. Hiramoto, Y. Sakagami, M. Yokoyama, *J. Chem. Soc. Jpn.*, 1180 (1992).
119. K. Suemori, T. Miyata, M. Hiramoto, M. Yokoyama, *Jpn. J. Appl. Phys.*, 43, L1014 (2004).
120. S. Suemori, T. Miyata, M. Yokoyama, M. Hiramoto, *Appl. Phys. Lett.*, 86, 063509 (2005).
121. J. M. Warman, A. M. van de Croats, *Mol. Cryst. Liq. Cryst.*, 396, 41 (2003).
122. R. G. Kepler, *Phys. Rev.*, 119, 1226 (1960).
123. B. A. Gregg, in S.-S. Sun and N. S. Sariciftci (eds), *Organic Photovoltaics, Mechanism, Materials, and Devices* (London/New York: Taylor and Francis, 2005), Chapter 6.
124. N. F. Mott, *Proc. Cambridge, Philos. Soc.*, 34, 568 (1938).
125. R. K. Ahrenkiel, *Phys. Rev.*, 180, 859 (1969).
126. A. Solbrand, H. Lindstrom, H. Rensmo, A. Hagfeldt, S-E. Lindquist, *J. Phys. Chem. B*, 101, 2514 (1997).
127. L. Forro, O. Chauvet, D. Emin, L. Zuppiroll, H. Berger, F. Levy, *J. Appl. Phys.*, 75(1), 633 (1994).
128. M. Hiramoto, private communication, 2007.
129. E. Frankevich, Y. Maruyama, H. Ogata, *Chem. Phys. Lett.*, 214, 39 (1993).
130. K. Itaka, M. Yamashiro, J. Yamaguchi, M. Haemori, S. Yaginuma, Y. Matsumoto, M. Kondo, H. Koinuma, *Adv. Mater.*, 18, 1713 (2006).
131. K. Sakai, M. Hiramoto, *Mol. Cryst. Liq. Cryst.*, 491, 284 (2008).
132. J. van Biesen, *J. Appl. Phys.*, 41, 1910 (1970).
133. T. A. Russell, R. M. Wilson, C. R. Sanford, "Multiple Coating Apparatus", US Patent 2,761,417 (1956).
134. J. Hens, W. van Abbenyen, in S. F. Kistler and P. M. Schweizer (eds), *Liquid Film Coating Scientific Principles and Their Technological Implications* (London: Chapman and Hall, 1997), Chapter 11b.
135. R. Schoenheimer, *The Dynamic State of Body Constituents* (New York, Hafner, 1964).
136. R. E. Olson, *J. Nutrition*, 127, 1017S (1997).

11

Models for Research and Development in Manufacturing Industry and University

11.1 Product Portfolio Management for the Photographic Industry

The present author has been involved in the research and development of photographic films for nearly a half of a century at a university and a photographic company, and has observed the rise and fall of the industry of silver halide (AgX) photographic films and papers.

According to Product Portfolio Management [1], industrial products are classified into four groups: "problem child", "star", "cash cow", and "dog" in a matrix of their market share and growth rate as shown in Figure 11.1. "Problem child" is a product with small market share and large growth rate, and needs to enhance additional investment for it to get proper market share and growth rate. "Star" is a product with large growth rate and large market share, and still needs the investment to maintain the present growth rate. "Cash cow" is a product with large market share and small growth rate, and does not need any more investment for its growth. "Dog" is a product with small share and small growth rate, and has no prospect for its growth even with investment.

The products of photographic films belonged to "problem child" when the present author was involved in the field of AgX photography, eventually became "star", reached and stayed as "cash cow" for many years, and later fell into "dog" [2,3]. In Japan, the shipment of color films peaked at 20 million m^2 in 1997, and declined to 2 million m^2 by 2008 [4]. During the period of "star" and "cash cow", the research and development organization for the photographic films industry was very powerful and efficient, and was regarded as a good model (i.e., the center of excellence; COE) for manufacturing industries [2,3].

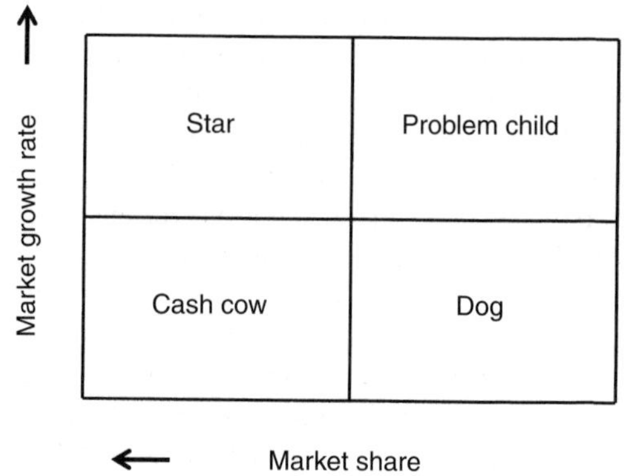

Figure 11.1 Product portfolio management.

The present author was involved in the research on photographic science at Kikuchi's Laboratory at the University of Tokyo for 5 years, and at the Ashigara Research Laboratories of Fuji Photo Film for more than 40 years. Based on these experiences, he has analyzed the research and development organization of the photographic films industry, and describes their results in this chapter.

11.2 Analysis of the Rise and Fall of the Photographic Industry

When the present author began to study photographic science in Kikuchi's Laboratory in 1963, the photographic films and papers industry in Japan was still regarded as "problem child". The industry made great progress for many years and became "star", eventually being able to enjoy "cash cow". Although it is a pity that the industry fell into "dog" after that, it was a unique experience for the present author to observe how one of the most typical fine-chemical industries moved through all the groups of the product portfolio.

The photographic films and papers industry at "cash cow" was characterized by high profitability with small investment, and can be regarded as the ideal condition for manufacturing industries to aim for. The "star" and "cash cow" conditions were based on the technologies, which were developed and accumulated without attracting attention from other industries for many years [2,3]. When the products of photographic films became "star" and "cash cow", many other industries became interested in the industry, but found it very difficult to join it because of the high barrier set by the above-stated technologies, which had been developed and accumulated for many years.

In other words, the photographic films and papers industry formed a strong "domain", based on its own technologies which allowed the companies in this domain to compete quite favorably with those outside of it. The domain allowed the companies within to have a very powerful and efficient organization for researching and developing its products [2,3].

In this organization, different areas represented by products such as color negative films, color reversal films, color papers, photographic films for printing industry, and those for medical uses interacted with each other intimately to enhance the creation and exchange of new ideas and technologies based on common knowledge and technologies such as the design and preparation of mono-dispersed AgX grains, the design and synthesis of various kinds of organic functional compounds, theory and mechanisms of latent image formation and dye sensitization, the design and evaluation of the quality of pictures taken by photographic films, the design and production of film bases for the above-stated photographic products, and the design and construction of the apparatuses for the preparation of photographic films and papers [2,3]. In addition, routine experiments for preparing and evaluating photographic films and papers were carried out precisely, efficiently, and powerfully based on abundant techniques and information that had been proposed and accumulated mainly by many technicians over many years [5].

Ashigara Research Laboratories were provided with all the facilities necessary for developing new products of photographic films and papers. Their structure was composed of a matrix of two different kinds of research laboratories. The row of the matrix, for example, was composed of laboratories for developing new products of the above-stated photographic films and papers. The line of the matrix was composed of laboratories for developing common technologies and knowledge necessary for developing the above-stated products. Table 11.1 shows an example of such a matrix structure. The research laboratories for developing new products included those in charge of color negative films, color reversal films, color papers, photographic films for printing industry, and photographic films for medical uses. The research laboratories for developing common technologies and knowledge included those in charge of organic synthesis, analyses, design and evaluation of image quality of pictures, design of equipment, and basic research [2,3].

It is said that the presence of a basic research laboratory in a company is proof of the fact that the corresponding company has become full-fledged. Since it usually takes a long time to get the results of a research activity, it is important that the basic research laboratory was already founded by Y. Koseki in Ashigara Research Laboratories when the product of photographic films and papers in Japan was still a "problem child".

It was obvious that the mission of Ashigara Research Laboratories was to develop and deliver new products of high-quality photographic films and papers. The development of new products can in most cases be achieved by improving their sensitivity and/or image quality. Since sensitivity and image quality are correlated, the sensitivity of photographic films can be the index for developing new products. Accordingly, the target of a new product and the plan to achieve it can be specific and quantitative, making it clear what each laboratory and its members need to do. As a result, the target of each product can be successfully achieved for a relatively short period thanks to the great effort of members of Ashigara Research Laboratories.

As to the index (i.e., photographic sensitivity) for the development of new products, the basic research laboratory collaborated with other laboratories to analyze the index to grasp the present state and future prospect of corresponding products as a guideline, which is exemplified in Table 6.1 and described in Section 6.2.2 [6]. The guideline includes an extraction of all the factors determining photographic sensitivity, an evaluation of the quantitative difference between the ultimate limit and the present state of each factor,

Table 11.1 An example of the matrix structure of research laboratories for production of AgX photographic materials

	Development of color films	Development of color papers	Development of films for medical use	Development of films for printing
Fundamental research				
Analytical research				
Organic Synthesis				
Research on polymers				
Research on image analysis				
Design of apparatus				

analysis of the origin for determining the difference between the present state and the ultimate limit, and proposals of technologies expected to contribute to the reduction of the above-stated difference. In light of the guideline, the progress of each project for developing new products can be grasped properly by evaluating the result of each research activity with respect to the present state and ultimate limit, and by recognizing the problems to be solved.

11.3 Analysis of Basic Research for Photographic Films and Papers

The present author has been involved in basic research of AgX photographic films and papers for nearly a half of a century, mostly at Ashigara Research Laboratories, as described in the previous section. For the first five years, he was, however, involved in the basic research at Kikuchi's Laboratory at the University of Tokyo, which was quite different from that at Ashigara Research Laboratories, as stated in the previous section.

It is said that the appropriate fraction of basic research out of a company's whole research and development organization is about 5%. Namely, the foundation of a basic research laboratory needs a large-scale research and development organization, as exemplified by a basic research laboratory of 50 members being part of a research and development organization of 1000 members.

In order to achieve the specific and quantitative targets of Ashigara Research Laboratories, not only each laboratory, but also each member of the laboratory, was required to be in good order to accomplish their tasks, which had been planned to contribute to the achievement of the targets, and to report the results regularly to confirm progress with respect to the plans. Since nearly all the members were involved in developing new products of AgX photographic films and papers from various aspects, they could exchange abundant information on AgX photographic films and papers among the many other members. On the other hand, they scarcely received advanced knowledge of non-silver photographic systems and methods such as xerography, photo-resists, diazonium compounds, and organic semiconductors for photoelectric conversion devices [3].

In 1963–8, the present author was a graduate student of Kikuchi's Laboratory, in which there were always about ten research members, all of whom were involved in basic research on one of the themes of "photographic chemistry". Actually, photographic chemistry was the domain of Kikuchi's Laboratory with a fundamental background of electrochemistry. The discovery of the Honda–Fujishima effect, by which water can be photo-decomposed in the presence of TiO_2 [7], was one of the most important achievements at Kikuchi's Laboratory, and proved that Kikuchi's Laboratory provided its members with an excellent environment for basic research in photographic chemistry. However, basic research at Kikuchi's Laboratory was quite different from that at Ashigara Research Laboratories. The former is considered a model of a basic research laboratory that produces exploratory results, while the latter is a model for a basic research laboratory in the organization to develop new products [3].

In photographic chemistry, there are various fields including the mechanisms of photographic sensitivity and dye sensitization of AgX systems, electrostatic copying systems such as xerography and electrofax, photo-resists with polyvinyl-cinnamate, etc., chemical copying systems such as those with diazonium compounds and dichromate salts, organic electrochemistry, and photo-electrochemistry, where each member of a laboratory can cover one of these areas [3]. Exchanges of information among these fields can be vividly made in a laboratory of about ten members, in contrast to Ashigara Research Laboratories where the number of members was two orders of magnitude larger that of Kikuchi's laboratory, and where most of the time only advanced knowledge of AgX photography was exchanged among its members. Although these conditions of Ashigara Research Laboratories were ideal for developing new products of AgX photographic films and papers, the conditions of Kikuchi's laboratory were considered much more suitable for exploratory research, since interactions among properly different areas in the domain of photographic chemistry could enhance the creation of new ideas.

In contrast to Ashigara Research Laboratories, where members needed to be in good order and devoted to achieving their specific and quantitative target within its deadline, each member of Kikuchi's Laboratory was given a wide theme that was neither specific nor quantitative, and was allowed to pursue it in any direction without planning [3]. Although the present author was given the theme to theoretically study the absorption spectra of AgX crystals, he left it shortly and became involved in the quantum-chemical study of absorption spectra of sensitizing dyes, and then in the study of the dye sensitization mechanism of AgX grains in photographic films. This change of research theme was rather common in Kikuchi's laboratory. Although this kind of behavior involving huge freedom was not necessarily appreciated in research laboratories where defined missions of developing new products by a fixed date had to be achieved, it was a very important factor in exploratory research.

Ashigara Research Laboratories' research and development activities were well planned with fixed dates of achievement in order to meet the demand that new fine-quality products should be developed in a short period. On the other hand, the Kikuchi's Laboratory's research activities were unplanned and carried out for a long period in some cases. It took more than twenty years for photo-catalysts with the photo-oxidation effect of TiO_2 [8] to appear on the market as the application of the Honda–Fujishima effect since its discovery.

Although Kikuchi's Laboratory's and Ashigara Research Laboratories' research activities for AgX photographic materials were able to achieve good results, they were quite different from each other, as described, and regarded as the research models for creating seeds and developing new products, respectively. It is important to note the difference between them and to properly apply them to research activities at various stages such as in creating a promising "problem child", and in developing new products under the conditions of "star" and "cash cow". The analyses of these laboratories are expected to serve as reference for the research and development organization of manufacturing industries.

11.4 Development of Photographic Science and Technology for the Future

Although AgX photographic films enjoyed "cash cow" by forming a strong domain, scientists had to create other promising "problem children" that were expected to grow up to become the "stars" and "cash cows" of the future, since AgX photographic films would not stay in the "cash cow" state forever. Based on the above-stated analyses, it was necessary to grasp the strong technical points of the "domain", in order to carry out exploratory research and to have great freedom when creating promising "problem children" by taking as long as necessary.

It was important to understand properly the essential technologies of the domain for AgX photographic films. Unexpectedly, it did not seem that "photography" or "imaging" was a correct answer. Although photography needed to include digital still cameras with CCD as their essential part, the knowledge and technologies cultivated for photographic films could hardly help in the development of CCD for digital still cameras, since the technologies for AgX photographic films and for CCD were so different to bring about synergetic interaction between them. On the other hand, the essential technologies in the domain of photographic films were considered functional materials that constituted photographic films as their indispensable parts.

Another important point in the strategy was to carry out exploratory research with great freedom for a long period of time in order to create promising "problem children" with the aid of the domain's essential technologies. Although it is not necessary to specify the goal of exploratory research at the start, some examples of the transfer of the results of photographic science and technology to other fields are being made apparently and are briefly described below.

Triacetate cellulose (TAC) film bases developed for photographic films are characterized by their physical and optical uniformity, and are now widely used for liquid crystal display. The preparation and characterization of well-defined AgX grains are unique and expected to serve as reference for those of nanoparticles of metals and semiconductors. Dye sensitization, invented and developed for AgX photographic films over many years, is now the central technology in dye-sensitized solar cells. The vacuum level shift between substrates and organic layers was observed for the first time in the basic research on dye sensitization of AgX photographic films, and is now indispensable knowledge for organic electronic devices. J-aggregates together with H-aggregates were discovered for the first time with cyanine dyes specifically used for dye sensitization of AgX photographic films and are expected to be applicable to the desirable self-alignment of molecules in organic devices. Photographic films with well-defined AgX nanocrystals still provide us with unique sensors, and are expected, for example, to serve as an indispensable tool for detecting dark matter, which is one of the most important subjects in space nuclear physics.

Although the industry of photographic films has declined, photographic science and technology as cultivated for AgX photographic films and papers for many years are based on various fundamentals and have been deeply investigated, as described in this book, and therefore are expected to be further extended in developing knowledge and technologies in other fields. It would be a great pleasure if this book, an updated edition of *Photographic Sensitivity*, is able to contribute to the development of the results of photographic science and technology for advanced knowledge and devices in the future.

References

1. B. Bensoussan, C. S. Fleisher, *Strategic and Competitive Analysis Methods and Techniques for Analyzing Business Competition* (Upper Saddle River, NJ: Pearson Education, 2004).
2. T. Tani, *Kenkyu Kaihatsu Leader*, 5, 33 (2008).

3. T. Tani, *Photochemistry*, 41, 1 (2010).
4. H. Taniguchi, *Photo Market*, 19 (2009).
5. A. Kumai, *J. Appl. Photogr. Eng.*, 7, 88 (1981).
6. T. Tani, *J. Imaging Sci. Tech.*, 51, 110 (2007).
7. A. Fujishima, K. Honda, *Nature*, 238, 37 (1972).
8. R. Wang, K. Hashimoto, A. Fujishima, T. Watanabe, et al., *Nature*, 388, 431 (1997).

Index